NARRATIVE

OF AN

EXPEDITION

INTO

CENTRAL AUSTRALIA

BY

CHARLES STURT

ISBN 1-876247-11-8

NARRATIVE OF AN EXPEDITION INTO
CENTRAL AUSTRALIA, PERFORMED UNDER THE
AUTHORITY OF HER MAJESTY'S GOVERNMENT
DURING THE YEARS 1844, 5, 6.
TOGETHER WITH A NOTICE OF THE
PROVINCE OF SOUTH AUSTRALIA, IN 1847.
BY CAPTAIN CHARLES STURT, F.L.S. AND F.R.G.S.

First published in London by T and W Boone, 1849.
First edition by Corkwood Press 2001

For a free brochure listing titles published by Corkwood Press
send your name and address (post free) to:

Corkwood Press
Reply Paid 1313
PO Box 237
North Adelaide
South Australia 5006

Telephone 08 8431 5784

e-mail corkwood@onaustralia.com.au

NARRATIVE

OF AN

EXPEDITION

INTO

CENTRAL AUSTRALIA

PERFORMED

Under the Authority of her Majesty's Government,

DURING

THE YEARS 1844, 5. AND 6.

TOGETHER WITH A NOTICE OF THE

PROVINCE OF SOUTH AUSTRALIA,

IN 1847.

BY

CAPTAIN CHARLES STURT, F.L.S. F.R.G.S.

ETC. ETC.

AUTHOR OF "TWO EXPEDITIONS INTO SOUTHERN AUSTRALIA."

CONTENTS

INTRODUCTION

ABOVE a doorway in the Family Hotel in Tibooburra, long the haunt of ringers, doggers, old-timers and bush characters, a lovely, elusive fresco has been painted. It shows a mirage, fading, losing its definition against the desert sky: "Sturt's Dream — Our Destination," reads the legend.

For most of the century and a half since his tortured expedition into Central Australia, the name of Captain Charles Sturt has been linked with a set of cherished, almost stereotypical traits of national character: endurance, pioneering persistence, a romantic dreaminess, a burning desire to see what lies ahead.

Sturt's last voyage had an almost mythical resonance and potency: his urge to find the Centre; his conviction that the red heart must be watered by an inland sea; his march with his party northward through the Barrier Ranges; the entrapment at Depôt Glen; the despairing sortie across the Stony Desert; the final failure on the margins of the Simpson, all routes forward blocked by endless dunes of blazing sand.

Once, these things would have been stamped into schoolchildren's minds. Now, though, all is changed. An epochal reassessment and revision of the truths of Australian history has been under way over the course of the last two decades. The standing of Sturt, and the reputations of his fellow explorers, have been among the casualties of the new accounting.

The broad lines of this revision will be familiar to anyone who breathes in the contemporary intellectual, political or moral atmosphere. If settlement was invasion, then exploration was expropriation, and discovery desecration. Even the image of the simple, noble-seeming Sturt has been subject to fresh interpretative effort. His skills as a bushman have been treated to withering critique; his casual assumptions of Britannic supremacy have been laid bare; the publication of private journals written by his expedition members suggest he may have been petty and vindictive as much as austere and Olympian. Should we, then, tear down the statues, now we realise, with all our subtle hindsight, that this 19th century man was blinkered by a 19th century set of perspectives? Let me put a brief, contrarian case in Sturt's favour.

There are several reasons to be willing to spend a few hours in the company of this engaging Captain. The first is one that stems purely from an accident of circumstance and time. It fell to Sturt to be the first European to see and describe some of the most extreme faces of the Australian inland. His route, from 1844-6, took him through new ranges and gibber country, through breakaways and unseen deserts. To retrace his path today, book in hand, to follow his narrative, is to experience Australia anew as it seemed — in all its beauty, its weirdness and its horror — before the friendly impasto of bush kitsch had been ladled onto the land and our perception of it.

Even at the start of the 21st Century, and in the air-conditioned cocoon of a four-wheel drive, the route from Mildura, through Menindee, on up the Silver City Highway to Tibooburra leaves the traveller on edge. The sky shimmers; the hills lead off to nowhere; mirage-lakes dance before the eye. Depôt Glen, and Mount Pole, just off the Tibooburra road, and Fort Grey, on the Cameron Corner track, have lost none of their primal charge of desolation. Examining Sturt's words onsite gives them a different quality: the party's efforts come to seem an attempt to withstand, to gain the acceptance of the landscape. Northward, Sturt's ride can be simply traced by a pair of drives. One road to Innamincka heads through Strzelecki Creek, still as weedy and disappointing as when Sturt crossed it, while the "Great Stony Desert" that so transfixed the Captain now forms a visual interlude along the neatly-graded Birdsville Track. Sturt's furthest north is less often visited. He penetrated into the Simpson and headed, beside the course of Eyre Creek, along the great dunes west of "Big Red". At the ruins of Annandale homestead, before a lovely water hole much frequented by budgerigars, a small plaque records his passage. On September the 8th, 1845, he halted a short distance further up the dune-way, at sunset, "in a country such as I firmly believe has no parallel on Earth's surface, and one which was terrible in its aspect." Sturt's dream was at its end; its despairing songline lies traced out across some of the most unyielding, and magnificent, terrain in all Australia. The principal appeal of his narrative, though, lies not in its value as a gazetteer or a telescope into the past, or even in its compelling tone of emotional turmoil and despair barely mastered, or in its kind-hearted and enlightened judgements of Aboriginal people.

No: the chief claim Sturt has on our attention is the one least often entered on his behalf by the critics, colonial theorists, subaltern historians and cultural geographers who take him as the archetype of his age's shortcomings. The *Expedition Into Central Australia* is, in its understated, innocent fashion, a grand and epic piece of literature — a distinction it shares with a handful of the early exploration journals, such as those of Grey, Eyre and Lort Stokes. Sturt's clean and transparent prose style was already evident in his narrative of his journey down the Murray, published in 1833. But in the record of his second voyage, even the initial note is much darker. It is clear from the outset, as the expedition sets off under the gaze of that gloomy romantic, Edward Eyre, that this will be an affair of endurance, endeavour and ultimate renunciation. The sublime futility of Sturt's quest, the perfect lunacy of his desire to sail on the inland sea — a desire he indulged by taking a boat with him in his baggage train (a sculpture at Tibooburra depicts it)—has somehow lent his mission a symbolic dimension. The biblical quality of his ordeal in the desert has led sensitive souls to see the journey as a kind of metaphysical rite of passage: a notion Sturt himself seems to have toyed with, since he refers regularly to the "veil" drawn across the heart of Australia, "that could neither be pierced or raised." But what underpins and makes possible these slightly fanciful higher readings is the starkness, the thematic pureness, the simplicity of the book: it is an exposed outcrop of writing, weathered and lashed into its shape by the elements of Australia.

We begin, as we shall end, with Sturt's watery fantasia of the inland; and throughout the tale, its mirage-like gleam is sighted, until each fresh range of flat-topped mountains, each sandy plain Sturt encounters becomes one more piece in the chain of evidence proving that a sea had indeed once lain at the heart of Australia: "All I can say is, would that I had discovered such a feature, for I could then have done more upon its waters ten-fold, than I was enabled to accomplish in the gloomy and burning deserts over which I wandered during more than 13 months." This dichotomy between what should be present in the new continent and what is actually found there runs through the tale: mountains look like ruined citadels; natives who should build cities roam the land, birds mimic human voices, or worse, become worrying and Hitchcockian enemies. Nature everywhere is upside down — the task of the explorer is to maintain human grace, form and structure in this mad wilderness. Accordingly, marooned at the Glen, and with his faithful assistant James Poole lying on his deathbed, Sturt decides to raise a pyramid upon a nearby hill "to give the men occupation".

"I little thought," he goes on, "that I was erecting Mr Poole's monument, but so it was, that rude structure looks out over his lonely grave, and will stand for ages as a record of all we suffered in the dreary region to which we were so long confined..." Tombs instead of survey cairns. Stillness instead of daily motion. The experience of Sturt and his party became, eventually, a form of living death, couched, in this narrative, in terms that are sharply reminiscent of purple passages from the romantic poets of the early 19th Century: If Coleridge's *Goddess Death-in-Life* or Keats' *Belle Dame sans Merci* are not actually present to Sturt's heat-infested brain as he writes, they are very much part of the cultural ambience the Captain brought with him in his packsaddle. This was a heroic journey into the unknown, made at a time when the mode of internal life, and of descriptive prose, was intensely emotive, self-conscious, vivid and unalloyed. Even so simple a soldier as Charles Sturt had been drenched in this affective climate: it is largely responsible for the extraordinary passages in which he describes his men's ordeal. (Intriguingly, John McDouall Stuart, who was on this expedition, and who later endured equal if not greater horrors on his own, successful, crossing of the continent, wrote them up in a far different fashion, being a silent Scotsman from a tighter cultural background.)

The forthcoming publication of the full text of Sturt's private expedition journal by the Hakluyt Society may provide further clues to his personal imaginings (already, in fact, the chief subject of an intriguing novel of the desert, James Baron's *Revolution by Night*). But even this volume's "official" presentation of Sturt's experiences portrays a world of violent extremes, full of depictions of ruin and decay, of fertility dashed and evanescent greenery returning to a scorched and dreadful bleakness. He was lost for words to convey "to the reader's mind an idea of the intense and oppressive nature of the heat." Even scientific measurement was left behind — the thermometer bulb had cracked in the hot wind blown from the Centre. "The lead dropped out of our pencils, our signal rockets were entirely spoilt; our hair, as well as the wool on the sheep, ceased to grow, and our nails had become as brittle as glass." Where was he? Not amongst the sand and spinifex, but in a hybrid, almost abstract landscape: "The stillness of death reigned around us, no living creature was to be heard; nothing visible inhabited that dreary desert but the ant, even the fly shunned it, and yet its yielding surface was marked all over with the tracks of native dogs." Instead of water, there was another element predominant: fire — the sun would "lick up" every pool; matches, falling to the ground, ignited on contact; the hot wind was "so terrific, I wondered the very grass did not take fire." Sturt stood, he felt, not just on the fringes of human experience and endurance, but on the borders of language — on the brink of paradox.

And this, of course, was the core of his journey: the reverse of his hope; the truth of his dream. "Solid waves" of sand, and a stony ocean, turned a mocking purple by the sunset's glow. Ridges that resembled "sea dunes, backed by storm clouds," and through which navigation was only possible, as at sea, by compass. From an epic beginning to his exploration career upon the ambiguous waters of the Murray, Sturt fought through to a strange end. Its mode was dictated by the Australian landscape, which scorned to give him tragedy: nothing so human as chaos, sin or even flaws of character undid him, but merely the silence, the indifference of his veiled, beloved Centre.

Sturt reached within one degree of the Tropic of Capricorn. He was turned back only 150 miles from the geographic heart of Australia. His story had, at last, a shape, but one quite different from his expectations. Together with his companion on his final northward probe, the surgeon John Harris Browne, he gazed out from a high ridge, across dunes that succeeded each other in mazy and monotonous prospect:

"… The sand was of a deep red colour, and a bright narrow line of it marked the top of each ridge, amidst the sickly pink and glaucous coloured vegetation around."

"Good heavens," exclaimed Browne: "did ever man see such country!" …It was the westward throw of the Simpson Desert — the true and only inland sea — that they beheld, in all its other-worldly beauty, at that moment. Many explorers would follow Sturt in time, with more success and less romance — but in this sight, at least, he was the first.

Nicolas Rothwell,
Sydney,
March 2001.

TO THE

RIGHT HONORABLE THE EARL GREY,

ETC., ETC., ETC.,

MY LORD,

ALTHOUGH the services recorded in the following pages, which your Lordship permits me to dedicate to you, have not resulted in the discovery of any country immediately available for the purposes of colonization, I would yet venture to hope that they have not been fruitlessly undertaken, but that, as on the occasion of my voyage down the Murray River, they will be the precursors of future advantage to my country and to the Australian colonies.

Under present disappointment it must be as gratifying to those who participated in my labours, as it is to myself to know that they are not the less appreciated by your Lordship, because they were expended in a desert.

I can only assure your Lordship, that it has been my desire to give a faithful description of the country that has been explored, and of the difficulties attending the task; nor can I refuse myself the anticipation that the perusal of these volumes will excite your Lordship's interest and sympathy.

I have the honour to be,
My Lord,
Your Lordship's
Most obedient humble servant,

CHARLES STURT.

London, November 21, 1848.

NOTICE

IT might have been expected that many specimens, both of Botany and Ornithology, would have been collected during such an Expedition as that which the present narrative describes, but the contrary happened to be the case.

I am proud in having to record the name of my esteemed friend, Mr. Brown, the companion of Flinders, and the learned author of the "Prodromus Novæ Hollandiæ," to whose kindness I am indebted for the Botanical Remarks in the Appendix.

To my warm-hearted friend, Mr. Gould, whose splendid works are before the Public, and whose ardent pursuits in furtherance of his ambition, I have personally witnessed, I owe the more perfect form in which my ornithological notice appears.

I have likewise to acknowledge, with very sincere feelings, the assistance I have received from Mr. Arrowsmith, in the construction of my Map, to whose anxious desire to ensure correctness and professional talent I am very greatly indebted.

I hope the gentlemen whose names I have mentioned will accept my best thanks for the assistance they have afforded me in my humble labours. It is not the least of the gratifications enjoyed by those who are employed on services similar to which I have been engaged, to be brought more immediately in connection with such men.

London, November 21, 1848.

TRAVELS IN AUSTRALIA.

CHAPTER I.

CHARACTER OF THE AUSTRALIAN CONTINENT—OF ITS RIVERS—PECULIARITY OF THE DARLING—SUDDEN FLOODS TO WHICH IT IS SUBJECT—CHARACTER OF THE MURRAY—ITS PERIODICAL RISE—BOUNTY OF PROVIDENCE—GEOLOGICAL POSITION OF THE TWO RIVERS—OBSERVATIONS—RESULTS—SIR THOMAS MITCHELL'S JOURNEY TO THE DARLING—ITS JUNCTION WITH THE MURRAY—ANECDOTE OF MR. SHANNON—CAPTAIN GREY'S EXPEDITION—CAPTAIN STURT'S JOURNEY—MR. EYRE'S SECOND EXPEDITION—VOYAGE OF THE BEAGLE—MR. OXLEY'S OPINIONS—STATE OF THE INTERIOR IN 1828—CHARACTER OF ITS PLAINS AND RIVERS—JUNCTION OF THE DARLING—FOSSIL BED OF THE MURRAY—FORMER STATE OF THE CONTINENT—THEORY OF THE INTERIOR.

THE AUSTRALIAN continent is not distinguished, as are many other continents of equal and even of less extent, by any prominent geographical feature. Its mountains seldom exceed four thousand feet in elevation, nor do any of its rivers, whether falling internally or externally, not even the Murray, bear any proportion to the size of the continent itself. There is no reason, however, why rivers of greater magnitude, than any which have hitherto been discovered in it, should not emanate from mountains of such limited altitude, as the known mountains of that immense and sea-girt territory. But, it appears to me, it is not in the height and character of its hilly regions, that we are to look for the causes why so few living streams issue from them. The true cause, I apprehend, lies in its climate, in its seldom experiencing other than partial rains, and in its being subject to severe and long continued droughts. Its streams descend rapidly into a country of uniform equality of surface, and into a region of intense heat, and are subject, even at a great distance from their sources, to sudden and terrific floods, which subside, as the cause which gave rise to them ceases to operate; the consequence is, that their springs become gradually weaker and weaker, all back impulse is lost, and whilst the rivers still continue to support a feeble current in the hills, they cease to flow in their lower branches, assume the character of a chain of ponds, in a few short weeks their deepest pools are exhausted by the joint effects of evaporation and absorption, and the traveller may run down their beds for miles, without finding a drop of water with which to slake his thirst.

In illustration of the above, I would observe that during the progress of the recent expedition up the banks of the Darling, and at a distance of more than 300 miles from its sources, that river rose from a state of complete exhaustion, until in four days it overflowed its banks. It was converted in a single night, from an almost dry channel, into a foaming and impetuous stream, rolling along its irresistible and turbid waters, to add to those of the Murray.

There can be no doubt, but, that this sudden rise in the river, was caused by heavy rains on the mountains, in which its tributaries are to be found, for the Darling does not receive any accession to its waters below their respective junctions, of sufficient magnitude to account for such an occurrence.*

When, on the return of the expedition homewards the following year, some two months later in the season than that of which I have just been speaking, Oct. 1844, there had been no recurrence of the flood of the previous year, but the Darling was at a still lower ebb than before, and every lagoon, and creek in its vicinity had long been exhausted and waterless.† Now, it is evident, as far as I can judge, that if the rains of

* The principal tributaries of the Darling, are the Kindur, the Keraula, the Namoy, and the Gwydir. They are beautiful mountain streams, and rise in the hilly country, behind Moreton Bay, in lat. 27°S., and in longitude 152°E.

† It may be necessary to warn my readers that a creek in the Australian colonies, is not always an arm of the sea. The same term is used to designate a watercourse, whether large or small, in which the winter torrents may or may not have left a chain of ponds. Such a watercourse could hardly be called a river, since it only flows during heavy rains, after which it entirely depends on the character of the soil, through which it runs, whether any water remains in it or not.

Australia were as regular as in other countries, its rivers would also be more regular in their flow, and would not present the anomaly they now do, of being in a state of rapid motion at one time, and motionless at another.

But, although I am making these general observations on the rivers, and to a certain extent of climate of Australia, I would not be understood to mean more than that its seasons are uncertain, and that its summers are of comparatively long duration.

In reference to its rivers also, the Murray is an exception to the other known rivers of this extensive continent. The basins of that fine stream are in the deepest recesses of the Australian Alps—which rise to an elevation of 7000 feet above the sea. The heads of its immediate tributaries, extend from the 36th to the 32nd parallel of latitude, and over two degrees of longitude, that is to say, from the 146° to the 148° meridian, but, independently of these, it receives the whole westerly drainage of the interior, from the Darling downwards. Supplied by the melting snows from the remote and cloud-capped chain in which its tributaries rise, the Murray supports a rapid current to the sea. Taking its windings into account, its length cannot be less than from 1300 to 1500 miles. Thus, then, this noble stream preserves its character throughout its whole line. Uninfluenced by the sudden floods to which the other rivers of which we have been speaking are subject, its rise and fall are equally gradual. Instead of stopping short in its course as they do, its never-failing fountains have given it strength to cleave a channel through the desert interior, and so it happened, that, instead of finding it terminate in a stagnant marsh, or gradually exhausting itself over extensive plains as the more northern streams do, I was successfully borne on its broad and transparent waters, during the progress of a former expedition, to the centre of the land in which I have since erected my dwelling.

A lagoon is a shallow lake, it generally constitutes the back water of some river, and is speedily dried up. In Australia, there is no surface water, properly so called, of a permanent description.

As I have had occasion to remark, the rise and fall of the Murray are both gradual. It receives the first addition to its waters from the eastward, in the month of July, and rises at the rate of an inch a day until December, in which month it attains a height of about seventeen feet above its lowest or winter level. As it rises it fills in succession all its lateral creeks and lagoons, and it ultimately lays many of its flats under water.

The natives look to this periodical overflow of their river, with as much anxiety as did ever or now do the Egyptians, to the overflowing of the Nile. To both they are the bountiful dispensation of a beneficent Creator, for as the sacred stream rewards the husbandman with a double harvest, so does the Murray replenish the exhausted reservoirs of the poor children of the desert, with numberless fish, and resuscitates myriads of crayfish that had laid dormant underground; without which supply of food, and the flocks of wild fowl that at the same time cover the creeks and lagoons, it is more than probable, the first navigators of the Murray would not have heard a human voice along its banks; but so it is, that in the wide field of nature, we see the hand of an overruling Providence, evidences of care and protection from some unseen quarter, which strike the mind with overwhelming conviction, that whether in the palace or in the cottage, in the garden, or in the desert, there is an eye upon us. Not to myself do I accord any credit in that I returned from my wanderings to my home. Assuredly, if it had not been for other guidance than the exercise of my own prudence, I should have perished: and I feel satisfied the reader of these humble pages, will think as I do when he shall have perused them.

An inspection of the accompanying chart, will shew that the course of the Murray, as far as the 138° meridian is to the W.N.W., but that, at that point, it turns suddenly to the south, and discharges itself into Lake Victoria, which again communicates with the ocean, in the bight of Encounter Bay. This outlet is called the "Sea mouth of the Murray," and immediately to the eastward of it, is the Sand Hill, now called Barker's Knoll—under which the excellent and amiable officer after whom it is named fell by the hands of the natives, in the cause of geographical research.

Running parallel with its course from the southerly bend, or great N.W. angle of the Murray, there is a line of hills, terminating southwards, at Cape Jarvis; but, extending northwards beyond the head of Spencer's Gulf. These hills contain the mineral wealth of South Australia, and immediately to the westward of them is the fair city of Adelaide.

On gaining the level interior, the Murray passes through a desert country to the 140° meridian, when it enters the great fossil formation, of which I shall have to speak hereafter. In lat. 34°, and in long. 142°, the Darling forms a junction with it; consequently, as that river rises in latitude 27°, and in long. 152°, its direct course will be about S.W. There is a distance of nine degrees of latitude, therefore, between their respective sources, and, as the Darling forms a considerable angle with the Murray at this junction, it necessarily follows, as I have had occasion to remark, that the two rivers must receive all the drainage from the eastward, falling into that angle. If I have been sufficiently clear in explaining the geographical position and character of these two rivers, which in truth almost make an island of the S.E. angle of the Australian continent, it will only remain for me to add in this place, that neither the Murray nor the Darling receive any tributary stream from the westward or northward, and at the time at which I commenced my last enterprise, the Darling was the boundary of inland discovery, if I except the journey of my gallant friend Eyre, to Lake Torrens, and the discovery by him of the country round Mount Serle. Sir Thomas Mitchell had traced the Darling, from the point at which I had been obliged from the want of good water to abandon it, in 1828, to lat. 32°26', and had marked down some hills to the westward of it. Still I do not think that I detract from his merit, and I am sure I do not wish to do so, when I say that his having so marked them can hardly be said to have given us any certain knowledge of the Cis-Darling interior.

More than sixteen years had elapsed from the period when I undertook the exploration of the Murray River, to that at which I commenced my preparations for an attempt to penetrate Central Australia. Desolate, however, as the country for the most part had been, through which I passed, my voyage down that river had been the forerunner of events I could neither have anticipated or foreseen. I returned indeed to Sydney, disheartened and dissatisfied at the result of my investigations. To all who were employed in that laborious undertaking, it had proved one of the severest trial and of the greatest privation; to myself individually it had been one of ceaseless anxiety. We had not, as it seemed, made any discovery to gild our enterprise, had found no approximate country likely to be of present or remote advantage to the Government by which we had been sent forth; the noble river on whose buoyant waters we were hurried along, seemed to have been misplaced, through such an extent of desert did it pass, as if it was destined thus never to be of service to civilized man, and for a short time the honour of a successful undertaking, as far as human exertion could ensure it, was all that remained to us after its fatigues and its dangers had terminated, as the reader will conclude from the tenour of the above passage; for, although at the termination of the Murray, we came upon a country, the aspect of which indicated more than usual richness and fertility, we were unable, from exhausted strength, to examine it as we could have wished, and thus the fruits of our labours appeared to have been taken from us just as we were about to gather them. But if, amidst difficulties and disappointments of no common description, I was led to doubt the wisdom of Providence, I was wrong. The course of events has abundantly shewn how presumptuous it is in man to question the arrangements of that Allwise Power whose operations and purposes are equally hidden from us, for in six short years from the time when I crossed the Lake Victoria, and landed on its shores, that country formed another link in the chain of settlements round the Australian continent, and in its occupation was found to realize the most sanguine expectations I had formed of it. Its rich and lovely valleys, which in a state of nature were seldom trodden by the foot of the savage, became the happy retreats of an industrious peasantry; its plains were studded over with cottages and cornfields; the very river which had appeared to me to have been so misplaced, was made the high road to connect the eastern and southern shores of a mighty continent; the superfluous stock of an old colony was poured down its banks into the new settlement to save it from the trials and vicissitudes to which colonies, less favourably situated, have been exposed; and England, throughout her wide domains, possessed not, for its extent, a fairer or a more promising dependency than the province of South Australia. Such, there can be no doubt, have been the results of an expedition from which human foresight could have anticipated no practical good.

During my progress down the Murray River I had passed, the junction of a very considerable stream with it, in lat.* 34°S' and long. 142°. Circumstances, however, prevented my examining it to any distance above its point of union with the main river. Yet, coming as it did, direct from the north, and similar as it was to the Darling in its upper branches, neither had I, nor any of the men then with me, and who had accompanied me when I discovered the Darling in 1828, the slightest doubt as to its identity. Still, the fact might reasonably be disputed by others, more especially as there was abundant space for the formation of another river, between the point where I first struck the Darling and this junction.

It was at all events a matter of curious speculation to the world at large, and was a point well worthy of further investigation. Such evidently was the opinion of her Majesty's Government at the time, for in accordance with it, in the year 1835, Sir Thomas Mitchell, the Surveyor-General of the colony of New South Wales, was directed to lead an expedition into the interior, to solve the question, by tracing the further course of the Darling. This officer left Sydney in May, 1835, and pushing to the N.W. gradually descended to the low country on which the Macquarie river all but terminates its short course. In due time he gained the Bogan river (the New Year's Creek of my first expedition, and so called by my friend, Mr. Hamilton Hume, who accompanied me as my assistant, because he crossed it on that day), and tracing it downwards to the N.W., Sir Thomas Mitchell ultimately gained the

* The Darling.

banks of the Darling, where I had before been upon it, in latitude 30°. He then traced it downwards to the S.S.W to latitude 32°26". At this point he determined to abandon all further pursuit of the river, and he accordingly returned to Sydney, in consequence, as he informs us, of his having ascertained that just below his camp a small stream joined the Darling from the westward. The Surveyor-General had noticed distant hills also to the west; and it is therefore to be presumed that he here gave up every hope of the Darling changing its course for the interior, and of proving that I was wrong and that he was right. The consequence, however, was, that he left the matter as much in doubt as before, and gained but little additional knowledge of the country to the westward of the river.

In the course of the following year Sir Thomas Mitchell was again sent into the interior to complete the survey of the Darling. On this occasion, instead of proceeding to the point at which he had abandoned it, the Surveyor-General followed the course of the Lachlan downwards, and crossing from that river to the Murrumbidgee, from it gained the banks of the Murray. In due time he came to the disputed junction, which he tells us he recognised from its resemblance to a drawing of it in my first work. As I have since been on the spot, I am sorry to say that it is not at all like the place, because it obliges me to reject the only praise Sir Thomas Mitchell ever gave me; but I mention the circumstance because it gives me the opportunity to relate an anecdote, connected with the drawing, in which my worthy and amiable friend, Mr. Shannon, a clergyman of Edinburgh, and a very popular preacher there, but who is now no more, took a chief part. I had lost the original drawing of the junction of the Murray, and having very imperfect vision at the time I was publishing, I was unable to sketch another. It so happened that Mr. Shannon, who sketched exceedingly well with the pen, came to pay me a visit, when I asked him to try and repair my loss, by drawing the junction of the Darling with the Murray from my description. This he did, and this is the view Sir Thomas Mitchell so much approved. I take no credit to myself for faithfulness of description, for the features of the scene are so broad, that I could not but view them on my memory; but

I give great credit to my poor friend, who delineated the spot, so as that it was so easily recognised. It only shews how exceedingly useful such things are in books, for if Sir Thomas Mitchell had not so recognised the view, he might have doubted whether that was really the junction of the Darling or not, for he had well nigh fallen into the mistake of thinking that he had discovered another river, when he came upon the Darling the year before, and had as much difficulty in finding a marked tree of Mr. Hume's upon its banks, as if it had been a needle in a bundle of straw. Fortunately, however, the Surveyor-General was enabled to satisfy himself as to this locality, and he accordingly left the Murray, and traced the junction upwards to the north for more than eight miles, when he was suddenly illuminated. A ray of light fell upon him, and he became convinced, as I had been, of the identity of this stream with the Darling, and suddenly turning his back upon it, left the question as much in the dark as before. Neither did he therefore on this occasion, throw any light on the nature and character of the distant interior. In the year 1837 the Royal Geographical Society, assisted by Her Majesty's Government, despatched an expedition under the command of Lieuts. afterwards Captains Grey and Lushington—the former of whom has since been Governor of South Australia, and is at the present moment Governor in Chief of New Zealand—to penetrate into the interior of the Australian continent from some point on the north-west or west coast; but those gentlemen were unable to effect such object. The difficulties of the country were very great, and their means of transport extremely limited; and in consequence of successive untoward events they were ultimately obliged to abandon the enterprise, without any satisfactory result. But I should be doing injustice to those officers, more particularly to Captain Grey, if I did not state that he shewed a degree of enthusiasm and courage that deserve the highest praise.

As, however, both Sir Thomas Mitchell and Capt. Grey† have published accounts of their respective expeditions, it may not be

† Journals of Expeditions of Discovery in North-West and Western Australia, during the years 1837-8-9, by Captain George Grey.

necessary for me to notice them, beyond that which may be required to connect my narrative and to keep unbroken the chain of geographical research upon the continent.

In the year 1838, I myself determined on leading a party overland from New South Wales to South Australia, along the banks of the Murray; a journey that had already been successfully performed by several of my friends, and among the rest by Mr. Eyre. They had, however, avoided the upper branches of the Murray, and particularly the Hume, by which name the Murray itself is known above the junction of the Murrumbidgee with it. Wishing therefore to combine geographical research with my private undertaking, I commenced my journey at the ford where the road crosses the Hume to Port Phillip, and in so doing connected the whole of the waters of the south-east angle of the Australian continent.

In this instance, however, as in those to which I have already alluded, no progress was made in advancing our knowledge of the more central part of the continent.

In the year 1839 Mr. Eyre, now Lieutenant-Governor of New Zealand, fitted out an expedition and under the influence of the most praiseworthy ambition, tried to penetrate into the interior from Mount Arden; but, having descended into the basin of Lake Torrens, he was baffled at every point. Turning, therefore, from that inhospitable region, he went to Port Lincoln, from whence he proceeded along the line of the south coast to Fowler's Bay, the western limit of the province of South Australia.

He then determined on one of those bold movements, which characterise all his enterprises, and leaving the coast, struck away to the N.E. for Mount Arden along the Gawler Range; but the view from the summit of that rugged line of hills, threw darkness only on the view he obtained of the distant interior, and he returned to Adelaide without having penetrated further north than 29°30', notwithstanding the unconquerable perseverance and energy he had displayed.

In the following year, the colonists of South Australia, with the assistance of the local government, raised funds to equip another expedition to penetrate to the centre of the continent, the command of which was entrusted to the same dauntless officer. On the morning on which he was to take his departure, from the fair city of Adelaide, Colonel Gawler, the Governor, gave a breakfast, to which he invited most of the public officers and a number of the colonists, that they might have the opportunity of thus collectively bidding adieu to one who had already exerted himself so much for the public good.

Few, who were present at that breakfast will ever forget it, and few who were there present, will refuse to Colonel Gawler the mead of praise due to him, for the display on that occasion of the most liberal and generous feelings. It was an occasion on which the best and noblest sympathies of the heart were roused into play, and a scene during which many a bright eye was dim through tears.

Some young ladies of the colony, amongst whom were Miss Hindmarsh and Miss Lepson, the one the daughter of the first Governor of the province, the other of the Harbour-master, had worked a silken union to present to My. Eyre, to be unfurled by him in the centre of the continent, if Providence should so far prosper his undertaking, and it fell to my lot, at the head of that fair company, to deliver it to him.

When that ceremony was ended, prayers were read by the Colonial Chaplain, after which Mr. Eyre mounted his horse, and escorted by a number of his friends, himself commenced a journey of almost unparalleled difficulty and privation*—a journey, which, although not successful in its primary objects, yet established the startling fact, that there is not a single watercourse to be found on the South coast of Australia, from Port Lincoln to King George's Sound, a distance of more than 1500 miles. To what point then, let me ask, does the drainage of the interior set? It is a question of deep interest to all—a question bearing strongly on my recent investigations, and one that, in connection with established facts, will, I think, enable the reader to draw a reasonable conclusion, as to the probable character of the country, which is hid

* Journals of Expeditions of Discovery into Central Australia, and Overland from Adelaide to King George's Sound, in the years 1840 and 41, by E. J. Eyre, Esq.

from our view by the adamantine wall which encircles the great Australian bight.

On this long and remarkable journey, Mr. Eyre again found it impossible to penetrate to the north, but steadily advancing to the westward, he ultimately reached the confines of Western Australia, with one native boy, and one horse only. Neither, however, did this tremendous undertaking throw any light on the distant interior, and thus it almost appeared that its recesses were never to be entered by civilized man.

From this time neither the government of South Australia, or that of New South Wales, made any further effort to push geographical inquiry, and all interest in it appeared to have past away.

It remains for me to observe, however, that, whilst these attempts were being made to prosecute inland discovery, Her Majesty's naval service was actively employed upon the coast. Captain Wickham, in command of the Beagle, was carrying on a minute survey of the intertropical shores of the continent, which led to the discovery of two considerable rivers, the Victoria and the Albert, the one situated in lat. 14°26'S. and long. 139°22'E, the other in lat. 17°35' and long. 139°54'; but in tracing these up to lat. 15°30' and 17°58', and long. 130°50' and 139°28" respectively, no elevated mountains were seen, nor was any opening discovered into the interior. Captain Wickham having retired, the command of the Beagle devolved on Lieut. now Captain Stokes, to whose searching eye the whole of the coast was more or less subjected, and who approached nearer to the centre than any one had ever done before,* but still no light was thrown on that hidden region; and the efforts which had been made both on land and by water, were, strictly speaking, unsuccessful, to push to any conclusive distance from the settled districts on the one hand, or from the coast into the interior on the other. Reasoning was lost in conjecture, and men, even those most interested in it, ceased to talk on the subject.

* Discoveries in Australia, and Expeditions into the Interior, surveyed during the Voyage of H.M.S. Beagle, between the years 1837 and 43, by Captain J. Lort Stokes.

It may not be of any moment to the public to be made acquainted with the cause which led me, after a repose of more than fourteen years, to seek the field of discovery once more. It will be readily admitted, that from the part, as I have observed in my preface, which I had ever taken in the progress of Geographical Discovery on the Australian continent, I must have been deeply interested in its further development.

I had adopted an impression, that this immense tract of land had formerly been an archipelago of islands, and that the apparently boundless plains into which I had descended on my former expeditions, were, or rather had been, the sea-beds of the channels, which at that time separated one island from the other; it was impossible, indeed, to traverse them as I had done, and not feel convinced that they had at one period or the other been covered by the waters of the sea. It naturally struck me, that if I was correct in this conjecture, the difficulty or facility with which the interior might be penetrated, would entirely depend on the breadth and extent of these once submarine plains, which in such case would now separate the available parts of the continent from each another, as when covered with water they formerly separated the islands. This hypothesis, if I may so call it, was based on observations which, however erroneous they may appear to be, were made with an earnest desire on my part to throw some light on the apparently anomalous structure of the Australian interior. No one could have watched the changes of the country through which he passed, with more attention than did I—not only from a natural curiosity, but from an anxious desire to acquit myself to the satisfaction of the Government by which I was employed.

When Mr. Oxley, the first Surveyor-General of New South Wales, a man of acknowledged ability and merit, pushed his investigations into the interior of that country, by tracing down the rivers Lachlan and Macquarie, he was checked in his progress westward by marshes of great extent, beyond which he could not see any land. He was therefore led to infer that the interior, to a certain extent, was occupied by a shoal sea, of which the marshes were the borders, and into which the rivers he had been tracing discharged themselves.

My friend, Mr. Allan Cunningham, who was for several years resident in New South Wales, and who made frequent journeys into the interior of the continent as botanist to his late Majesty King George IV, and who also accompanied Captain P. P. King, during his survey of its intertropical regions, if he did not accompany Mr. Oxley also on one of his expeditions, strongly advocated the hypothesis of that last-mentioned officer; but as Mr. Cunningham kept on high ground on his subsequent excursions, he could not on such occasions form a correct opinion as to the nature of the country below him. His impressions were however much influenced by the observations made by Captain King in Cambridge Gulf, the water of which was so much discoloured, as to lead that intelligent and careful officer to conclude, that it might prove to be the outlet of the waters of the interior, and hence a strong opinion obtained, that the dip of the continent was in the direction of that great inlet, or to the W.N.W. I therefore commenced my investigations, under an impression that I should be led to that point, in tracing down any river I might discover, and that sooner or later I should be stopped by a large body of inland waters. I descended rapidly from the Blue Mountains, into a level and depressed interior, so level indeed, that an altitude of the sun, taken on the horizon, on several occasions, approximated very nearly to the truth. The circumference of that horizon was unbroken, save where an isolated hill rose above it, and looked like an island in the ocean.

When I reached the point at which Mr. Oxley had been checked, I found the Macquarie, not "running bank high," as he describes it, but almost dry; and although ten years had passed since his visit to this distant spot, the grass had not yet grown over the foot-path, leading from his camp to the river; nor had a horse-shoe that was found by one of the men lost its polish. In this locality there are two hills, to which Mr. Oxley gave the names of Mount Harris and Mount Foster, distant from each other about five miles, on a bearing of 45° to the west of south. Of these two hills Mount Foster is the highest and the nearest, and as the Macquarie runs between them to the westward, it must also be closer than Mount Harris to the marshes. I therefore naturally looked for any discovery that was to be made from Mount Foster, and I accordingly ascended that hill just as the sun was setting. I looked in vain however for the region of reeds and of water, which Mr. Oxley had seen to the westward; so different in character were the seasons, and the state of the country at the different periods in which the Surveyor-General and I visited it. From the highest point I could gain I watched the sun descend; but I looked in vain for the glittering of a sea beneath him, nor did the sky assume that glare from reflected light which would have accompanied his setting behind a mass of waters, I could discover nothing to intercept me in my course. I saw, it is true, a depressed and dark region in the line of the direction in which I was about to go. The terrestrial line met the horizon with a sharp and even edge, but I saw nothing to stay my progress, or to damp my hopes. As I had observed the country from Mount Foster, so I found it to be when I advanced into it. I experienced little difficulty therefore in passing the marshes of the Macquarie, and in pursuing my course to the N.W. traversed plains of great extent, until at length I gained the banks of the Darling, in lat. 30°S. and in long. 146°E. This river, instead of flowing to the N.W. led me to the S.W.; but I was ultimately obliged to abandon it in consequence of the saltness of its waters, I could not, however, fail to observe that the plains over which I had wandered were wholly deficient in timber of any magnitude or apparently of any age, excepting the trees which grew along the line of the rivers; that the soil of the plains was sandy, and the productions almost exclusively salsolaceous. Their extreme depression, indeed their general level, since they were not more than 250 or 300 feet above the level of the sea, together with their general aspect, instinctively, as it were, led the mind to the conviction that they had, at a comparatively recent period, been covered by the ocean. On my return to the Blue Mountains, and on a closer examination of the streams falling from them into the interior, I observed that at a certain point, and that too nearly on the same meridian, they lost their character as rivers, and soon after gaining the level interior, terminated in marshes of greater or less extent; and I further remarked that at certain points, and that too where the channels of the rivers seemed to change, certain trees, as the swamp oak, casuarina, and others ceased, or were sparingly to be found on the lower country—a fact that may not be of any great importance in itself, but which it is still as well to record. The field, however, over which I wandered on this occasion was too limited to enable me to draw any conclusions applicable to so large a tract of land as the Australian continent. On this, my first expedition, I struck the Darling River twice, 1st, as I have stated in latitude 30°S. and in long. 146°; and 2ndly, in lat. 30°10'0"S., and in long. 147°30'E. From neither of these points was any elevation visible to the westward of that river, but plains similar to those by which I had approached it continued beyond the range of vision or telescope from the highest trees we could ascend; beyond the Darling, therefore, all was conjecture.

At the close of the year 1829, I was again sent into the interior to trace its streams and to ascertain the further course of the Darling. I proceeded on this occasion to the south of Sydney, and intersecting the Murrumbidgee, a river at that time but little known, but which Mr. Hume had crossed, in lat. 35°10', and long. 147°28'30", on his journey to the south coast, at a very early period of discovery, and which thereabouts is a clear, rapid and beautiful stream. I traced it downwards to the west to lat. 34°44', and to long. 143°5'00"E. or thereabouts, having taken to my boats a few miles above the junction of the Lachlan with it, in lat. 34°25'00" and in long. 144°03'E.; having at that point left all high lands 200 miles behind me, and being then in a low and depressed country, precisely similar to that over which I had crossed the previous year. As on the first expedition, so on the present one, I descended rapidly into a country of general equality of surface; reeds grew in extensive patches along the line of the river, but beyond them sandy plains extended, covered with salsolæ of various kinds. From the Murrumbidgee, I passed into the Murray, the largest known river in Australia, unless one of greater magnitude has recently been discovered by Sir Thomas Mitchell to the north.

In lat. 34° and in long. 142°, I arrived, (as I have already had occasion to inform my readers), at the junction of a very considerable stream with the Murray. At this point, being then 200 miles distant from the south coast in a direct line, I was less than 100 feet above the level of the sea; circumstances prevented my examining this new river however for many miles above its junction with the main stream, but coming, as I have elsewhere remarked, direct from the north, and possessing, as it did, all the character and appearance of the Upper Darling, I had no doubt as to its identity; in which case no stronger fact could have been adduced to prove the southerly fall or dip of the interior as far as it had been explored. Proceeding down the Murray, I reached at length the commencement of the great fossil formation, through which that river flows. This immense bed rose gradually before me as I pushed to the westward, and it gained an elevation of from 2 to 250 feet, but on my turning southward, it presented an horizontal and undulating surface, until at the point at which the river enters the Lake Victoria, it suddenly dipped and ceased. The lower part of this formation was entirely composed of Serritullæ, but every description of shell with the bones and teeth of sharks and other animals, have subsequently been found in the upper parts of the bed, the summit of which is in many places covered with oyster shells so little changed by time, as to appear as if they had only just been thrown in a heap on the ground they occupy.

The general appearance of the country through which I had passed, and the numerous deposits of fine sand upon the face of it, like sea dunes, still more convinced me, that, when the events which had produced such a change in the physical structure of the continent took place, a current of some description or other must have swept over the interior from the northward; and that this current had deposited the great fossil bed where it now rests; for I cannot conceive that such a mass and mixture of animal remains could have been heaped together in any other way. From the outline of this bed, it struck me that some natural obstacle or other had checked the detritus, brought down by the current, as sand and gravel are checked and accumulated against a log or other impediment athwart a stream, presenting a gradual ascent on the side next the current and a sudden fall on the other. Such, in truth, is the apparent form of the great fossil bed of the Murray. This idea, which struck me as I journeyed down the river, was strengthened, when at a lower part of it I observed a ridge of coarse red granite, running across the channel of the river, and disappearing under the fossil formation on either side of it. It appeared to me to be probable that this ridge of granite might rise higher in other places, and that stretching across the current as it did, that is to say from west to east, the great accumulation of fossil and other remains had been gradually deposited against it, forming a gradual ascent on the northern side of the ridge, and a precipitous fall upon the other.

I have already observed that at a particular point the rivers of the interior, which I had traced on my first expedition, appeared to lose their character as such, and that they soon afterwards ceased in some extensive marsh, the evaporation and absorption over such extensive surfaces being greater than the supply of water they received. This point is about 250 or 300 feet above the level of the sea, and if we draw a little eastward, from the summit of the fossil formation, and prolong it to the western base of the Blue Mountains, we shall find that it will pass over the marshes of the several rivers falling into the interior, and will strike these rivers where their channels appear to fail, as if that had been the former sea-level.

The impressions I have on this interesting subject are clear enough in my own mind, but they are difficult to explain, and I fear I have but ill expressed myself so as to be understood by my readers. I only wish however to record my own ideas, and if I am in error in any particular, I shall thank any one of the many who are better versed in these matters than myself to correct me.

I have stated in a former part of this chapter, that I undertook a journey to South Australia in 1838. I advert to the circumstance again because it is connected with the present inquiry. After I had turned the north-west angle of the Murray, and had proceeded southwards to latitude 34°26' (Moorundi), where Mr. Eyre has built a residence, I turned from the river to the westward, alone the summit of the fossil formation, which, at the distance of a few miles, was succeeded by sandstone, and this rock again, as we gained the hills, by a fine slate, and this again, as we crossed the Mount Barker and Mount Lofty ranges, by a succession of igneous rocks, of a character and form such as could not but betray to a less experienced geologist even than myself the abundant mineral veins they contained. On descending to the plains of Adelaide I again crossed sandstone, and to my surprise discovered that the city of Adelaide stood on the same kind of fossil formation I had left behind me on the banks of the Murray, and it was on the discovery of this fact that the probability of the Australian continent having once been an archipelago of islands first occurred to me.

A more intimate acquaintance with the opinions of Flinders, as to the probable character of the interior of the continent, from the character and appearance of the coast along the Great Australian Bight; the information I have collected as to the extent of the fossil bed, and my own past experience, have led me to the following general conclusions. That the continent of Australia has been subjected to great changes from subigneous agency, and that it has been bodily raised, if I may so express myself, to its present level above the sea; that, as far as we can judge, the north and N.E. portions of the continent are higher than the southern or S.W. parts of it, and that there has consequently been a current or rush of waters, from the one point to the other—that this current was divided in its progress into branches, by hills, or some other intervening obstacle and that one branch of it, following the line of the Darling, discharged itself into the sea, through the opening between the western shores of Encounter Bay and Cape Bernouilli; that the other, taking a more westerly direction, escaped through the Great Australian Bight. From what I could judge, the desert I traversed is about the breadth of that remarkable line of coast, and I am inclined to think that it (the desert) retains its breadth the whole way, as it comes gradually round to the south, thus forming a double curve, from the Gulf of Carpentaria, on the N.E angle of the continent, to the Great Bight on its south-west coast; but my readers will, as they advance into my narrative, see the grounds upon which I have rested these ideas. If such an hypothesis is correct, it necessarily follows, that the north and north-west coasts of the continent were once separated from the south and east coasts by water; and as I have stated my impression that the current from the north, passed through vast openings, both to the eastward and westward of the province of South Australia, it as necessarily follows, that that province must also have been an island. I hope it will be understood that I started with the supposition that the continent of Australia was formerly an archipelago of islands, but that some convulsion, by which the central land has been raised, has caused the changes I have suggested. It was still a matter of conjecture what the real character of Central Australia really was, for its depths had been but superficially explored before my recent attempt. My own opinion, when I commenced my last expedition, inclined me to the belief, and perhaps this opinion was fostered by the hope that such would prove to be the case, as well as by the reports of the distant natives, which invariably went to confirm it, that the interior was occupied by a sea of greater or less extent, and very probably by large tracts of desert country.

With such a conviction I commenced my recent labours, although I was not prepared for the extent of desert I encountered—with such a conviction I returned to the abodes of civilized man. I am still of opinion that there is more than one sea in the interior of the Australian continent, but such may not be the case. All I can say is, Would that I had discovered such a feature, for I could then have done more upon its waters tenfold, than I was enabled to accomplish in the gloomy and burning deserts over which I wandered during more than thirteen months. My readers, however, will judge for themselves as to the probable correctness of my views, and also as to the probable character of the yet unexplored interior, from the data the following pages will supply. I have recorded my own impressions with great diffidence, claiming no more credit than may attach to an earnest desire to make myself useful, and to further geographical research. My desire is faithfully to record my own feelings and impulses under peculiar embarrassments, and as faithfully to describe the country over which I wandered.

My career as an explorer has probably terminated for ever, and only in the cause of humanity, had any untoward event called for my exertions, would I again have left my home. I wish not to hide from my readers the disappointment, if such a word can express the feeling, with which I turned my back upon the centre of Australia, after having so nearly gained it; but that was an achievement I was not permitted to accomplish.

CHAPTER II.

PREPARATIONS FOR DEPARTURE—ARRIVAL AT MOORUNDI—NATIVE GUIDES—NAMES OF THE PARTY—SIR JOHN BARROW'S MINUTE—REPORTS OF LAIDLEY'S PONDS—CLIMATE OF THE MURRAY—PROGRESS UP THE RIVER—ARRIVAL AT LAKE BONNEY—GRASSY PLAINS—CAMBOLI'S HOME—TRAGICAL EVENTS IN THAT NEIGHBOURHOOD—PULCANTI—ARRIVAL AT THE RUFUS—VISIT TO THE NATIVE FAMILIES—RETURN OF MR. EYRE TO MOORUNDI—DEPARTURE OF MR. BROWNE TO THE EASTWARD.

ENTERTAINING the views I have explained in my last chapter, I wrote in January, 1843, to Lord Stanley, at that time Her Majesty's principal Secretary of State for the Colonies, tendering my services to lead an expedition from South Australia into the interior of the Australian continent. As I was personally unknown to Lord Stanley, I wrote at the same time to Sir Ralph Darling, under whose auspices I had first commenced my career as an explorer, to ask his advice on so important an occasion. Immediately on the receipt of my letter, Sir Ralph addressed a communication to the Secretary of State, in terms that induced his Lordship to avail himself of my offer.

In May, 1844, Captain Grey, the Governor of South Australia, received a private letter from Lord Stanley, referring to a despatch his Lordship had already written to him, to authorise the fitting out of an expedition to proceed under my command into the interior. This despatch, however, did not come to hand until the end of June, but on the receipt of it Captain Grey empowered me to organise an expedition, on the modified plan on which Lord Stanley had determined.

Aware as I was of the importance of the season in such a climate as that of Australia, I had written both to the Secretary of State, and to Sir Ralph Darling, so that I might have time after the receipt of replies from Europe, in the event of my proposals being favourably entertained, to make my preparations, and commence my journey at the most propitious season of the year, but my letter to Sir Ralph Darling unfortunately miscarried, and did not reach him until three months after its arrival in England. The further delay which took place in the receipt of Lord Stanley's despatch, necessarily threw it late in the season before I commenced my preparations for the long and trying task that was before me. By the end of July, however, my arrangements were completed, and my party organised, and only awaited the decision of Mr. John Browne, the younger of two brothers who were independent settlers in the province, whose services I was anxious to secure as the medical officer to the expedition, to fix on the day when it should leave Adelaide.

On the 4th of the month (August), I saw Mr. W. Browne, who informed me that his brother had determined to accept my proposals, and that he would join me with the least possible delay; upon which I felt myself at liberty to make definitive arrangements, and to direct that the main body of the expedition should commence its journey on Saturday, the 10th. On the morning of that day I attended a public breakfast, to which I had been invited by the colonists, at the conclusion of which the party, under the charge of Mr. L. Piesse (who subsequently acted as storekeeper) proceeded to the Dry Creek, a small station about five miles from Adelaide. At that place he halted for the night. Mr. Browne not having yet joined me, I kept Davenport, one of the men, who was to attend on the officers, with a riding horse for his use, and the spring cart (in which the instruments were to be carried), for the purpose of forwarding his baggage to the Murray, on the banks of which the party was to muster.

I have said that on the 10th of August I attended a public breakfast, to which I and my party had been invited by the colonists, on the occasion of our quitting the capital. I may be permitted in these humble pages to express my gratitude to them for the kind and generous sympathy they have ever evinced in my success in life, as well as the delicacy and consideration which has invariably marked the expression of their sentiments towards me. If, indeed, I have been an instrument, in the hands of Providence, in bringing about the speedier establishment of the province of South Australia, I am thankful that I have been permitted to witness the happiness of thousands whose prosperity I have unconsciously promoted. Wherever I may go, to whatever part of the world my destinies may lead me, I shall yet hope one day to return to my adopted home, and make it my resting-place between this world and the next. When I went into the interior I left the province with storm-clouds overhanging it and sunk in adversity. When I returned the sun of prosperity was shining on it, and every heart was glad. Providence had rewarded a people who had borne their reverses with singular firmness and magnanimity. Their harvest fields were bowed down by the weight of grain; their pastoral pursuits were prosperous; the hills were yielding forth their mineral wealth, and peace and prosperity prevailed over the land. May the inhabitants of South Australia continue to deserve and to receive the protection of that Almighty power, on whose will the existence of nations as well as that of individuals depends!

Not having had time as yet to attend to my own private affairs, I was unable to leave Adelaide for a few days after the departure of Mr. Piesse. A similar cause prevented Mr. James Poole, who was to act as my assistant, from accompanying the drays. On the 12th Mr. Browne arrived in Adelaide, when he informed me that he had remained in the country to give over his stock, and to arrange his affairs, to prevent the necessity of again returning to his station. He had now, therefore, nothing to do but to equip himself, when he would be ready to accompany me. When I wrote to Mr. Browne, offering him the appointment of medical officer to the expedition, I was personally unacquainted with him, but I was aware that he enjoyed the respect and esteem of every one who knew him, and that he was in every way qualified for the enterprise in which I had invited him to join. Being an independent settler, however, I doubted whether he could, consistently with his own interests, leave his homestead on a journey of such doubtful length as that which I was about to commence. The spirit of enterprise, however, outweighed any personal consideration in the breast of that resolute and intelligent officer, and I had every reason to congratulate myself in having secured the services of one whose value, under privation, trial, and sickness, can only be appreciated by myself.

The little business still remaining for us to do was soon concluded, and as Mr. Browne assured me that it would not take more than two or three days to enable him to complete his arrangements, I decided on our final departure from Adelaide on the 15th of the month; for having received my instructions I should then have nothing further to detain me. That day, therefore, was fixed upon as the day on which we should start to overtake the party on its road to Moorundi. The sun rose bright and clear over my home on the morning of that day. It was indeed a morning such as is only known in a southern climate; but I had to bid adieu to my wife and family, and could but feebly enter into the harmony of Nature, as everything seemed joyous around me.

I took breakfast with my warm-hearted friend, Mr. Torrens, and his wife, who had kindly invited a small party of friends to witness my departure; but although this was nominally a breakfast, it was six in the afternoon before I mounted my horse to commence my journey. My valued friend, Mr. Cooper, the Judge, had returned to Adelaide early in the day, but those friends who remained accompanied us across the plain lying to the north of St. Clare, to the Gawler Town road, where we shook hands and parted.

We reached Gawler Town late at night, and there obtained intelligence that the expedition had passed Angus Park all well. I also learnt that Mr. Calton, the master of the hotel, had given the men a sumptuous breakfast as they passed through the town, and that they had been cheered with much enthusiasm by the people.

On the 16th we availed ourselves of the hospitality of Mrs. Bagot, whose husband was absent on his legislative duties in Adelaide, to stay at her residence for a night. Nothing however could exceed the kindness of the reception we met from Mrs. Bagot and the fair inmates of her house.

On the 17th we turned to the eastward for the Murray, under the guidance of Mr. James Hawker, who had a station on the river. At the White Hut, Mr. Browne, who had left me at Gawler Town, to see his sister at Lyndoch Valley, rejoined me; and at a short distance beyond it, we overtook the party in its slow but certain progress towards the river.

At the Dust Hole, another deserted sheep station on the eastern slope of the mountains, I learnt that Flood, an old and faithful follower of mine, whom I had added to the strength of the expedition at the eleventh hour, was at the station. He was one of the most experienced stockmen in the colonies, and intimately acquainted with the country. I had sent him to receive over 200 sheep I had purchased from Mr. Dutton, which I proposed taking with me instead of salt meat. He had got to the Dust Hole in safety with his flock, and was feeding them on the hills when I passed. The experiment I was about to make with these animals was one of some risk; but I felt assured, that under good management, they would be of great advantage. Not however to be entirely dependent on the sheep, I purchased four cwt. of bacon from Mr. Johnson of the Reed Beds, near Adelaide, by whom it had been cured; and some of that bacon I brought back with me as sweet and fresh as when it was packed, after an exposure of eighteen months to an extreme of heat that was enough to try its best qualities. I was aware that the sheep might be lost by negligence, or scattered in the event of any hostile collision with the natives; but I preferred trusting to the watchfulness of my men, and to past experience in my treatment of the natives, rather than to overload my drays. The sequel proved that I was right. Of the 200 sheep I lost only one by *coup de soleil.* They proved a very valuable supply, and most probably prevented the men from suffering, as their officers did, from that fearful malady the scurvy.

I had them shorn before delivery, to prepare them for the warmer climate into which I was going. And I may here remark, although I shall again have to allude to it, that their wool did not grow afterwards to any length. It ceased indeed to grow altogether for many months, nor had they half fleeces after having been so long as a year and a half unshorn. I did not see Flood at the Dust Hole; but continuing my journey, entered the belt of the Murray at 1 P.M., and reached Moorundi just as the sun set, after a ride of four hours through those dreary and stunted brushes.

My excellent friend, Mr. Eyre, had been long and anxiously expecting us. Altogether superior to any unworthy feeling of jealousy that my services had been accepted on a field in which he had so much distinguished himself, and on which he so ardently desired to venture again, his efforts to assist us were as ceaseless as they were disinterested. Whatever there was of use in his private store, whether publicly beneficial or for our individual comfort, he insisted on our taking. He had had great trouble in retaining at Moorundi two of the most influential natives on the river to accompany us to Williorara (Laidley's Ponds). Mr. Eyre was quite aware of the importance of such attachées, and had spared no trouble in securing their services. Their patience however had almost given way, and they had threatened to leave the settlement when fortunately we made our appearance, and all their doubts as to our arrival vanished. Nothing but jimbucks (sheep) and flour danced before their eyes, and they looked with eager impatience to the approach of the drays.

These two natives, Camboli and Nadbuck, were men superior to their fellows, both in intellect and in authority. They were in truth two fine specimens of Australian aborigines, stern, impetuous, and determined, active, muscular, and energetic. Camboli was the younger of the two, and a native of one of the most celebrated localities on the Murray. It bears about N.N.E. from Lake Bonney, where the flats are very extensive, and are intersected by numerous creeks and lagoons. There, consequently, the population has always been greater than elsewhere on the Murray, and the scenes of violence more frequent. Camboli was active, light-hearted, and confiding, and even for the short time he remained with us gained the hearts of all the party.

Nadbuck was a man of different temperament, but with many good qualities, and capable of strong attachments. He was a native of Lake Victoria, and had probably taken an active part in the conflicts between the natives and overlanders in that populous part of the Murray river. He had somewhat sedate habits, was restless, and exceedingly fond of the *fair* sex. He was a perfect politician in his way, and of essential service to us. I am quite sure, that so long as he remained with the party, he would have sacrificed his life rather than an individual should have been injured. I shall frequently have to speak of this our old friend Nadbuck, and will not therefore disturb the thread of my narrative by relating any anecdote of him here. It may be enough to state that he accompanied us to Williorara, even as he had attended Mr. Eyre to the same place only a few weeks before, and that when he left us he had the good wishes of all hands.

In the afternoon of the day following that of our arrival at Moorundi, Mr. Piesse arrived with the drays, and drew them up under the fine natural avenue that occupies the back of the river to the south of Mr. Eyre's residence. Shortly afterwards Davenport arrived with the light cart, having the instruments and Mr. Browne's baggage. Flood also came up with the sheep, so that the expedition was now complete, and mustered in its full force for the first time, and consisted as follows of officers, men, and animals:—

Captain Sturt, *Leader.*
Mr. James Poole, *Assistant.*
Mr. John Harris Browne, *Surgeon.*
Mr. M'Dougate Stuart, *Draftsman.*
Mr. Louis Piesse, *Storekeeper.*
Daniel Brock, *Collector.*
George Davenport,
Joseph Cowley, *Servants.*
Robert Flood, *Stockman.*
David Morgan, *with horses.*
Hugh Foulkes,
John Jones
John Turpin, *Bullock drivers.*
William Lewis, *sailor,*
John Mack,
John Kerby, *with sheep.*
11 horses; 30 bullocks; 1 boat and boat carriage; 1 horse dray; 1 spring cart; 3 drays, 200 sheep; 4 kangaroo dogs; 2 sheep dogs.

The box of instruments sent from England for the use of the expedition had been received, and opened in Adelaide. The most important of them were two sextants, three prismatic compasses, two false horizons, and a barometer. One of the sextants was a very good instrument, but the glasses of the other were not clear, and unfortunately the barometer was broken and useless, since it had the syphon tube, which could not be replaced in the colony. I exceedingly regretted this accident, for I had been particularly anxious to carry on a series of observations, to determine the level of the interior. I manufactured a barometer, for the tube of which I was indebted to Captain Frome, the Surveyor-General, and I took with me an excellent house barometer, together with two brewer's thermometers, for ascertaining the boiling point of water on Sykes' principle. The first of the barometers was unfortunately broken on the way up to Moorundi, so that I was a second time disappointed.

It appears to me that the tubes of these delicate instruments are not secured with sufficient care in the case, that the corks placed to steady them are at too great intervals, and that the elasticity of the tube is consequently too great for the weight of mercury it contains. The thermometers sent from England, graduated to 127° only, were too low for the temperature into which I went, and consequently useless at times, when the temperature in the shade exceeded that number of degrees. One of them was found broken in its case, the other burst when set to try the temperature, by the over expansion of mercury in the bulb.

The party had left Adelaide in such haste that it became necessary before we should again move, to rearrange the loads. On Monday, the 18th, therefore I desired Mr. Piesse to attend to this necessary duty, and not only to equalize the loads on the drays, and ascertain what stores we had, but to put everything in its place, so as to be procured at a moment's notice.

The avenue at Moorundi presented a busy scene, whilst the men were thus employed reloading the drays and weighing the provisions. Morgan, who had the charge of the horse cart, had managed to snap one of the shafts in his descent into the Moorundi Flat, and was busy replacing it. Brock, a gunsmith by trade, was cleaning the arms. Others of the men were variously occupied, whilst the natives looked with curiosity and astonishment on all they saw. At this time, however, there were not many natives at the settlement, since numbers of them had gone over the Nile, to make their harvest on the settlers.

On Monday I sent Flood into Adelaide with despatches for the Governor, and with letters for my family, as well as to bring out some few trifling things we had overlooked, and as Mr. Piesse reported to me on that day that the drays were reloaded, I directed him, after I had inspected them, to lash down the tarpaulines, and to warn the men to hold themselves in readiness to proceed on their journey at 8 A.M. on the following morning—for, as I purposed remaining at Moorundi with Mr. Eyre until Flood should return, I was unwilling that the party should lose any time, and I therefore thought it advisable to send the drays on, under Mr. Poole's charge, until such time as I should overtake him. The spirit which at this time animated the men ensured punctuality to any orders that were given to them. Accordingly the bullocks were yoked up, and all hands were at their posts at early dawn. As, however, I was about to remain behind for a few days, it struck me that this would be a favourable opportunity on which to address the men. I accordingly directed Mr. Poole to assemble them, and with Mr. Eyre and Mr. Browne went to join him in the flat, a little below the avenue. I then explained to them that I proposed remaining at Moorundi for a few days after their departure. I thought it necessary, in giving them over into Mr. Poole's charge, to point out some of the duties I expected from them.

That in the first place I had instructed Mr. Poole to mount a guard of two men every evening at sunset, who were to remain on duty until sun-rise; that I expected the utmost vigilance from this guard, and that as the safety of the camp would depend on their attention, I should punish any neglect with the utmost severity. I then adverted to the natives, and interdicted all intercourse with them, excepting with my permission. That as I attributed many of the acts of violence that had been committed on the river to this irritating source, so I would strike the name of any man who should disobey my orders in this respect off the strength of the party from that moment, and prevent his receiving a farthing of pay; or whoever I should discover encouraging any of the natives, but more particularly the native women, to the camp. I next drew the attention of the men to themselves, and pointed out to them the ill effects of discord, expressing my hope that they would be cheerful and ready to assist one another, and that harmony would exist in the camp; that I expected the most ready obedience from all to their superiors; and that, in such case, they would on their part always find me alive to their comforts, and to their interests. I then confirmed Mr. Piesse in his post as store-keeper; gave to Flood the general superintendence of the stock; to Morgan the charge of the horses, and to each bullock-driver the charge of his own particular team. To Brock I committed the sheep, with Kirby and Sullivan to assist, and to Davenport and Cowley (Joseph) the charge of the officers' tents. I then said, that as they might now be said to commence a journey, from which none of them could tell who would be permitted to return, it was a duty they owed themselves to ask the blessing and protection of that Power which alone could conduct them in safety through it; and having read a few appropriate prayers to the men as they stood uncovered before me, I dismissed them, and told Mr. Poole he might move off as soon as he pleased. The scene was at once changed. The silence which had prevailed was broken by the cracks of whips, and the loud voices of the bullock-drivers. The teams descended one after the other from the bank on which they had been drawn up, and filed past myself and Mr. Eyre, who stood near me, in the most regular order. The long line reached almost across the Moorundi flat, and looked extremely well. I watched it with an anxiety that made me forgetful of everything else, and I naturally turned my thoughts to the future. How many of those who had just passed me so full of hope, and in such exuberant spirits, would be permitted to return to their homes? Should I, their leader, be one of those destined to remain in the desert, or should I be more fortunate in treading it than the persevering and adventurous officer whose guest I was, and who shrank from the task I had undertaken. My eyes followed the party as it ascended the gully on the opposite side of the flat, and turned northwards, the two officers leading, until the whole were lost to my view in the low scrub into which it entered. I was unconscious of what was passing around me, but when I turned to address my companions, I found that I was alone. Mr. Eyre, and the other gentlemen who had been present, had left me to my meditations. In the afternoon Kusick, one of the mounted police, arrived with despatches from the Governor, and letters from my family. He had met Flood at Gawler Town, whose return, therefore, we might reasonably expect on the Friday.

Amongst the first purchases that had been made was a horse for the service of the expedition, which had not very long before been brought in from Lake Victoria, Nadbuck's location, distant nearly 200 miles from Adelaide, where he had been running wild for some time. This horse was put into the government paddock at Adelaide when bought, but he took the fence some time during the night and disappeared, nor could he be traced anywhere. Luckily, however, Kusick had passed the horses belonging to the settlers at Moorundi, feeding at the edge of the scrub upon the cliffs, and amongst them had recognised this animal, which had thus got more than 90 miles back to his old haunt. He had, however, fallen into a trap, from which I took care he should not again escape; but we had some difficulty in running him in and securing him. Prior to the departure of the expedition from Adelaide, a considerable quantity of rain had fallen there. Since our arrival at Moorundi also we could see heavy rain on the hills, although no shower fell in the valley of the Murray. Kusick informed us that he had been in constant rain, and it was evident, from the dense and heavy clouds hanging upon them, that it was still pouring in torrents on the ranges. We feared, therefore, and it eventually proved to be the case, that Flood would not be able to cross the Gawler on his return to us. He was, in fact, detained a day in consequence of the swollen state of that little river, but swam his horse over on the following day, at considerable risk both to himself and his animal. He did not, in consequence, reach us until Saturday. In anticipation, however, of his return on that day, we had sent Kenny, the policeman stationed at Moorundi who was to accompany Mr. Eyre, up the river in advance of us at noon, with Tampawang, the black boy I intended taking with me, and had everything in readiness to follow them, as soon as Flood should arrive. He did not, however, reach Moorundi until 5 P.M. It took me some little time to reply to the communications he had brought, but at seven we mounted our horses, and leaving Flood to rest himself, and to exchange his wearied animal for the one we had recovered, with Tenbury in front, left the settlement. The night was cold and frosty, but the moon shone clear in a cloudless sky, so that we were enabled to ride along the cliffs, from which we descended to one of the river flats at 1 A.M. and, making a roaring fire, composed ourselves to rest.

It may here be necessary, before I enter on any detail of thc proceedings of the expedition, to explain the general nature of my instructions, the object of the expedition, and the reasons why, in some measure, contrary to the opinion of the Secretary of State, I preferred trying the interior by the line of the Darling, rather than by a direct northerly route from Mount Arden.

As the reader will have understood, I wrote, in the year 1843, to Lord Stanley, the then colonial minister, volunteering my services to conduct an expedition into Central Australia. It appeared to his Lordship as well as to Sir John Barrow, to whom Lord Stanley referred my report, that the plan I had proposed was too extensive, and it was therefore determined to adopt a more modified one, and to limit the resources of the expedition and the objects it was to keep in view, to a certain time, and to the investigation of certain facts. After expressing his opinion as to the magnitude of the undertaking I had contemplated, "There is, however," says Sir J. Barrow, in a minute to the Secretary of State, "a portion of the continent of Australia, to which he (Captain Sturt) adverts, that may be accomplished, and in a reasonable time and at a moderate expense.

"He says, if a line be drawn from lat. 29°30' and long. 146°, N.W., and another from Mount Arden due north, they will meet a little to the northward of the tropic, and there, I will be bound to say, a fine country will be discovered. On what data he pledges himself to the discovery of this fine country is not stated. It may, however, be advisable to allow Mr. Sturt to realize the state of this fine country.

"This, however, is not to be done by pursuing the line of the Darling to the latitude of Moreton Bay, which would lead him not far from the eastern coast, where there is nothing of interest to be discovered, nor does it appear advisable to pursue the Darling to the point to which he and Major Mitchell have already been, for this reason. His preparation will, no doubt, be made at Adelaide; from thence to the point in question is about 600 miles, and from this point to the fine country, a little beyond the tropic, is 700 miles, which together make a journey of 1300 miles. Now a line directly north from Adelaide, through Mount Arden, to the point where it crosses the former in the fine country, is only 80 miles, making a saving, therefore, of 500 miles, which is of no little importance in such a country as Australia.

"But Mr. Sturt assigns reasons for supposing that a range of mountains will be found about the 29th parallel of latitude, and Mr. Eyre, whilst exploring the Lake he discovered to the northward of the Gulf of St. Vincent, Adelaide, notices mountains to the N.E., in about the latitude of 28°. Supposing, then, a range of mountains to exist about that parallel their direction will probably be found to run from N.E. to S.W., which is that generally of the river Darling and its branches; and in this case it may reasonably be concluded that these mountains form the division of the waters and that all the branches of the several rivers (some of them of considerable magnitude) which have been known to fall into the bays and gulfs on the W. and N.W. coasts, between the parallels of 14° and 21°, have their sources on the northern side of this range of mountains; but, even if no such range exists, it is pretty evident, from what we know of the southern rivers, adjuncts chiefly of the Darling, that somewhere about the latitudes of 28° or 29° the surface rises to a sufficient height to cause a division of the waters, those on the northern side taking a northerly direction, and those on the southern side a southerly one.

"To ascertain this point is worthy of a practical experiment in a geographical point of view, as the knowledge of the direction that mountains and rivers lake, the bones and blood vessels of bodies terrestrial give us at least a picture of the body of that skeleton. To these

Mr. Sturt will no doubt direct his particular attention, as constituting the main object of such an expedition, and these, with the great features of the country, its principal productions in the animal and vegetable part of the creation, the state and condition of the original inhabitants, will render a great service to the geography of the southern part of Australia."

On this memorandum the Secretary of State observes, in a private letter to Captain Grey, that came to hand before the receipt of Lord Stanley's public despatch:—

"In considering Sir John Barrow's memorandum, enclosed in my public despatch, you will see that a strong opinion is expressed against ascending the Darling in the first instance, and in favour of making a direct northerly course from Adelaide to Mount Arden. I do not wish this to be taken as an absolute injunction, because I am aware that there may be local causes why the apparently circuitous route may after all be the easiest for the transport of provisions, and may really facilitate the objects of the expedition. In like manner I do not wish to be understood as absolutely prohibiting a return by Moreton Bay, extensive as that deviation would be, if it should turn out that the exploration of the mountain chain led the party so far to the eastward as to be able to reach that point by a route previously known to Captain Sturt or to Major Mitchell, more easily than they could return on their steps down the Darling. What Captain Sturt will understand as absolutely prohibited, is any attempt to conduct his party through the tropical regions to the northward, so as to reach the mouths of any of the great rivers. The present expedition will be limited in its object, to ascertaining the existence and the character of a supposed chain of hills, or a succession of separate hills, trending down from N.E. to S.W., and forming a great natural division of the continent; to examining what rivers take their source in those mountains, and what appears to be their course; to the general lie of the country to the N.W. of the supposed chain; and to the character of the soil and forests, as far as can be ascertained by such an investigation as shall not draw the party away from their resources, and shall make the south the constant base of their operations."

I presume, from the tenor of Sir John Barrow's memorandum, that he was not fully aware of the insurmountable difficulties the course he recommends presented. Valuing his judgment as I did on such an occasion, and anxious as I was to act on the suggestions of the Secretary of State, the strongest grounds could alone have made me pursue a course different to that which had been recommended to me. Certainly the fear of any ordinary difficulty would not have influenced me to reject the line pointed out, but I felt satisfied that if Lord Stanley and Sir John Barrow could be made aware of the nature of the country to the north of Mount Arden, and the reasons why I considered it would be more advantageous to take the line of the Darling, they would have concurred in opinion with me. I would myself much rather have taken the line by Mount Arden, since it would have been a greater novelty, and I would have precluded the chance of any collision with the natives of the Darling, more especially at that point to which I proposed to go, and at which Sir Thomas Mitchell had had a rupture with them in 1836. The journeys of Mr. Eyre had, however, proved the impracticability of a direct northerly course from Mount Arden. Such a course would have led me into the horseshoe of Lake Torrens; and although I might have passed to the westward of it, I could hope for no advantage in a country such as that which lies to the north of the Gawler Range. On the other hand, the Surveyor-General of South Australia had attempted a descent into the interior from the eastward, and had encountered great difficulties from the want of water. Local inquiry and experience both went to prove the little likelihood of that indispensable element being found to the north of Spencer's Gulf. It appeared to me also that Sir John Barrow had mistaken the point on the Darling to which I proposed going. It was not, as he seems to have conjectured, to any point to which I had previously been, but to an intermediate one. It is very true that if I had contemplated pushing up the Darling to Fort Bourke, the distance would have been 600 miles, and that, too, in a direction contrary to the one in which I was instructed to proceed; but to Laidley's Ponds, in lat. 32°26'0"S. and long. 142°30'W., (the point to which I proposed to go) the distance would have been a little more than 300 miles. It was from this point that Sir Thomas Mitchell retreated after his rupture with the natives in 1836; because, as he himself informs us, he just then ascertained that a small stream joined the Darling from the westward a little below his camp, and he likewise saw hills in the same direction.

In consequence of the inhospitable character of the country to the north, I had turned my attention to the above locality, and had been assured by the natives, both of the Murray and the Darling, that the Williorara (Laidley's Ponds) was a hill stream, that it came far from the N.W., that it had large fish in it, and that its banks were grassy. It struck me, therefore, that it would be a much more eligible line for the expedition to run up the Darling to lat. 32°26', and then to trace the Williorara upwards into the hills, with the chance of meeting the opposite fall of waters, rather than to entangle myself and waste my first energies amidst scrub and salt lagoons. As I understood my instructions and the wishes of the Secretary of State, I was to keep on the 138th meridian (that of Mount Arden) until I should reach the supposed chain of mountains, the existence of which it was the object of Lord Stanley to ascertain, or until I was turned aside from it by some impracticable object. Lake Torrens being due north of Mount Arden would, if I had taken that line, have been direct in my way, and I should have had to turn either its eastern or its western flank. The Surveyor-General, Captain Frome, had tried the former, but although he went considerably to the eastward into the low and desert interior before he turned northwards, he still found himself entangled in that sandy basin, so that it appeared to me that I should do little more than clear it on the course I proposed to take.

As the reader, however, will learn in the perusal of these pages, I was wholly disappointed in the character of the Williorara. Where that channel joins the Darling, the upward course of that river is to the north-east; and as that was a course directly opposite to the cue I felt myself bound to take, I abandoned it and took at once to the hills. At my Depôt Prison, in lat. 29°40', and in long. 141°30'E., I hoped that we had sufficiently cleared the north-east limit of Lake Torrens; but when on the fall of rain we resumed our labours, we measured 131¾ miles with the chain before we arrived on the shore of a vast sandy basin, which I would not cross, and to the northward of which I could not penetrate. Thus disappointed in my attempt to gain the 138th meridian on a westerly course, as well as in my anticipation of finding Lake Torrens connected with some more central feature, it appeared to me that I could not follow out my instructions better than by attempting to penetrate towards the centre of the continent on a north-west course, for it was clear that if there were any ranges or any mountain chains traversing the interior from north-east to south-west I should undoubtedly strike them; but that if no such chains existed the proposed course would take me to the Tropic on the meridian of 138°, and would enable me to determine the character of the interior, and more central regions of the continent. In this attempt I succeeded in gaining the desired meridian, but failed in reaching the Tropic. My position was about 500 miles north of Mount Arden, 60 miles from the Tropic, and somewhat less than 150 to the eastward of the centre of the Australian continent. Forced back to my depôt a second time, from the total failure both of water and grass, in the quarter to which I had penetrated with the above objects in view, having passed the centre in point of latitude, I again left it on a due north course to ascertain if there were any ranges or hills between my position and the Gulf of Carpentaria, as well as to satisfy myself as to the character and extent of a stony desert I had crossed on my last excursion. That iron region however again stopped me in my progress northwards, and obliged me to fall back on a place of safety. For fourteen months I kept my position in a country which never changed but for the worse and from which it was with difficulty that I ultimately escaped; but as the minuter details of the expedition will be given in the subsequent pages of this work, any mention of them here would be superfluous. I shall only express my regret that we were unable to make the centre or to gain the Tropic. As regards the objects for which the expedition was fitted out, I hope it will be granted that they were accomplished, and that little doubt can now be entertained as to the non-existence of the mountain chains, the supposed existence of which I was sent to ascertain. It would, however, have gratified me exceedingly to have crossed into the Tropic, to have decided my own hypothesis as to the fine country I ventured to predict would be found to exist beyond it. My reasons for supposing which I thought I had explained in my first letter to the Secretary of State, but as it would appear from an observation in Sir John Barrow's memorandum, that I had not done so, I deem it right briefly to record them here.

I had observed on my first expedition to the Darling, in 1828, when in about lat. 29°30'S. that the migration of the different kinds of birds which visit the country east of the Darling during the summer, was invariably to the W.N.W. Cockatoos and parrots that whilst staying in the colony were known to frequent elevated land, and to select the richest and best watered valleys for their temporary location, passed in flights of countless number to the above-mentioned point. I had also observed, during my residence in South Australia, that several of the same kind of birds annually visited it, and that they came directly from the north. I had seen the *Psytacus Novæ Hollandiæ* and the *Shell Parroquet* following the line of the shore of St. Vincent Gulf like flights of starlings in England, and although intervals of more than a quarter of an hour elapsed between the passing of one flight and that of another, they all came from the north and followed in the same direction. Now, although I am quite ready to admit that the casual appearance of a few strange birds should not influence the judgment, yet I think that a reasonable inference may be drawn from the regular and systematic migration of the feathered races. Now, if we were to draw a line from Fort Bourke to the W.N.W., and from Mount Arden to the north, we should find that they would meet a little to the northward of the Tropic, and as I felt assured of two lines of migration thus tending to the same point, there could be little doubt but that the feathered races migrating upon them rested at that point, for a time, so I was led to conclude that the country to which they went would in a great measure resemble that which they had left—that birds which delighted in rich valleys or kept on lofty hills, surely would not go into deserts and into a flat country; and therefore it was that I was led to hope, that as the fact of large migrations from various parts of the continent to one particular part, seemed to indicate the existence either of deserts or of water to a certain distance, so the point at which migration might be presumed to terminate would be found a richer country than any which intervened. On the late expedition, I accidentally fell into the line of migration to the north-west, and birds that I was aware visited Van Diemen's Land passed us, after watering, to that point of the compass. Cockatoos would frequently perch in our trees at night, and wing their way to the north-west after a few hours of rest; and to the same point wild fowl, bitterns, pigeons, parrots, and parroquets winged their way, pursued by numerous birds of the Accipitrine class. From these indications I was led still more to conclude that I might hope for the realization of my anticipations if I could force my own way to the necessary distance.

During our stay at Moorundi, the weather had been beautifully fine, although it rained so much in the hills. A light frost generally covered the ground, and a mist rose from the valley of the Murray at early dawn; but both soon disappeared before the sun, and the noon-day temperature was delicious—nothing indeed could exceed the luxury of the climate of that low region at that season of the year, August.

We had directed Kenny, the policeman, and Tampawang, to bivouac in the valley in which we ourselves intended to sleep, but we saw nothing of them on our arrival there. The night was bitter cold, insomuch that we would hardly keep ourselves warm, notwithstanding that we laid under shelter of a blazing log. As dawn broke upon us, we prepared for our departure, being anxious to escape from the misty valley to the clearer atmosphere on the higher ground. At eight A.M. we passed the Great Bend of the Murray, and I once more found myself riding over ground every inch of which was familiar to me, since not only on my several journeys down and up the river had I particularly noticed this spot, but I had visited it in 1840 with Colonel Gawler, the then Governor of South Australia; who, finding that he required relaxation from his duties, invited me to accompany him on an excursion he proposed taking to the eastward of the Mount Lofty Range, for the purpose of

N.W. angle of the Murray. Colonel Gawler's Camp.

examining the country along the shores of Lake Victoria and the River Murray, as far as the Great Bend. It was a part of the province at that time but little known save by the overlanders, and the Governor thought that by personally ascertaining the capabilities of the country contiguous to the Murray, he might throw open certain parts of it for location. Being at that time Surveyor-General of the Province, I was glad of such an opportunity to extend my own knowledge of the province to the north and north-east of Adelaide, more especially as this journey gave me an opportunity to cross from the river to the hills westward of the Great Bend. Not only was the land on the Murray soon afterwards occupied to that point, but Colonel Gawler and I also visited the more distant country on that occasion. Since my return, indeed, from my recent labours, the line of the Murray is occupied to within a short distance of the remoter stations of the colony of New South Wales, and there can be no doubt but that in the course of a few years the stock stations from the respective colonies will meet. I was afraid, when I came the second time down the Murray, that I had exaggerated the number of acres in the valley, but on further examination, it appears to me that I did not do so; for as the traveller approaches Lake Victoria the flats are very extensive, but more liable to inundation than those on the higher points of the river, for being so little elevated above the level of the water, especially those covered with reeds, the smallest rise in the stream affects them. Lake Victoria, although it looks like a clear and open sea, as you look from the point of Pomundi, which projects into it to the south, is after all exceedingly shallow, and is rapidly filling up from the decay of seaweed and the deposits brought into it yearly by the floods of the Murray. No doubt but that future generations will see that fine sheet of water confined to a comparatively narrow bed, and pursuing its course through a rich and extensive plain. When such shall be the case, and that the strength of the Murray shall be brought to bear in one point only, it is probable its sea mouth will be navigable, and that the scenery on this river will be enlivened by the white sails of vessels on its ample bosom. I can fancy that nothing would be more beautiful than the prospect of vessels, however small they might be, coming with swelling sails along its reaches. It may, however, be said, that it will be a distant day when such things shall be realized. There is both reason and truth in the remark; but Time, with his silent work, has already raised the flats in the valley of the Murray, and as we are now benefiting by his labours, so it is to be hoped will our posterity. However that may be, for it is a matter only of curious speculation, nothing will stay the progress of improvement in a colony which has received such an impulse as the province of South Australia. As men retain their peculiarities, so, I believe, do communities; and where a desirable object is to be gained, I shall be mistaken if it is lost from a want of spirit in that colony. Purposing, however, to devote a few pages to the more particular notice of the state of South Australia, and the prospects it holds out to those who may desire to seek in other lands more comforts and a better fortune than they could command in their native country, I shall not here make any further observation.

The morning, which had been so cold, gradually became more genial as the sun rose above us, and both Mr. Eyre and myself forgot that we had so lately been shivering, under the influence of the more agreeable temperature which then prevailed.

As we turned the Great Bend of the Murray, and pursued an easterly course, we rode along the base of some low hills of tertiary fossil formation, the summits of which form the table land of the interior. We were on an upper flat, and consequently considerably above the level of the water as it then was. In riding along, Tenbury pointed out a line of rubbish and sticks, such as is left to mark the line of any inundation, and he told us, that, when he was a boy, he recollected the floods having risen so high in the valley as to wash the foot of these hills. He stated, that there had been no previous warning; that the weather was beautifully fine, and that no rain had fallen; and he added that the natives were ignorant whence the water came, but that it came from a long way off. According to Tenbury's account, the river must have been fully five and twenty feet higher than it usually runs; and judging from his age, this occurrence might have taken place some twenty years before. As we proceed up the Darling, we shall see a clue to this phenomenon. But why, it may be asked, do not such floods more frequently occur? Is it that the climate is drier than it once was, and that the rains are less frequent? There are vestiges of floods over every part of the continent; but the decay of debris and other rubbish is so slow, that one cannot safely calculate how long it may have been deposited where they are so universally to be found.

After passing the Great Bend, as I have already stated, we turned to the eastward and overtook Mr. Poole at noon, not more than eight miles distant. Some of the bullocks had strayed, and he had consequently been prevented from starting so early as he would otherwise have done. The animals had, however, been recovered before we reached the party, and were yoked up; we pushed on therefore to a distance of nine miles, cutting across from angle to angle of the river, but ultimately turned into one of the flats and encamped for the night. We passed, during the day through some low bushes of cypresses and other stunted shrubs, but they were not so thick as to impede our heavy drays, by the weight of which every tree they came in contact with was brought to the ground. A meridian altitude of Vega placed us in lat. 34°4'20"S., by which it appeared that we had made four miles of southing, the Great Bend being in lat. 34°. Kenny and Tampawang had joined the party before we overtook it, and Flood arrived in the course of the afternoon. The cattle had an abundance of feed round our tents, and near a lagoon at the upper end of the flat. The thermometer stood at 40° at 7 P.M., with the wind at west.

On the morning of the 26th we availed ourselves of the first favourable point to ascend from the river flats to the higher ground, since it prevented our following the windings of the river and shortened our day's journey. In doing this we sometimes travelled at a considerable distance from the Murray—the surface of the country was undulating and sandy, with clumps of stunted cypress trees, and eucalyptus dumosa scattered over it. Low bushes of rhagodia, at great distances apart, were growing on the more open ground; the soil, consisting of a red clay and sand, only superficially covering the fossil formation beneath it. At 11 A.M. we entered a dense brush of cypress and eucalypti growing in pure sand. Fortunately for us the overlanders had cut a passage through it, so that we had a clear road before us, but the drays sunk deep into the loose sand in which these trees were growing, and the bullocks had a constant strain on the yoke for six miles. We then broke into more open ground, and ultimately reached the river in sufficient time to arrange the camp before sunset, although we had 2 1/2 miles to travel on a S.W. course before we found a convenient place to stop at. Our course during the day having been S.S.E., we had thus been obliged to turn back upon it, but this was owing to the direction the river here takes and was unavoidable. At 6 P.M. the thermometer stood at 55° of Farenheit, the barometer at 30.000, and the boiling point of water by two thermometers with a difference of 2°212' and 214', respectively, our distance from the sea coast being about 180 miles as the crow flies.

It was generally thought in Adelaide that having started so late in the season, I should experience some difficulty in getting feed for the cattle. From my experience, however, of the seasons in the low region through which the Murray flows, I had no such anticipation. The only fear I had, was, that we should be shut out from flats of the river by the floods, as I knew it would be on the rise at the time we should be upon it. To this point, however (and I may add, with few exceptions), we found an abundance of feed, both along the line of the Murray and the Darling, but at our present encampment our animals fared very indifferently, in consequence of the poor nature of the soil. Our tents were pitched at the northern extremity of a long flat, between the river and a serpentine lagoon, which left but a narrow embankment between itself and the stream. The soil of the flat was a cold white clay, on which there was scarcely any vegetation, so that the cattle wandered and kept us about an hour after our appointed hour of starting. There had been a sharp frost during the night, and the morning was bitterly cold. At sunrise the thermometer stood at 29° the dew point being 43° and the barometer at 29.700.

When we left this place, our course, for the first three miles, was along the embankment separating the river from the lagoon, and I remarked that although there was so little vegetation on the ground, there were some magnificent trees on the bank of the river itself, which gradually came up to the north-east. At three miles, however, our farther course along the flats was checked by the hills of fossil formation, which approached the river so closely as to leave no passage for the drays between it and them. We were, therefore, obliged to ascend to the upper levels, in doing so we were also obliged to put two teams, or sixteen bullocks, to each dray, and even then found it difficult to master the ascent.

Referring back to a previous remark, I would observe that the Murray river is characterised by bold and perpendicular cliffs of different shades of yellow colour, varying from a light hue to a deep ochre. These cliffs rise abruptly from the water to the height of 250 and occasionally 300 feet. They occur first on one side of the river, and then on the other, there being an open or a lightly-timbered flat on the opposite side, with a line of trees almost invariably round it, especially along the river. These flats are backed, at uncertain distances, by the fossil formation, as by a natural inclosure—sometimes it rises perpendicularly from the flats, but more generally assumes the character of sloping hills. The cliffs occasionally extend, like a wall, along the river for two or three miles, and look exceedingly well; but their constant recurrence, at length fatigues the eye. At the point at which we had now arrived this remarkable formation ceases, or, as we are going up the river, I should perhaps be more correct if I said, begins. Above it a long line of hills, broken by deep and rugged stony gullies, and with steep sides, extends to the eastward (that also being the upward course of the river). On gaining the crest of these hills we found ourselves, as usual, on a flat table land, notwithstanding the broken faces of the hills themselves. There was only a narrow space between them, and a low thick brush of eucalyptus to the north. The soil was, as usual, a mixture of clay and sand, with small rounded nodules of limestone. From this ground, the view to the south as a medium point, was over as dark and monotonous a country as could well be described. There was not a single break in its sombre hue, nor was there the slightest rise on the visible horizon; both to the eastward and westward we caught glimpses of the Murray glittering amid the dark foliage beneath us, but it made no change in the character of the landscape.

We kept on the open ground, just cutting the heads of the gullies, and advanced eight miles before we found a convenient spot at which to drive the cattle down to water, and feed in the flats below, and into which it appeared impracticable to get our drays. I halted, therefore, on the crest of the hills, and sent Flood and three other men to watch the animals, and to head them back if they attempted to wander. In the afternoon we went down to the river, and on crossing the flat came upon the dray tracks of some overland party, the leader of which had taken his drays down the hills, notwithstanding the apparent difficulty of the attempt. But what is there of daring or enterprise that these bold and high-spirited adventurers will shrink from?

I had hoped that the more elevated ground we here occupied, would have been warmer than the flats on which we had hitherto pitched our tents, but in this I was disappointed. The night was just as cold as if we had been in the valley of the Murray. At sunrise the thermometer stood at 27°, and we had thick ice in our pails.

At five miles from this place, having left the river about a mile to our right, we arrived at the termination of this line of hills. They gradually fell away to the eastward and disappeared; nor does the fossil formation extend higher up the Murray. It here commences or terminates, as the traveller is proceeding up or down the stream. A meridian altitude on the hill just before we descended, placed it in lat. 34°9'56", so that we had still been going gradually to the south. At the termination of the hills, the Murray forms an angle in turning sharp round to that point, and after an extensive sweep comes up again, so as to form an opposite angle; the distance between the two being 14 or 15 miles, and from the ground on which we stood the head of Lake Bonney bore E.5°S., distant six miles.

On descending from these hills we fell into the overland road, but were soon turned from it by reason of the floods, and obliged to travel along a sandy ridge, forming the left bank of a lagoon, running parallel to the river, into which the waters were fast flowing; but finding a favourable place to cross, at a mile distant, we availed ourselves of it, and encamped on the river side. In the afternoon we had heavy rain from the west. During it, Mr. James Hawker, a resident at Moorundi, joined us, and took shelter in our tents. He had, indeed, kept pace with us all the way from the settlement in his boat, and supplied us with wild fowl on several occasions.

We had showers during the night, but the morning, though cloudy, did not prevent our moving on to Lake Bonney, distant, according to our calculation, between four and five miles. To determine this correctly, however, I ordered Mr. Poole to run the chain from the river to the lake. We had seen few or no natives as yet; but expecting to find a large party of them assembled at Lake Bonney, Mr. Eyre went before us with Kenny and Tenbury, leaving Nadbuck and Camboli to shew us the most direct line to the mouth of the little channel which connects Lake Bonney with the Murray, at which I purposed halting. The greater part of our way was through deep sandy cypress brushes, so that the cattle had a heavy pull of it. We reached our destination at 1 P.M., where we found Mr. Eyre, with eight or nine natives, all, who were then in the neighbourhood.

The back-water of the Murray was fast flowing into the lake, which already presented a broad expanse of water to the eye. It was covered with wild fowl of various kinds, and there were several patches of reeds in which they were feeding. As I purposed stopping for a day or two, to rest the bullocks, I directed Mr. Poole to survey the lake, whilst I undertook to lay down the creek or channel connecting it with the river, in which service I enlisted Mr. Hawker, who had formerly been on the survey, and whose name I gave to the creek on the completion of our work.

Lake Bonney is a shallow sandy basin, which is annually filled by the Murray; and as it rises, so, to a certain extent, it falls with the river, until at length, being left very shallow, it is soon dried up. The Hawker being too small to discharge the water equally with the fall of the river, has a current in it after the river has lowered considerably, for which reason I thought, when I passed it on my second expedition, that it had been a tributary; but such is not the case—Lake Bonney receiving no water save from the Murray. To the south of it, or next the river, the ground is low, grassy, and wooded; but on every other side the lake is confined by a low sand hill, of about fifteen feet in height, behind which there is a barren flat covered with salsolaceous plants, and exactly resembling a dry sea marsh, if I may say so. The more distant interior is alternate brush and plain, and exceedingly barren. The day after we arrived, however, Tenbury, with the dogs, killed four large kangaroos and as he saw many more, it is to be presumed that thereabouts they are pretty numerous. The lake is ten miles in circumference. Hawker's Creek, taking its windings, is nearly six in length. The latitude of our camp was 34°13'42"S.; its longitude 140°26'16". On September 1st. the thermometer, at 8 A.M. and at noon, stood at 48° and 60° respectively; the barometer at 29.750, and the boiling point was 212° nearly, thus indicating that we had risen but a few feet above the level of the sea. We left Lake Bonney on the 3rd of

September, and crossing the bank of sand by which it is confined, traversed the flat behind it for about three miles, when we ascended some feet and entered a low brush that continued for nearly nine miles, with occasional openings in it to that angle of the river which is opposite to the one at the end of the fossil formation.

Our camp at this place was on one of the prettiest spots on the Murray. Our tents were pitched on some sloping ground, sheltered from the S.W. wind. The feed was excellent, and the soil of better quality than usual. We had a splendid view of the river, which here is very broad and flanked on the right by a dark clay cliff, which is exceedingly picturesque. On the opposite side of the stream there is an extensive, well wooded and grassy flat of beautiful and park-like appearance. Altogether it was a cheerful and pleasant locality, and we were sorry to leave it so soon. Our observations placed us in lat. 34°11'12"S. and in long. 140°39'42"E. From this point the general course of the Murray is much more to the north than heretofore, so that on leaving it we had more of northing in our course than anything else. Some strange natives brought up our cattle for us, to whom I made presents; but although so kindly disposed, they did not follow us. Indeed, the natives generally, seemed to regard our progress with suspicion. and could not imagine why we were going up the Darling with so many drays and cattle. Our sheep had now become exceedingly tame and tractable; they followed the party like dogs, and I therefore felt satisfied that I had not done wrong in bringing them with me. We travelled on the 4th, over harder and more open ground than usual, having extensive polygonium flats to our right. There were belts of brush however on the plains, the soil and productions of which were sandy and salsolaceous. At 4½ miles we struck a lagoon, and coming upon a creek at 13 miles, we halted, although the feed was bad, as the cattle were unable to get to the river flats in consequence of the flooded state of the creek itself.

On the 5th we travelled through a country that consisted almost entirely of scrub on the poorest soil. However, we were now approaching that part of the river at which the flats (extensive enough) are intersected by numerous creeks and lagoons, so that our approach to the Murray was likely to be cut off altogether. At 3 1/2 miles we again struck the creek on the banks of which we had slept, and as it was the point at which the native path from Lake Bonney also strikes it, I halted to take a meridian altitude, which placed it in 34°4'5"S. We had allowed our horses to go and feed with their bridles through the stirrups, and were sitting on the ground when we heard a shot, and a general alarm amongst them, insomuch that we had some difficulty in quieting them, more especially Mr. Poole's horse. It was at length discovered that one of that gentleman's pistols had accidentally gone off in the holster, to the dismay of the poor animal. Fortunately no damage was done.

After noon, we pushed on, and at a mile crossed a creek, where we found a small tribe of scrub natives, one of whom had a child of unusual fatness: its flesh really hung about it; a solitary instance of the kind as far as I am aware. We then traversed good grassy plains for about two miles, when we fell in with another small tribe up a second creek: our introduction to which was more than ordinarily ceremonious. The natives remained seated on the ground, with the women and children behind them, and for a long time preserved that silence and reserve which is peculiar to these people when meeting strangers; however, we soon became more intimate, and several of them joined our train. Our friend Nadbuck was very officious (not disagreeably so, however), on the occasion, and shewed himself a most able tactician since he paid more attention to the fair than his own sex, and his explanation of our movements seemed to have its due weight.

We soon passed from the grassy plains I have mentioned, to plains of still greater extent, and still finer herbage. Nothing indeed could exceed the luxuriance of the grass on these water meadows, for we found on crossing that the floods were beginning to incroach upon them. These were marked all over with cattle tracks, many of them so fresh that they could only have been made the night before, but independently of these there were others of older date. The immense number of these tracks led me to inquire from the natives if there were any cattle in the neighbourhood, when they informed me that there were numbers of wild cattle in the brushes to the westward of the flats, and that they came down at night to the river for water and food. The grass upon the plain over which we were travelling was so inviting, that I determined to give the horses and bullocks a good feed, and turning towards the river with Mr. Eyre, I directed Mr. Poole and Mr. Browne to try the brushes with Flood and Mack, for a wild bullock, whilst we arranged the camp. We scarcely had time to do this, however, when Mr. Browne returned to inform me that soon after gaining the brush they had fallen in with a herd of about fifty cattle, out of which they had singled and shot a fine animal, and that on his way back to the camp the dogs had killed a large kangaroo. Upon this I sent Morgan with the cart to fetch in the quarters of the animal, and desired the natives to go with him to benefit by what might be left behind, and to feast on the kangaroo. The beast the party had killed fully justified Mr. Browne's account of it, and its fine condition proved the excellent nature of the pastures on which it had fed. We had not killed many of the sheep, as I was anxious to preserve them, since they had given us little or no trouble, so that I was led to hope that by ordinary care they would prove a most valuable and important stock.

We were here unable to approach the river, and therefore encamped near a creek, the banks of which were barren enough; however, as we had stopped for the benefit of the cattle it was of no consequence. But although on this occasion they were absolutely up to their middles in the finest grass, the bullocks were not satisfied, but with a spirit of contradiction common to animals as well as men they separated into mobs and wandered away; the difficulty of recovering them being the greater, because of the numerous tracks of other cattle in every direction around us. We recovered them, however, although too late to move that day, and it is somewhat remarkable to record, that this was the only occasion on which during this long journey we were delayed for so long a time by our animals wandering. Had it not been for Tampawang, whose keen eye soon detected the fresher tracks, we might have been detained for several days.

As Mr. Browne had been on horseback the greater part of the day, I left him in the camp with Mr. Poole, both having been after the cattle, and in the afternoon walked out with Mr. Eyre, to try if we could get to the river, but failed, for the creeks were full of water, and our approach to it or to the nearer flats was entirely cut off. So intersected indeed was this neighbourhood, that we got to a point at which five creeks joined. The scene was a very pretty one, since they formed a sheet of water of tolerable size shaded by large trees. The native name of this place was "Chouraknarup," a name by no means so harmonious as the names of their places generally are. We had not commenced any collection at this time, there being nothing new either in the animals or plants, but I observed that everything was much more forward on this part of the river than near Lake Bonney, although there was no material difference between the two places in point of latitude. A meridian altitude of the sun gave our latitude 34°1'33"S, and one of Altair 34°2'2"S.

The night of the 6th Sept. was frosty and cold, and we had thick ice in the buckets. We left our camp on a N. by E. course, at 8 o'clock on the morning of the 7th, and at 4 miles struck the river, where its breadth was considerable, and it looked exceedingly well. The flooded state of the creeks however prevented our again approaching it for several days. Shortly after leaving the river we turned more to the eastward, having gained its most northern reach. About noon we fell in with a few natives, who did not trouble themselves much about us, but we found that their backwardness was rather the result of timidity at seeing such a party than anything else. We traversed huge and well-grassed flats almost all day long, and ultimately encamped on the banks of a creek of some size, opposite to our tents the floods had made an island, on which we put our cattle for security during the night.

Mr. Eyre and I were again disappointed in an attempt to gain the banks of the Murray, but we returned to the camp with a numerous retinue of men, women, and children, who treated us to a corrobori at night. The several descriptions which have been given by others of these scenes, might render it unnecessary for me to give my account of such here; but as my ideas of these ceremonies may differ from that of other travellers I shall trespass on the patience of my readers for a few moments to describe them. However rude and savage a corrobori may appear to those to whom they are new, they are, in truth, plays or rather dramas, which it takes both time and practice to excel in. Distant tribes visiting any other teach them their corrobori, and the natives think as much of them as we should do of the finest play at Covent Garden. Although there is a great sameness in these performances they nevertheless differ. There is always a great bustle when a corrobori is to be performed, and the men screw themselves up to the acting point, as our actors do by other means than these poor creatures possess. On the present occasion there was not time for excitement; our's was as it were a family corrobori, or private theatricals, in which we were let into the secrets of what takes place behind the scenes. A party of the Darling natives had lately visited the Murray, and had taught our friends their corrobori, in which, however, they were not perfect; and there was consequently want of that excitement which is exhibited when they have their lesson at their fingers' ends, and are free to give impulse to those feelings, which are the heart and soul of a corrobori.

We had some difficulty in persuading our friends to exhibit, and we owed success rather to Mr. Eyre's influence than any anxiety on the part of the natives themselves. However, at last we persuaded the men to go and paint themselves, whilst the women prepared the ground. It was pitch dark, and ranging themselves in a line near a large tree, they each lit a small fire, and had a supply of dry leaves to give effect to the acting. On their commencing their chanting, the men came forward, emerging from the darkness into the obscure light shed by the yet uncherished fires, like spectres. After some performance, at a given signal, a handful of dry leaves was thrown on each fire, which instantly blazing up lighted the whole scene, and shewed the dusky figures of the performers painted and agitated with admirable effect, but the fires gradually lowering, all were soon again left in obscurity.

But, as I have observed, for some reason or other the thing was not carried on with spirit, and we soon retired from it; nevertheless, it is a ceremony well worth seeing, and which in truth requires some little nerve to witness for the first time.

We had now arrived at Camboli's haunt, and were introduced by him to his wife and children, of whom he seemed very proud; but a more ugly partner, or more ugly brats, a poor Benedict could not have been blessed with. Whether it was that he wished to remain behind, for he had not been very active on the road, or taken that interest in our proceedings which Nadbuck had done; or that our praises of his wife and pickaninnies had had any effect I know not, but he would not leave his family, and so remained with them when we left on the following morning. The neighbourhood of our camp was, however, one of great celebrity—since in it some of the most remarkable and most tragical events had taken place. It was near it that the volunteers who went out to rescue Mr. Inman's sheep, which had been seized by the natives to the number of 4,000, were driven back and forced to retreat; not, I would beg to be understood, from want of spirit, but because they were fairly overpowered and caught in a trap. The whole of the party, indeed, behaved with admirable coolness, and one of them, Mr. Charles Hawker, as well as their leader, Mr. Fidd, shewed a degree of moderation and forbearance on the occasion that was highly to their credit. Here also was the Hornet's Nest, where the natives offered battle to my gallant friend, Major O'Halloran, whose instructions forbade his striking the first blow. I can fancy that his warm blood was up at seeing himself defied by the self-confident natives; but they were too wise to commence an attack, and the parties, therefore, separated without coming to blows. Here, or near this spot also, the old white-headed native, who used to attend the overland parties, was shot by Miller, a discharged soldier, I am sorry to say, of my own regiment This old man had accompanied me for several days in my boat, when I went down the Murray to the sea coast in 1830, and I had made him a present which he had preserved, and shewed to the first overland party that came down the river, and thenceforward he became the guide of the parties that followed along that line. He attended me when I came overland from Sydney, in 1838, on which occasion he recognised me, and would sleep no where but at my tent door. He was shot by Miller in cold blood, whilst talking to one of the men of the party of which unfortunately he had the charge; but retribution soon followed. Miller was shortly afterwards severely wounded by the natives; and, having aneurism of the heart, was cautioned by his medical attendant never to use violent exercise; but, disregarding this, when he had nearly recovered, he went one day to visit a friend at the gaol in which he ought to have been confined, and in springing over a ditch near it, fell dead on the other side, and wholly unprepared to appear before that tribunal, to which he will one day or other be summoned, to answer for this and other similar crimes.

About a dozen natives followed us from our camp, on the morning of the 8th. We again struck the creek, on which we had rested, and which had turned to our right at 2½ miles on an east by south course, and followed along its banks, until it again trended too much to the south. We crossed alluvial flats of considerable extent, on which there was an abundance of grass. Just at the point at which we turned from the creek, we ascended a small sand hill, covered with the amaryllis, then beautifully in flower. The latitude of this little hill, from which the cliffs on the most northern reach of the Murray bore N.170°E. distant four miles, was 33°57'11"; so that the Murray does not extend northwards beyond latitude 34°1' or thereabouts. We again struck the creek, the course of which had been marked by gum-trees, at six miles, and were forced by it to the N.E., but ultimately turned it and descended southwards to the river; but as we were cut off from it we encamped on a lagoon of great length, backed by hills of a yellow and white colour, the rock being a soft and friable sandstone, slightly encrusted with salt. We had, shortly before we halted, passed a salt lagoon in the centre of one of the grassy flats, but such anomalies are not uncommon in the valley of the Murray. That part of the river which I have described, from the point where we shot the bullock to this lagoon, appeared to me admirably adapted for a cattle station, and has since been occupied as such.

As I have observed, the lagoon on which we encamped was backed by hills of 150 or 200 feet elevation, which were covered with thick brushwood. The flat between us and these hills was unusually barren, and all the trees at the side of the lagoon were dead. Whether this was owing to there being salt in the ground or to some other cause, there was here but little grass for the cattle to eat, so that, although they were watched, twenty of them managed to crawl away, and we were consequently delayed above an hour and a half after our usual hour of starting, and commenced our day's journey wanting two of our complement, but we stumbled upon them in passing through the brush, in which they were very comfortably lying down. We travelled for about six miles through a miserable undulating country of sand and scrub. At noon we were abreast of a little sandy peak that was visible from our camp, and is a prominent feature hereabouts. This peak Mr. Browne and I ascended, though very little to our gratification, for the view from it was as usual over a sea of scrub to whatever quarter we turned. The peak itself was nothing more than a sandy eminence on which neither tree or shrub was growing, and the whole locality was so much in unison with it, that we called it "Mount Misery." After passing this hill, and forcing through some stunted brush, we debouched on open plains and got once more on the overland road, which was distinctly marked by a line of bright green grass, that was springing up in the furrows the drays had left. This road took us to the edge of a precipitous embankment, from which we overlooked the river flowing beneath it. This embankment was 60 or 70 feet high, and presented a steep wall to the river; for although the Murray had lost the fossil cliffs it was still flanked by high level plains on both sides, and cliffs of 100 or 120 feet in height, composed of clay and sand, rose above the stream, the faces of which presented the appearance of fretwork, so deeply and delicately had they been grooved out by rains. The soil of this upper table land was a bright red ferruginous clay and sand. The vegetation was chiefly salsolaceous, but there was, notwithstanding, no want of grass upon it, though the tufts were very far apart. If our cattle had fared badly at our last camp, they had no reason to complain at this; for we encamped on a beautifully green flat, about seven miles short of the Rufus, and about eight from the nearest point of lake Victoria. There were now seventeen natives in our train, amongst whom was one of remarkable character. This was "Pulcanti," who was engaged in, wounded and taken prisoner at an affair on the Rufus, to which I shall again have to allude.

Whilst the police were conveying this man handcuffed to Adelaide, he threw himself off the lofty cliffs at the Great Bend into the river beneath, and attempted to escape by swimming across it, but he was recaptured and taken safe to Adelaide, where subsequent kind treatment had considerable influence on his savage disposition. His attempt to escape was of the boldest kind, and was spoken of with astonishment by those who witnessed it, but so desperate an act only proved how much more these people value liberty than life. I am sure that bold savage would have submitted to torture without a groan; he was the most repulsive native in aspect that I ever saw, and had a most ferocious countenance. The thick lip and white teeth, the lowering brow, and deep set but sharp eye, with the rapidly retiring forehead all betrayed the savage with the least intellect, but his demeanour was now quiet and inoffensive.

Mr. Eyre again preceded us to the Rufus, with Kenny and Tenbury; for although we had been disappointed in seeing any natives at Lake Bonney, it was hardly to be doubted but that we should find a considerable number at Lake Victoria.

We joined Mr. Eyre about noon at the junction of the Rufus with the Murray, and which serves like Hawker's Creek as a channel of communication between that river and the Murray. Here Mr. Eyre had collected 69 natives, who were about to go out kangarooing when he arrived. They had their hunting spears and a few waddies, but no other weapons.

We had now arrived at Nadbuck's native place, and he left us to join his family, promising still to accompany us up the Darling. A principal object Mr. Eyre had in joining me had been to distribute some blankets to those natives who, living in the distance, seldom came to Moorundi to benefit by the distribution of food and clothing there. In the position we now occupied we were flanked by the Rufus to our left, and had the Murray in front of us. The ground in our rear and to our right was rather bushy, and numerous Fusani, covered with fruit, were growing there; Lake Victoria being about four miles to our rear also. Considering the spirit of the natives on this part of the Murray, the position was not very secure, as we were too confined; but I had no apprehension of any attack from them, they having for some time shewn a more pacific disposition, and against whom we were otherwise always well prepared. As soon, therefore, as the tents were pitched, we walked together along the bank of the Rufus to its junction with the lake, but not seeing any of the native families we turned back, until observing some young men on the opposite side of the channel we called to them, and one of them ferried us over in a canoe. We had then a long round of visits to make to the different families of the natives, since they were all encamped on the eastern or opposite side of the Rufus.

The first huts to which we went happened to be that of our friend Nadbuck, and he introduced us, as Camboli had done, to his wives and children, of whom the old gentleman was very proud. We then visited eleven other huts in succession, after which we returned to the place where the canoe had been left, with twelve patriarchs, to whom Mr. Eyre (wisely selecting the oldest) intended making some presents. We were again ferried across the Rufus, the current setting strong into Lake Victoria at the time, and had well nigh gone down in our frail bark, to the infinite amusement of our Charon. We had just time, however, to reach the bank and to get out of her when she went down.

It was at this particular spot that the natives sustained so severe a loss when Pulcanti was taken. They got between two fires, that of Mr. Robinson's party of overlanders, with whom they had been fighting for three days; and a party of police who, providentially for Mr. Robinson, came up just in time to save him from being overwhelmed by numbers. Astonished at finding themselves taken in flank, the blacks threw themselves into the Rufus, and some effected their escape, but about forty fell, whose grave we passed on our way back to the camp.

The natives who accompanied us pointed out the mound to Mr. Eyre and myself as we walked along, and informed us that thirty of their relatives laid underneath; but they did not seem to entertain any feelings of revenge for the loss they had sustained.

On the morrow, my worthy friend left me, on his return to Moorundi, together with Kenny and Tenbury, and a young native of the Rufus. We all saw them depart with feelings of deep regret; but Mr. Eyre had important business to attend to which did not admit of delay.

A little before Mr. Eyre mounted his horse, I had sent Mr. Browne, with Flood and Pulcanti, to the eastward, to ascertain how high the backwaters of the Murray had gone up the Ana-branch of the Darling, since that ancient channel laid right in our way, and I was anxious if possible to run up it, rather than proceed to the river itself, as being a much nearer line. In the afternoon Mr. Poole and I moved the camp over to the lake, and on the following day I directed him to ascertain its circumference, as we should be detained a day or two awaiting the return of Mr. Browne.

CHAPTER III.

MR. BROWNE'S RETURN—HIS ACCOUNT OF THE COUNTRY—CHANGE OF SCENE—CONTINUED RAIN—TOONDA JOINS THE PARTY—STORY OF THE MASSACRE—LEAVE LAKE VICTORIA—ACCIDENT TO FLOOD—TURN NORTHWARDS—CROSS TO THE DARLING—MEET NATIVES—TOONDA'S NAUGHTY MANNER—NADBUCK'S CUNNING—ABUNDANCE OF FEED—SUDDEN FLOODS—BAD COUNTRY—ARRIVAL AT WILLIORARA—CONSEQUENT DISAPPOINTMENT—PERPLEXITY—MR. POOLE GOES TO THE RANGES—MR. BROWNE'S RETURN—FOOD OF THE NATIVES—POSITION OF WILLIORARA

LAKE VICTORIA is a very pretty sheet of water, 24 miles in diameter, very shallow, and at times nearly dry. As I have previously observed of Lake Bonney, it is connected with the Murray by the Rufus, and by this distribution of its waters, the floods of the Murray are prevented from being excessive, or rising above a certain height. The southern shore of Lake Victoria is very picturesque, as well as the line of the Rufus. The latter however is much wooded, whereas the S.W. shore of the lake is low and grassy, and beautiful umbrageous trees adorn it, in number not more than two or three to the acre. As Mr. Poole was engaged near me, I remained stationary on the 13th, but on the following day moved the camp seven miles to the westward, for his convenience. On the 15th I again moved so as to keep pace with him, and was highly delighted at the really park-like appearance of the scenery. This pretty locality is now occupied as a cattle run, and must be a place of amusement as well as profit.

We met Mr. Browne and Flood on their return to the camp from the journey on which I had sent them, about an hour before we halted.

Mr. Browne informed me that the day he left me he rode for some miles along the shore of the lake, and that after leaving it he encamped in the scrub, having travelled about seventeen miles. The brush was very dense, although there were open intervals; it consisted of trees and shrubs of the usual kind, the soil was very sandy, and there was a good deal of spinifex upon it.

The next day, still on a due east course (that on which he had travelled from the lake), and at five miles from where he had slept, Mr. Browne came on a salt lake, about 800 yards in circumference. A third of the bed was under water, and half of the remainder was white with crystallized salt, that glittered in the sun's rays, and looked like water at a distance. At about five miles farther on there were two other lakes of the same kind, but both were dry and without any salt deposits in their beds. At five miles beyond these lakes Mr. Browne intersected the Ana-branch of the Darling, which I had detached him to examine. To within a short distance of the Ana-branch the country was similar to that through which he had passed the day before, but on nearing it he crossed an open plain. This old channel of the Darling had been crossed by Mr. Eyre on a recent journey to the north, but at that time was dry. Where Mr. Browne struck it the banks were rather high, and its course was N.W. by W. It was about eighty yards wide, with a strong current running upwards, caused by the back waters of the Murray. Its general course for 12 miles was N. by E. The country was very open, and high banks, similar to those on the Murray, occurred alternately on either Side. The channel maintained the same appearance as far as Mr. Browne; rode and as he found the waters still running upwards, he considered that the object of his journey was attained, and that we should find no difficulty in pursuing our route northwards along this new line. It may be necessary for me to inform the reader that no water ever flows down the Ana-branch from the north. When Mr. Eyre first arrived on its banks it was dry, and he was consequently obliged to cross the country to the Darling itself, a distance of between 40 and 50 miles. Pulcanti, the native I sent with Mr. Browne, however, made a rough sketch of the two channels, by which it appeared that the Ana-branch held very much to the eastward, in proof of which he pointed to a high line of trees, at a great distance, as being the line of the river Darling. Considering from this that, even if water failed us in the Ana-branch, we should have no difficulty in crossing to the main stream, and that however short our progress might be, it would greatly curtail our journey to Laidley's Ponds, I decided on trying the new route.

Mr. Browne saw a great many red kangaroos (foxy), some very young, others very large; and he chased a jerboa, which escaped him. He also saw a new bird with a black crest, about the size of a thrush.

The morning of the 14th had been cloudy, but the day was beautifully fine; so that we had really enjoyed our march, if so it might be called. From our tents there was a green and grassy slope to the shore of the lake, with a group of two or three immense trees, at distances of several hundred yards apart, and the tranquil waters lay backed by low blue hills.

On the morning of the 15th the barometer fell to 27.672, the thermometer standing at 56°, at 8 A.M.The air was heavy, the sky dull, and the flies exceedingly troublesome. All these indications of an approaching change in the weather might have determined me to remain stationary, but I was anxious to push on. I therefore directed Mr. Poole to complete the survey of the lake, and at eleven moved the whole party forward.

The picturesque scenery which had, up to this point, adorned the shores of Lake Victoria ceased at two miles, when we suddenly and at once found ourselves travelling on sand, at the same time amidst reeds. The rich soil disappeared, the trees becoming stunted and low. As the travelling was also bad, we went along the margin of the lake, where the sand was firm, although marked with ripples like those left on the sea-shore by the tide, between the water and a line of rubbish and weeds inside of us, so that it appeared the lake had not yet risen so high as the former year. We had moved round to its eastern side, which being its lea side also, the accumulation of rubbish and sand was easily accounted for. We traversed about eight miles of as dreary a shore as can be imagined, backed, like Lake Bonney, by bare sand hills and barren flats, and encamped, after a journey of thirteen miles, on a small plain, separated from the lake by a low continuous sand ridge, on which the oat-grass was most luxuriant. The indications of the barometer did not deceive us, for soon after we started it began to rain, and did not cease for the rest of the day, the wind being in the N.E. quarter.

It continued showery all night, nor on the morning of the 16th was there any appearance of a favourable change. At nine a steady and heavy rain setting in we remained stationary.

The floods in the Rufus had obliged us to make a complete circuit of the lake, so that we had now approached that little stream to within six miles from the eastward. Our friend Nadbuck, therefore, thinking that we were about to leave the neighbourhood, rejoined the party. With him about eighty natives came to see us, and encamped close to our tents; forty-five men, sixteen women, and twenty-six children. I sent some of the former out to hunt, but they were not successful.

Amongst the natives there were two strangers from Laidley's Ponds, the place to which we were bound. The one was on his way to Moorundi, the other on his return home. Pulcanti had given us a glowing account of Laidley's Ponds, and had assured us that we should not only find water, but plenty of grass beyond the hills to the N.W. of that place. This account the strangers confirmed; and the one who was on his way home expressing a wish to join us, I permitted him to do so; in the hope that, what with him and old Nadbuck, we should be the less likely to have any rupture with the Darling natives, who were looked upon by us with some suspicion. I was, in truth, very glad to take a native of Williorara up with me, because I entertained great doubts as to the reception we should meet with from the tribe, on our arrival there, in consequence of the unhappy occurrence that took place between them and Sir Thomas Mitchell, during a former expedition; and I hoped also to glean from this native some information as to the distant interior. Both the Darling natives were fine specimens of their race. One in particular. Toonda, was a good-looking fellow, with sinews as tough as a rope. It also appeared to me that they had a darker shade of colour than the natives of the Murray.

Nadbuck turned out to be a merry old man, and a perfect politician in his way, very fond of women and jimbuck (sheep), and exceedingly good-humoured with all. He here brought Davenport a large quantity of the fruit of the Fusanus, of which he made an excellent jam, too good indeed to keep; but if we could have anticipated the disease by which we were afterwards attacked, its preservation would have been above all price. The natives do not eat this fruit in any quantity, nor do I think that in its raw state it is wholesome. They appeared to me to live chiefly on vegetables during the season of the year that we passed up the Murray, herbs and roots certainly constituted their principal food.

I had hoped that the weather would have cleared during the night, but in this I was disappointed. On the 17th we had again continued rain until sunset, when the sky cleared to windward and the glass rose. We were however unable to stir, and so lost another day. About noon Nadbuck came to inform me that the young native from Laidley's Ponds, who was on his way to Moorundi, had just told him that only a few days before he commenced his journey, the Darling natives had attacked an overland party coming down the river, and had killed them all, in number fifteen. I therefore sent for the lad, and with Mr. Browne's assistance examined him. He was perfectly consistent in his story; mentioned the number of drays, and said that the white fellows were all asleep when the natives attacked them amongst the lagoons, and that only one native, a woman, was killed; the blacks, he added, had plenty of shirts and jackets. Doubtful as I was of this story, and equally puzzled to guess what party could have been coming down the Darling, it was impossible not to give some little credit to the tale of this young cub; for he neither varied in his account or hesitated in his reply to any question. I certainly feared that some sad scene of butchery had taken place, and became the more anxious to push my way up to the supposed spot, where it was stated to have occurred, to save any one who might have escaped. I felt it my duty also before leaving Lake Victoria to report what I had heard to the Governor.

As the barometer fell before the rain, so it indicated a cessation of it, by gradually rising. The weather had indeed cleared up the evening before, but the morning of the 18th was beautifully fine and cool; we therefore yoked up the cattle and took our departure from Lake Victoria at 9 A.M. At first the ground was soft, but it soon hardened again.

Shortly after starting we struck a little creek, which trended to the south, so that we were obliged to leave it, but we could trace the line of trees on its banks to a considerable distance. We traversed plains of great extent, keeping on the overland road until at length we gained the river, and encamped on a small neck of land leading to a fine grassy enclosure, into which we put our cattle. One side of this enclosure was flanked by the river, the other by a beautiful lagoon, that looked more like a scene on Virginia Water than one in the wilds of Australia.

As we crossed the plains we again observed numerous cattle tracks, and regularly beaten paths leading from the brushes to the river, to the very point indeed where we encamped. The natives had previously informed us, as far back as the place where we shot the first bullock, that we should fall in with other cattle hereabouts; we did not however see any of them during the day. Our tents were pitched on the narrow neck of land leading to an enclosure into which we had turned our animals. It was so narrow indeed that nothing could pass either in or out of it without being observed by the guard, so that neither could our cattle escape or the wild ones join them. It was clear, however, that we had cut off the latter from their favourite pasture, for at night they were bellowing all round us, and frequently approached close up to our fires. We had no difficulty in distinguishing the lowing of the heifers from that of the bullocks; of which last there appeared to be a large proportion in the herd.

Some of our cattle were getting very sore necks, and our loads at this time were too heavy for me to relieve them. Flood therefore suggested our trying to secure two or three of the bullocks running in the bush. We therefore arranged that a party should go out in the morning to scour the wood, and drive any cattle they might find towards the river, at which I was to be prepared to entice them to our animals. Accordingly Mr. Poole and Mr. Browne, with Flood and Mack, started at sunrise. It was near twelve, however, when Mr. Browne returned with Flood, who had met with a sad accident, and had three of the first joints of the fingers of his right hand carried off by the discharge of his fusee whilst loading. He had incautiously put on the cap and was galloping at the time, but kept his seat. Mr. Browne informed me they had seen a great many cattle, but that they were exceedingly wild, and started off the moment the horsemen appeared, insomuch that they could not turn them, and it was with a view to drive them towards the river that Flood fired at them. However none approached the camp. Mr. Poole returned late in the afternoon equally unsuccessful. Mr. Browne dressed Flood's hand, who bore it exceedingly well, and only expressed his regret that he should be of no use on the Darling in the event of any rupture with the natives. I remained stationary, as Mr. Browne thought it would be necessary to keep Flood quiet for a day or two. On the following day we resumed our journey, and reached the junction of the ancient channel of the Darling with the Murray about 11. The floods were running into it with great velocity, and the water had risen to a considerable height, so that many trees were standing in it. I remained here until noon, when a meridian altitude placed us in lat. 34°4'3". We then bade adieu to the Murray, and turned northwards to overtake the party, which under Nadbuck's guidance had cut off the angle into which we had gone. With the Murray we lost its fine trees and grassy flats. The Ana-branch had a broad channel and long reaches of water; but was wholly wanting in pasture or timber of any size. The plains of the interior formed the banks, and nothing but salsolæ grew on them. We encamped at eight miles from the junction, where there happened to be a little grass, but were obliged to keep the cattle in yoke and the horses tethered to prevent their wandering. As we advanced up the Ana-branch on the following day, its channel sensibly diminished in breadth, and at eleven miles we reached a hollow, beyond which the floods had not worked their way. Here we found. a tribe of natives, thirty-seven in number, by whom the account we had heard of the massacre of the overlanders at the lagoons of the Darling was confirmed. Nadbuck now informed me that we should have to cross the Ana-branch and go to the eastward, and that it would be necessary to start by dawn, as we should not reach the Darling before sunset. Nadbuck had now become a great favourite, and there was a dry kind of humour about him that was exceedingly amusing, at the same time that his services were really valuable.

Toonda, on the other hand, was a man of singular temperament. He was good-looking and more intelligent than any native I had ever before seen. His habit was spare, but his muscles were firm, and his sinews like whipcord. He must indeed have had great confidence in his own powers to have undertaken a journey of more than 200 miles from his own home. He was very taciturn, and would rather remain at the officers' fire than join his fellows.

The country we had passed through during the day had been miserable. Plains of great extent flanked the Ana-branch on either side, on which there were sandy undulations covered with stunted cypress trees or low brush.

Flood had from the time of his accident suffered great pain; but as he did not otherwise complain, Mr. Browne did not entertain any apprehension as to his having any attack of fever.

On the morning of the 24th, the natives paid us an early visit with their boys, and remained at the camp until we started. At the head of the water they had made a weir, through the boughs of which the current was running like a sluice; but the further progress of the floods was stopped by a bank that had been gradually thrown up athwart the channel. Crossing the Ana-branch at this point, we struck across barren sandy plains, on a N.N.E. course. From them we entered a low brush, in which there were more dead than living trees. At four miles this brush terminated, and we had again to traverse open barren plains. At their termination we had to force our way through a second brush, consisting for the most part of fusani, acaciæ, hakeæ, and other low shrubs, but there were no cypresses here as in the first brush. On gaining more open ground, the country gradually rose before us, and a ferruginous conglomerate cropped out in places. We at length began our descent towards the valley of the Darling. The country became better wooded: the box-tree was growing on partially flooded land, and there was no deficiency of grass. Mr. Browne went on ahead with Toonda and Flood, whilst I and Mr. Poole remained with the party.

From the appearance of the country, however, I momentarily expected to come on the river; but the approach to it from the westward is extremely deceptive, and we had several miles of box-tree flats to traverse before the gum-trees shewed their white bark in the distance. We reached the Darling at half-past five, as the sun's almost level beams were illuminating the flats, and every blade of grass and every reed appeared of that light and brilliant green which they assume when held up to the light. The change from barrenness and sterility to richness and verdure was sudden and striking, and nothing certainly could have been more cheering or cheerful than our first camp on the Darling River. The scene itself was very pretty. Beautiful and drooping trees shaded its banks, and the grass in its channel was green to the water's edge. Evening's mildest radiance seemed to linger on a scene so fair, and there was a mellow haze in the distance that softened every object. The cattle and horses were up to their flanks in grass and young reeds, and plants indicative of a better soil, such as the sowthistle, the mallow, peppermint, and indigofera were growing in profusion around us. Close to our tents there was a large and hollow gum-tree, in which a new fishing net had been deposited, but where the owner intended to use it was a puzzle to us, for it was impossible that any fish could remain in the shallow and muddy waters of the Darling; which was at its lowest ebb, and the current was so feeble that I doubted if it really flowed at all. Whether the natives anticipated the flood which shortly afterwards swelled it I cannot say, although I am led to believe they did, either from habit or experience.

So abundant had been the feed that none of the cattle stirred out of sight of the camp, and we should have started at an early hour, but for the visit of an old native, the owner of the net we had discovered. It was with some hesitation that he crossed the river to us, but he did so; and as soon as he saw me here recognised me as having been in the boat on the Murray in 1830, though fourteen years had passed since that time, and he could only have seen me for an hour or two. He was not, however, singular in his recollection of me, since one of the natives of the Ana-branch also recollected me; and Tenbury, the native constable at Moorundi, not only knew me the moment he saw me, but observed that a little white man sat by my side in the stern of the boat, and that I had something before me, which was a compass. There was a suspicious manner about our visitor, for which we could not very well account; but it arose from doubts he entertained as to the safety of his net, for after he had seen that it had not been taken away, his demeanour changed, and he expressed great satisfaction that we had not touched it. We commenced our journey up the Darling at nine o'clock, on a course somewhat to the eastward of north. We passed flat after flat of the most vivid green, ornamented by clumps of trees, sufficiently apart to give a most picturesque finish to the landscape. Trees of denser foliage and deeper shade drooped over the river, forming long dark avenues, and the banks of the river, grassed to the water, had the appearance of having been made so by art.

We halted, after a journey of fourteen miles, on a flat little inferior to that we had left, and again turned the cattle out to feed on the luxuriant herbage around them.

The Darling must have been in the state in which we found it for a great length of time, and I am led to infer, from the very grassy nature of its bed, that it seldom contains water to any depth, or length of time, since in such case the grass would be killed. Its flats, like those of the Murray, are backed by lagoons, but they had long been dry, and the trees growing round them were either dead or dying.

With the exception of the tribe at the Ana-branch, and the old man, we had seen no natives since leaving the Murray; but, from the reports we had beard of the recent massacre of the overland party at Williorara, and the character of the Darling blacks, I was induced to take double precautions as I journeyed up the river, and had the camp so formed that it could not be surprised. Two drays were ranged close to each other on either side, the boat carriage formed a face to the rear, and the tents occupied the front; thus leaving sufficient room in the centre to fold the sheep in netting. The guard, augmented to six men, occupied a tent at one angle. My own tent was in the centre of the front, and another tent at the angle opposite the guard tent. So that it would have been difficult for the natives to have got at the sheep (which they most coveted), without alarming us. Still, although we had no apprehension of the natives, both Nadbuck and Toonda were constantly on the watch, and it was evident the former considered himself in no mean capacity at this time. He put on an air of great importance, and shewed great anxiety about our next interview with the natives; but Toonda took everything quietly, and there was a haughty bearing about him, that contrasted strangely with the bustling importance of his companion.

We here heard that there was a large encampment of natives about three miles above us, but none of them ventured to our camp; nor, it is more than probable, were the people aware of our being in the neighbourhood; but our friend Nadbuck, as I have stated, was in a great bustle, and shewed infinite anxiety on the occasion. Neither were his apprehensions allayed on the following morning when we started. He went in advance to prepare the natives for our approach, and to ask permission for us to pass through their territory, but returned without having found them. Not long afterwards it was reported that the natives were in front.

On hearing this the old gentleman begged of me to stop the party, and away he went, full of bustle and importance, to satisfy himself. In a few minutes he returned and said we might go on. We had halted close to the brow of a gentle descent into a small creek junction at this particular spot, and on advancing a few paces came in view of the natives, assembled on the bank of the river below. Men only were present, but they appeared to have been taken by surprise, and were in great alarm. They had their spears for hunting, and a few hostile weapons, but not many; and certainly had not met together with any hostile intention.

Some of the men were very good looking and well made, but I think the natives of the Darling generally are so. They looked with astonishment on the drays, which passed close to them; and I observed that several of them trembled greatly. At this time Nadbuck had walked to some little distance with two old men, holding each by the hand in the most affectionate manner, and he was apparently in deep and earnest conversation with them. Toonda, on the other hand, had remained seated on one of the drays, until it descended into the creek. He then got off, and walking up to the natives, folded his blanket round him with a haughty air, and eyed the whole of them with a look of stern and unbending pride, if not of ferocity. Whether it was that his firmness produced any effect I cannot say, but after one of the natives had whispered to another, he walked up to Toonda and saluted him, by putting his hands on his shoulders and bending his head until it touched his breast. This Toonda coldly returned, and then stood as frigid as before, until the drays moved on, when he again resumed his seat and left them without uttering a word. Nadbuck had separated from his friends, after having as it seemed imparted to them some important information, and coming up to myself and Mr. Browne, whispered to us, "Bloody rogue that fellow, you look after jimbuck." The contrast between these two men was remarkable: the crafty duplicity of the one, and the haughty bearing of the other. But I am led to believe that there was some latent cause for Toonda's conduct, since he asked me to shoot the natives, and was so excited that he pushed his blanket into his mouth, and bit it violently in his anger. On this I offered him a pistol to shoot them himself, but he returned it to me with a smile. Of course it will be understood that I should not have allowed him to fire it.

Two of the old men followed when we left the other natives, to whom I made presents in the afternoon; but it is remarkable that many of them trembled whilst we staid with them, and although their women were not present, they hovered on the opposite bank of the Darling all the time. We kept wide of the river almost all day, travelling between the scrub and lagoons, but we had occasionally to ascend and cross ridges of loose sand, over which the bullock-drivers were obliged to help each other with their teams. There was not the slightest change in the character of the distant interior, but the vicinity of the Darling was thickly timbered for more than three-quarters of a mile from its banks, but the wood was valueless for building purposes.

I was exceedingly surprised at the course of the river at this point. We had gone a good deal to the eastward the day before, but on this day we sometimes travelled on a course to the southward of east, and never for the whole day came higher up than east by north. The consequence was, that we proceeded into a deep bight, and made no progress northwards up the river. At our camp it had dwindled to a mere thread, so narrow was the line of water in its bed. Its banks were as even and as smooth as those of a fortification, and covered with a thick, even sward. There was no perceptible current and the water was all muddy; but the scenery in its precincts was still verdant and picturesque, grassy flats with ornamental trees succeeding each other at every bend of the stream. The dogs killed a large kangaroo on the plains, the greater part of which we gave to the natives, all indeed but a leg, which Jones, whose duty it was to feed them, reserved for the dogs. Yet this appropriation excited Toonda's anger, "Kangaroo mine, sheep yours," said he, threatening Jones with his waddy; but he soon recovered his temper, and carried off his share of the animal, subduing his feelings with as much apparent facility as he had given vent to them. About this time the weather had become much warmer, although we had occasional cold winds. We started early on the morning of the 27th, without the intention of making a long journey, because the bullocks had been kept in yoke all night. We travelled for six miles over firm and even plains, but soon afterwards got upon deep sand, through which the teams fairly ploughed their way. I therefore turned towards the river, and encamped on the first flat we reached, having run about ten miles on an east-north-east course.

We here found the Darling so diminished in size, and so still, that I began to doubt whether or not we should find water higher up. Its channel, however preserved the appearance of a canal, with sloping grassy sides, shaded by trees of drooping habit and umbrageous foliage, but the soil of the flats had become sandy, and they appeared to be more subject to inundation than usual.

About this time I regretted to observe that many of the bullocks had sore necks, and I was in consequence obliged to make a different distribution of them; an alternative always better if possible to avoid, as men become attached to their animals, and part even with bad ones reluctantly.

On counting our sheep at this camp, I found that we had 186 remaining. Toonda came as usual to take his share of one that had just been killed; but I said, No! that, according to his own shewing, he had no claim to any—thinking this the best way of speaking to his reason. He seemed much astonished at the view I took of the matter, but on his acknowledging himself in error, I forgave his recent ebullition and allowed him his wonted meal; for, although I was always disposed to be kind to the natives, I still felt it right to shew them that they were not to be unruly. Neither is it without great satisfaction that I look back to the intercourse I have had with these people, from the fact of my never having had occasion to raise my arm in hostility against them.

The cattle fared well on the luxuriant grass into which they had been turned when we halted, and as they had no inducement to wander, so they were close to the camp at daybreak, and we started at 7 on an east-north-east course, which at a mile we changed to a northerly one; but soon afterwards finding that a pine ridge crossed our course, and extended to the banks of the river, I turned to the north-west to avoid it, but the country becoming generally sandy I again turned towards the stream, and by going round the sandy points instead of over them, lessened the labour to the cattle, although I increased the distance. We were glad to find that the Darling held a general northerly course, or one somewhat to the westward of that point, for we had during the last three or four days made a great deal of easting, and I had thus been prevented making the rapid progress I anticipated to Laidley's Ponds. I had observed for more than twenty miles below us that the immediate precincts of the river were not so rich in soil, or the flats so extensive as at first; they now however began to open out and assumed the character and size of those of the Murray. The state of the two rivers however was very different, for the Darling still continued without breadth or current, (I speak of its appearance in lat. 33°43'S.) whilst the Murray ever presents its bright and expanded waters to the view.

We had communicated with a native tribe the day before that of which I am now speaking, and again this day fell in with another, which we evidently took by surprise. All the men had their spears, but on seeing us approach they quietly deposited them under a tree. Amongst these people there was another native who recognised me as an old acquaintance of fourteen years' standing; but I began to doubt these patriarchs, to whom I generally made a present for old acquaintance sake. This tribe numbered forty-eight. All of them were handsome and well-made men, though short in stature, and their lower extremities bore some proportion to their busts.

For the first time this day we observed a ferruginous sandstone in the bed of the Darling, and saw it cropping out from under the sand hills on the western extremity of the flats.

Shortly after leaving the natives we arrived at a small plain, where they could only just have killed a kangaroo that was lying on the ground partly prepared for cooking. On seeing it I ordered the dogs to be tied up, and left it untouched. Indeed if I had been fortunate enough to kill a kangaroo at this place, I would have given it to these poor people. Three of them. who afterwards came to our camp, mentioned the circumstance, and seemed to be sensible of our feelings towards them. There can he no doubt but that the Australian aboriginal is strongly susceptible of kindness, as has been abundantly proved to me, and to the influence of such feeling I doubtlessly owe my life, for if I had treated the natives harshly, and had thrown myself into their power afterwards, as under a kind but firm system I have ever done without the slightest apprehension, they would most assuredly have slain me; and when I assure the reader that I have traversed the country in every direction, meeting numerous tribes of natives, with two men only, and with horses so jaded that it would have been impossible to have escaped, he will believe that I speak my real sentiments. Equally so the old native, (to whom the net we discovered in the hollow of a tree where we first struck the Darling belonged), evinced the greatest astonishment and gratification, when he found that his treasure had been untouched by us.

The flats of the Darling are certainly of great extent, but their verdure reached no farther than the immediate precincts of the river at this part of its course. Beyond its immediate neighbourhood they are perfectly bare, but lightly wooded, having low and useless box-trees (the *Gobero* of Sir Thomas Mitchell), growing on them. Their soil is a tenacious clay, blistered and rotten. These flats extend to uncertain distances from the river, and vary in breadth from a quarter of a mile to two miles or more. Beyond them the country is sandy, desolate, and scrubby. Pine ridges, generally lying parallel to the stream, render travelling almost impracticable where they exist, whilst the deep fissures and holes on the flats, into which it is impossible to prevent the drays from falling, give but little room for selection. Our animals were fairly worn out by hard pulling on the one, and being shaken to pieces on the other.

Some days prior to the 29th, Mr. Browne and I, on examining the waters of the river, thought that we observed a more than usual current in it; grass and bark were floating on its surface, and it appeared as if the water was pushed forward by some back impulse. On the 28th it was still as low as ever; but on the morning of the 29th, when we got up it was wholly changed. In a few hours it had been converted into a noble river, and had risen more than five feet above its recent level. It was now pouring along its muddy waters with foaming impetuosity, and carrying away everything before it. Whence, it may be asked, come these floods? and was it from the same cause that the Murray, as Tenbury stated, rose so suddenly? Such were the questions that occurred to me. From the natives I could gather nothing satisfactory. We were at this time between three and four hundred miles from the sources of the Darling, and I could hardly think that this fresh had come from such a distance. I was the more disposed to believe, perhaps, because I hoped such would be the case, that it was caused by heavy rains in the hills to the north-west of Laidley's Ponds, and that it was pouring into the river through that rivulet.

The natives who had accompanied us from the last tribe left at sunset, as is their custom, after having received two blankets and some knives. Being anxious to get to Laidley's Ponds, I started early, with the intention of making a long journey, but circumstances obliged me to halt at six miles. We crossed extensive and rich flats the whole of the way, and found as usual an abundance of feed for our cattle. It would perhaps be hazardous to give an opinion as to the probable availability of the flats of the Darling: those next the stream had numerous herbs, as spinach, indigoferæ, clover, etc., all indicative of a better soil; but the out flats were bare of vegetation, although there was no apparent difference in their soil. One peculiarity is observable in the Darling, that neither are there any reeds growing in its channel or on the flats.

Our journey on the last day of September terminated at noon, as we arrived at a point from which it was evident the river takes a great sweep to the eastward; and Nadbuck informed me that by going direct to the opposite point, where, after coming up again, it turned to the north, we should cut off many miles, but that it would take a whole day to perform the journey. I determined therefore to follow his advice, and to commence our Journey across the bight at an early hour the following morning, the 1st of October. I availed myself of the remainder of the day to examine the country for some miles to the westward, but there was no perceptible change in it. The same barren plains, covered sparingly with salsolæ and atriplex, characterised this distant part of the interior; and sandy ridges covered with stunted cypress trees, acaciæ, hakeæ, and other similar shrubs, proved to me that the productions of it were as unchanged as the soil.

As we had arranged, we broke up our camp earlier than usual on the 1st of October, for, from what Nadbuck had stated, I imagined that we had a long journey before us; but after going fifteen miles, we gained the river, and found that it was again trending to the north. It had now risen more than bank high, and some of its flats were partly covered with water. We had kept a N.N.W. course the whole day, and crossed hard plains without any impediment; but, although we kept at a great distance from the stream, we did not observe any improvement in the aspect of the country.

Our specimens, both of natural history and botany, were as yet very scanty; but we found a new and beautiful shrub in blossom, on some of the plains as we crossed the bight; and Mr. Browne discovered three nests of a peculiar rat, that have been partially described by Sir Thomas Mitchell.

Mr. Browne was fortunate enough to secure one of these animals, which is here figured. The nests they construct are made of sticks, varying in length from three inches to three feet, and in thickness from the size of a quill to the size of the thumb. They were arranged in a most systematic manner, so as to form a compact cone like a bee-hive, four feet in diameter at the base, and three feet high. This fabric is so firmly built, as to be pulled to pieces with difficulty. One of these nests had five holes or entrances from the bottom, nearly equi-distant from each other, with passages leading to a hole in the ground, beneath which I am led to conclude they had their store. There were two nests of grass in the centre of the pyramid, and passages running up to them diagonally from the bottom. The sticks, which served for the foundations of the nests, were not more than two or three inches long, and so disposed as to form a compact flooring, whilst the roofs were arched. The nests were close together, but in separate compartments, with passages communicating from the one to the other.

In a pyramid that we subsequently opened, there was a nest nearly at the top; so that it would appear that these singular structures are common to many families, and that the animals live in communities. The heap of sticks, thus piled up, would fill four large-sized wheel-barrows, and must require infinite labour. This ingenious little animal measures six inches from the tip of the nose to the tail, which is six inches long. The length of the head is two and a half inches, of the ears one and a quarter, and one inch in breadth. Its fur is of a light brown colour, and of exceedingly fine texture. It differs very little in appearance from the common rat, if I except the length of its ears, and an apparent disproportion in the size of the hind feet, which were large. The one figured is a male, which I obtained from one of the natives who followed us to the camp.

At this period of our journey the weather was exceedingly cold, and the winds high. We were about 45 miles from Laidley's Ponds; but could not, from the most elevated point, catch a glimpse of the ranges in its neighbourhood. It appeared to me that the river flats were getting smaller on both sides of it, the river still continuing to rise. It was now pouring down a vast body of water into the Murray. There was, however, an abundance of luxuriant pasture along its banks. Late in the afternoon the lubras (wives) of the natives, at our camp, made their appearance on the opposite side of the river, and Nadbuck, who was a perfect gallant, wanted to invite them over; but I told him that I would cut off the head of the first who came over with my lonely knife—my sword. The old gentleman went off to Mr. Browne, to whom he made a long complaint, asking him if he really thought I should execute my threat. Mr. Browne assured him that he was quite certain I should not only cut of the lubra's head, but his too. On this Nadbuck expressed his indignation; but however much he might have ventured to risk the lubra's necks he had no idea of risking his own.

One of the natives who visited us at this place was very old, with hair as white as snow. To this man I gave a blanket, feeling assured it would be well bestowed; although a circumstance occurred that had well nigh prevented my behaving with my usual liberality to the natives who were here with us. The butcher had been killing a sheep, and carelessly left the steel, an implement we could ill spare, under the tree in which he had slung the animal: and it was instantly taken by the natives. On hearing this, I sent for Nadbuck and Toonda, and told them that I should not stir until the steel was brought back, or make any more presents on the river. On this there was a grand consultation between the two. Toonda at length went to the natives, who had retired to some little distance, and, after some earnest remonstrances, he walked to the tree near which the sheep had been killed, and, after looking at the ground for a moment, began to root up the ground with his toes, when he soon discovered the stolen article, and brought it to me. The thief was subsequently brought forward, and we made him thoroughly ashamed of himself; although I have no doubt the whole tribe would have applauded his dexterity if he had succeeded.

The day was exceedingly cold, as the two or three previous ones had been, but still the temperature was delightful. We travelled, on this day, across the river flats, which again opened out to a distance of two or three miles; the ground, however. was of a most distressing character, and we had to cross several sandy points projecting into them, so that the poor animals were much jaded. This, however, was only the beginning of their troubles, for we were, in like manner, obliged to travel for several successive days over the same kind of ground—land on which floods have gradually subsided, and which has been blistered and cracked by solar heat. Travelling on this kind of ground was, indeed, more distressing to the cattle than even the hard pull over sand; for it was impossible for the bullock-drivers to steer clear of the many fissures and holes on these flats, and the shock, when the drays fell into any of them, was so great, that it shook the poor brutes almost to pieces.

From this period to the 9th there was a sameness in our progress up the Darling. On the 3rd we crossed a small creek, into which the waters of the river were flowing fast; and which both Nadbuck and Toonda informed us joined Yertello Lake, and that the Ana-branch was on the other side of the lake. This explanation accounted to us for a statement made by Toonda, shortly after he first joined us, that the Ana-branch hereabouts formed a great lake. On the 4th a little rain fell, but not in such quantity as to interfere with our travelling. On the 5th we passed a tribe of natives, in number about thirty-four. We were again led by Nadbuck across the country, to avoid the more circuitous route along the river. We passed through a more pleasing country than usual, and one that was better timbered and better grassed than it had been at any distance from the river.

I have mentioned that Toonda was attended by a young lad, his nephew, who, with another young lad, joined us at Lake Victoria. These two young lads used to keep in front with myself or Mr. Poole, or Mr. Browne, and were quite an amusement to us. This day both of them disappeared, not very long after we passed the last tribe. On making inquiries I ascertained, to my surprise, that they had been forcibly taken back by three men from the last tribe, and that both cried most bitterly at leaving the party. The loss of his nephew greatly afflicted poor Toonda, who sobbed over it for a long time. We could not understand why the natives had thus detained the boys; but, I believe, they were members of that tribe, between which and a tribe higher up the river some ground of quarrel existed. After the departure of these boys we had only three natives with us, who had been with the party from Lake Victoria, i.e. Nadbuck, Toonda, and Munducki, a young man who had attached himself to Kirby, who cooked for the men. The latter turned out to be a son of old Boocolo, a chief of the Williorara tribe, whom I shall, ere long, have occasion to introduce to the reader. Mr. Browne, with the assistance of Nadbuck, gathered a good deal of information from the natives then with us, as to the inhospitable character of the country to the north-west of the Williorara, or Laidley's Ponds, that agreed very little with the accounts we had previously heard. They stated that we should not be able to cross the ranges, as they were covered with sharp pointed stones and great rocks, that would fall on and crush us to death; but that if we did get across them to the low country on the other side, the heat would kill us all. That we should find neither water or grass, or wood to light a fire with. That the native wells were very deep, and that the cattle would be unable to drink out of them; and, finally, that the water was salt, and that the natives let down bundles of rushes to soak it up.

Such was the account the natives gave of the region into which we were going. We were of course aware that a great deal was fiction, but I was fully prepared to find it bad enough. From the opinion I had formed of the distant interior, and. from my knowledge of the country, both to the eastward and westward of me, I had no hope of finding it good within any reasonable distance.

Prepared, however, as I was for a bad country, I was not prepared for such as the natives described.

It was somewhat strange, that as we neared the supposed scene of the slaughter of the overlanders, we should fail in obtaining intelligence regarding it; neither were the natives, who must have participated in it, so high up the river as we now were, afraid of approaching us, as they undoubtedly would have been if they had been parties to it. I began, therefore, to suspect that it was one of those reports which the natives are, unaccountably, so fond of spreading without any apparent object in view.

As we approached Williorara the course of the river upwards was somewhat to the westward of north. The country had an improved appearance as we ascended it, and grass seemed to be more generally distributed over the flats. We passed several large lagoons, which had already been filled from the river, and were much pleased with the picturesque scenery round them.

On the 7th Jones broke the pole of his dray, and Morgan again broke his shaft, but we managed to repair both without the loss of much time—and made about ten miles of northing during the day. We hereabouts shot several new birds; and the dogs killed a very fine specimen of the Dipus of Mitchell, but, unfortunately, in the scuffle, they mangled it so much that we could not preserve it.

On the 8th the weather was oppressively hot, but we managed to get on some fifteen miles before we halted. Our journey up the Darling had been of greater length than I had anticipated, and it appeared to me that I could not do better than reduce the ration of flour at this early stage of the expedition to provide the more certainly for the future. I accordingly reduced it to eight pounds a week, still continuing to the men their full allowance of meat and other things.

Nadbuck had assured me on the 9th that if the bullocks did not put out their tongues we should get to Laidley's Ponds that day, but I hardly anticipated it myself, although I was aware we could not be many miles from them.

We had a great many natives in the neighbourhood at our encampment of the 5th, but they did not approach the tents. Their families generally were on the opposite side of the river, but one man had his lubra and two children on our side of it. My attention was drawn to him, from his perseverance in cutting a bark canoe, at which he laboured for more than an hour without success. Mr. Browne walked with me to the tree at which he was working, and I found that his only tool was a stone tomahawk, and that with such an implement he would hardly finish his work before dark. I therefore sent for an iron tomahawk, which I gave to him, and with which he soon had the bark cut and detached. He then prepared it for launching by puddling up its ends, and putting it into the water, placed his lubra and an infant child in it, and giving her a rude spear as a paddle pushed her away from the bank. She was immediately followed by a little urchin who was sitting on the bank, the canoe being too fragile to receive him; but he evidently doubted his ability to gain the opposite bank of the river, and it was most interesting to mark the anxiety of both parents as the little fellow struck across the foaming current. The mother kept close beside him in the canoe, and the father stood on the bank encouraging his little son. At length they all landed in safety, when the native came to return the tomahawk, which he understood to have been only lent to him. However I was too much pleased with the scene I had witnessed to deprive him of it, nor did I ever see a man more delighted than he was when he found that the tomahawk, the value and superiority of which he had so lately proved was indeed his own. He thanked me for it, he eyed it with infinite satisfaction, and then turning round plunged into the stream and joined his family on the opposite bank.

We journeyed as usual over the river flats, and occasionally crossed narrow sandy parts projecting into them. From one of these Mr. Poole was the first to catch a glimpse of the hills for which we had been looking out so long and anxiously. They apparently formed part of a low range, and bore N.N.W. from him, but his view was very indistinct, and a small cone was the only marked object he could distinguish. He observed a line of gum-trees extending to the westward, and a solitary signal fire bore due west from him, and threw up a dark column of smoke high into the sky above that depressed interior. A meridian altitude placed us in latitude 32°33'00"S., from which it appeared that we were not more than eight or ten miles from Laidley's Ponds, but we halted short of them, and received visits from a great many of the natives during the afternoon, who came to us with their families, a circumstance which led me to hope that we should get on very well with them. Poor Toonda here heard of the death of some relative during his absence, and had a great cry over it. He and the native who communicated the news sat down opposite to one another with crossed legs, and their hands on each other's shoulders. They then inclined their heads forward, so as to rest on each other's breasts and wept violently. This overflow of grief, however, did not last long, and Toonda shortly afterwards came to me for some flour for his friend, who he said was very hungry.

As it appeared to me that we should have to remain for some time in the neighbourhood of Laidley's Ponds, I had directed my inquiries to the state of the country near them, and learnt both from Nadbuck and Toonda, that we should find an abundance of grass for the cattle. I was not however very well satisfied with the change that had taken place within a few miles, in the appearance of the river, and the size of the flats, these latter having greatly diminished, and become less verdant. On the 10th we started on a west course, but at about a mile changed it for a due north one, which we kept for about five miles over plains rather more than usually elevated above the river flats. From these plains the range was distinctly visible, now bearing N.10°E., and N.26° and 38°W, distant 35 miles. It still appeared low, nor could we make out its character; three cones marked its southern extremity, and I concluded that it was a part of Scrope's Range. With the exception of these hills there were none other visible from Laidley's Ponds.

The ground whereon we now travelled was hard and firm, so that we progressed rapidly, and at five miles descended into a bare flat of whitish clay, on which a few bushes of polygonum were alone growing under box-trees. At about two hundred yards we were stopped by a watercourse, into which the floods of the Darling were flowing with great velocity. It was about fifty yards broad, had low muddy banks, and was decidedly the poorest spot we had seen of the kind. This, Nadbuck informed me, was the Williorara or Laidley's Ponds, a piece of intelligence at which I was utterly confounded. I could not but reproach both him and Toonda for having so deceived me; but the latter said he had been away a long time and that there was plenty of grass when he left. Nadbuck, on the other hand, said he derived his information from others, and only told me what they told him. Be that as it may, it was impossible for me to remain in such a place, and I therefore turned back towards the Darling, and pitched my tents at its junction with the Williorara.

For three or four days prior to our arrival at Laidley's Ponds, the upward course of the river had been somewhat to the west of north. The course of Laidley's Ponds was exceedingly tortuous, but almost due west. The natives explained to us that it served as a channel of communication between two lakes that were on either side of it, called Minandichi and Cawndilla. They stated that the former extended between the Darling and the ranges, but that Cawndilla was to the westward, at the termination of Laidley's Ponds, by means of which it is filled with water every time the Darling rose; but they assured me that the waters had not yet reached the lake. It was nevertheless evident that we were in an angle, and our position was anything but a favourable one. From the point where we had now arrived the upward course of the Darling for 300 miles is to the N.E., that which I was anxious to take, was to the W.N.W. It was evident, therefore, that until every attempt to penetrate the interior in that direction had proved impracticable, I should not have been justified in pushing farther up the river. My hopes of finding the Williorara a mountain stream had been wholly disappointed, and the intelligence both Mr. Eyre and I had received of it from the Murray natives had turned out to be false, for instead of finding it a medium by which to gain the hills, I now ascertained that it had not a course of more than nine or ten miles, and that it stood directly in my way. We were as yet ignorant what the conduct of the natives towards us would be, having seen none or very few who could have taken part in the dispute between Sir Thomas Mitchell and the Williorara tribe in 1836. Expecting that they might be hostilely disposed towards us, I hesitated leaving the camp, lest any rupture should take place between my men and the natives during my absence; much less could I think of fortifying the party in a position from which, in the event of an attack, they would find it difficult to retreat. I thought it best therefore to move the camp to a more distant situation with as little delay as possible, and send Mr. Poole to visit the ranges, and ascertain from their summit the probable character of the N.W. interior.

Having come to this decision, I procured a guide to accompany that officer to the hills, who accordingly started for them, with Mr. Stuart, my draftsman, the morning after our arrival at the ponds. Some of the natives had informed us that there was plenty of feed at the head of Cawndilla Lake, a distance of seven or eight miles to the W.S.W.; but we could not understand from them how far the waters of the Darling had passed up the creek, although it was clear from what they said that they had not yet reached Cawndilla. My instructions to Mr. Poole were framed with it a view to our removal from our present position nearer to the ranges, and I therefore told him to cross the creek at the head of the water, and if he should find grass there, to return to the camp, if not, to continue his journey to the hills, and use every effort to find water and feed. We had had a good deal of rain during the night of the 10th; the morning of the 11th was hazy, with the wind at S.W., and there appeared to be every prospect of continued wet. Under less urgent circumstances, therefore, I should have detained Mr. Poole until the weather cleared, but our movements at this time were involved in too much uncertainty to admit of delay. I had hoped that the morning would have cleared, but a light rain set in and continued for several days.

We had seen fewer natives on the line of the Darling than we had expected; but as we approached Williorara they were in greater numbers. Our tents were hardly pitched at that place, when, as I have observed, we were visited by the local tribe, with their women and children, who sat down at some little distance from the drays, and contented themselves with watching our motions. I had tea made for the ladies, of which they seemed to approve highly, and gave the youngsters two or three lumps of sugar a-piece. The circumstance of the women and children thus venturing to us, satisfied me that no present hostile movement was contemplated by the men; but, notwithstanding that there was a seeming friendly feeling towards us, there was a suspicious manner about them, which placed me doubly on my guard, and caused me to doubt the issue of our protracted sojourn in the neighbourhood.

I had several of the natives in my tent, and with Mr. Browne's assistance questioned them closely as to the character of the country to the north west, but we could gather nothing from what they said. They spoke of it in terror, as a region into which they did not dare to venture, and gave me dreadful accounts of the rocks and difficulties against which I should have to contend. They agreed, however, in saying that there was both water and grass at the lake; in consequence, I sent Mr. Browne with Nadbuck to examine the locality on the morning of the 12th, as the distance was not greater than from six to seven miles. He returned about one P.M., and informed me that there was plenty of feed for the cattle, and water also; but that the water was at least a mile and a half from the grass, which was growing in tufts round the edge of the lake. It appeared that the Williorara made a circuitous and extensive sweep and entered Cawndilla on the opposite side to that of the river, so that he had to cross a portion of the lake and thus found that the floods had not reached it. Mr. Browne also stated that the extent of the lake was equal to that of Lake Victoria, but that it could at no time be more than eighteen inches deep. It was indeed nothing more than a shallow basin filled by river floods, and retaining them for a short time only. Immense numbers of fish, however, pass into these temporary reservoirs, which may thus be considered as a providential provision for the natives, whose food changes with the season. At this period they subsisted on the barilla root, a species of rush which they pound and make into cakes, and some other vegetables; their greatest delicacy being the large caterpillar (laabka), producing the gum-tree moth, an insect they procure out of the ground at the foot of those trees, with long twigs like osiers, having a small hook at the end. The twigs are sometimes from eight to ten feet long, so deep do these insects bury themselves in the ground.

Mr. Browne communicated with a tribe of natives, one of whom, a very tall woman, as well as her child, was of a copper colour.

From the information he gave me of the neighbourhood of Cawndilla, I determined, on the return of Mr. Poole, and in the event of his not having found a better position, to move to that place; for it was evident from his continued absence that he must have crossed the creek at a distance from the lake, and not seeing any grass in its neighbourhood, had pushed on to the hills. I was now anxious for his return, for we had had almost ceaseless though not heavy rain since he left us. On the 12th, the day he started, we had thunder; on the 13th it was showery, with wind at N. W., and the thermometer at 62° at 3 P.M., and the barometer at 29.742; the boiling point of water being 211.25.

Assuming Sir Thomas Mitchell's data to be correct, my position here was in long. 142°5'E, and in lat. 32°25'S.

CHAPTER IV.

TOONDA'S TRIBE—DISPOSITION OF THE NATIVES—ARRIVAL OF CAMBOLI—HIS ENERGY OF CHARACTER—MR. POOLE'S RETURN—LEAVE THE DARLING—REMARKS ON THAT RIVER—CAWNDILLA—THE OLD BOOCOLO—LEAVE THE CAMP FOR THE HILLS—REACH A CREEK—WELLS—TOPAR'S MISCONDUCT—ASCEND THE RANGES—RETURN HOMEWARDS—LEAVE CAWNDILLA WITH A PARTY—REACH PARNARI—MOVE TO THE HILLS—JOURNEY TO N. WEST—HEAVY RAINS—RETURN TO CAMP—MR. POOLE LEAVES—LEAVE THE RANGES—DESCENT TO THE PLAINS—MR. POOLE'S RETURN—MR. POOLE'S REPORT—FLOOD'S CREEK—AQUATIC BIRDS—RANGES DIMINISH IN HEIGHT.

TOONDA left us on our arrival at this place, to go to his tribe at Cawndilla, but returned the day Mr. Poole left us, with the lubras and children belonging to it, and the natives now mustered round us to the number of sixty-six. Nadbuck, who the reader will have observed was a perfect lady's man, made fires for the women, and they were all treated as our first visitors had been with a cup of tea and a lump of sugar. These people could not have shewn a greater mark of confidence in us than by this visit; but the circumstances under which we arrived amongst them, the protection we had given to some of their tribe, and the kind treatment we had adopted towards the natives generally, in some measure accounted for this, nevertheless there was a certain restlessness amongst the men that satisfied me they would not have hesitated in the gratification of revenge if they could have mustered sufficiently strong, or could have caught us unprepared.

It was clear that the natives still remembered the first visit the Europeans had made to them, and its consequences, and that they were very well disposed to retaliate. It was in this matter that Nadbuck's conduct and representations were of essential service, for he did not hesitate to tell them what they might expect if they appeared in arms.

Mr. Poole was short and stout like Sir Thomas Mitchell, and personally very much resembled him; moreover, he wore a blue foraging cap, as, I believe, Sir Thomas did; be that as it may, they took Mr. Poole for that officer, and were exceedingly sulky, and Nadbuck informed us that they would certainly spear him. It was necessary, therefore, to explain to them that he was not the individual for whom they took him, and we could only allay their feelings by the strongest assurances to that effect; for some time, indeed, they were inclined to doubt what we said, but at length they expressed great satisfaction, and to secure himself still more Mr. Poole put on a straw hat. Nevertheless, there were manifestations of turbulence amongst the younger men on several occasions, and they certainly meditated, even though, for particular reasons, they refrained from any act of violence.

The constant rain had made the ground in a sad state. There was scarcely any stirring out of the tents into the tenacious clay of the flat in which they were pitched; and the Darling, continuing to rise, overflowed its banks, drove our cattle from their feed, and obliged us to send them to a more distant point. In the midst of all this we were, on the 13th, most agreeably surprised by the appearance of our friend Camboli, with two other natives from Lake Victoria. Camboli brought despatches and letters in reply to those I had sent from the lake. It is impossible to describe the unaffected joy this poor native evinced on seeing us again. He had travelled hard to overtake us, and his condition when he arrived, as well as that of his companions proved that they had not spared themselves; but neither of them shewed the same symptoms of fatigue as Camboli. His thighs and ancles, and the calves of his legs were much swollen, and he complained of severe pain in his back and loins; but he was excited beyond measure, and sprang about with surprising activity whilst his comrades fell fast asleep. "Papung," he exclaimed, meaning paper or letters. "I bring papung to Boocolo," meaning me; "to Sacoback," meaning Doctor Browne; "and Mr. Poole, from Gobbernor," the Governor; "Hugomattin," Mr. Eyre; "Merilli," Mr. Scott of Moorundi; "and Bullocky Bob. Papung Gobbernor, Boocolo, Hugomattin." Nothing could stop him, nor would he sit still for a moment. There were, at the fire near the tents, a number of the young men of the Williorara tribe; and it would appear, from what occurred, that they were talking about us in no friendly strain. Certain it is that they made some remark which highly offended our lately-arrived envoy, for he suddenly sprang upon his feet, and, seizing a carbine, shook it at them in defiance, and, pointing to the tents, again shook it with all the energy and fearlessness of a savage, and he afterwards told us that the natives were "murry saucy." The scene was of a kind that is seldom if ever witnessed in civilized life.

The reader may be assured we took good care of him and his companions; but his excitement continued, even after he had laid down to sleep; yet, he was the first man up on the following morning, to cut a canoe for Mr. Browne, who wished to cross the river with a young lad of the name of Topar, a native f the place, who had been recommended to me by Mr. Eyre, a fine handsome young man, about eighteen years of age, and exceedingly prepossessing in appearance; but I am sorry to say with very few good qualities. He was a boy about eight when Sir Thomas Mitchell visited the neighbourhood, and, with his mother, was present at the unfortunate misunderstanding between his men and the natives on that occasion.

The bark was not in a fit state to be stripped from the tree, so that Camboli had a fatiguing task, but he got the canoe ready in sufficient time for Mr. Browne to cross the river and visit Sir Thomas Mitchell's last camp, which I had intended doing myself, in order to connect it with my own, if circumstances had not, at that time, prevented me.

Mr. Poole returned on the 15th, after an absence of four days and a half. He informed me that he had crossed the creek, as I had imagined, where there was little or no vegetation in its vicinity. He then took up a north-west course for the hills, and rode over flats of polygonum for nine miles, when he crossed the bed of a large lagoon; arriving at a round hill, somewhat detached from the main range, at half-past one, and searched about for water, but found none, neither could the native point out any to him. He therefore descended to the plains, and encamped.

On the following morning Mr. Poole again crossed the hill he had ascended the day before, but at half-past one changed his course for a high peak on the same range, on the summit of which he arrived at 2 P.M.; but the day was unfavourable, and the bearings from it consequently uncertain. The following morning being clear he again ascended the hill, and took the following bearings:—To the point of a distant range N.54°W.; to a very distant cone, 00 or due north; to a peak in a distant range, S.40°W.; to a lake, S.20°W.; and to another distant range, S.65°W. The country between the ranges Mr. Poole had ascended and the more distant ones, appeared to be flat, and covered with brush and spear grass. There was an appearance of water between the ranges, and they looked like islands in an immense lake. He did not think he could have been deceived by the effect of mirage; but felt satisfied, according to his own judgment, that he had seen a large body of water to the N.W. Mr. Poole did not succeed in finding any convenient place to which to remove the party, and his guide persisting in his statement that there was no water in the hills, he thought it better to return, to the camp.

However doubtful I might have been as to the reality of the existence of water in the direction to which Mr. Poole referred, it was clear that the were other and loftier ranges beyond those visible from the river. Taking everything into consideration, I determined on moving the camp to Cawndilla, and on proceeding myself to the north-west as soon as I should have established it in a secure place.

I was employed on the 16th in reporting our progress to the Governor, as Nadbuck and Cambol, were to leave us in the afternoon on their return to Lake Victoria. Both were exceedingly impatient to commence their journey, but when I came out with the bag old Nadbuck evinced great emotion and sorrow, nor could we look on the departure of our old and tried guide without regret. He had really served us well and faithfully, and if he had anything to do in propagating the several reports by which we had been deceived in our progress up the Darling, I believe it was with a view to prevent our going into a country from which he thought we should never return. We rewarded him as he deserved, and sent both him and his companions away with provisions sufficient to last them during the greater part of their journey, but we afterwards learnt that with the improvident generosity of the savage, they had appointed to meet a number of their friends in the bush, and consumed their whole supply before sunset.

The weather had cleared, and as we were enabled to connect the Darling with the hilly country, I directed Mr. Poole to measure a base line from a point at the back of our camp to the westward. This base line ran along the sandy ridge above the flats of Laidley's Ponds towards Cawndilla, so that we had no detention, but left the Darling on the 17th. The drays started early in the forenoon, but I remained until two, to take some lunars with Mr. Browne. At that hour we rode along the dray tracks, and at six miles descended into the bed of the lake, and crossing a portion of it arrived at the camp at half-past five. The floods were just crossing the dray tracks as we passed, and gradually advancing into the basin. The ground was cracked and marked with narrow but deep fissures into which the waters fell as they rolled onwards, and it was really surprising to see the immense quantity these chasms required to fill them.

Having taken leave of the Darling, it may be as well that I should make a few general remarks upon it. The reader will have observed from my description, that the scenery on the banks is picturesque and cheerful, that its trees though of smaller size than those on the Murray, are more graceful and have a denser foliage and more drooping habit, and that the flats contiguous to the stream abundantly grassy. I have described the river as I found it, but I would not have the reader suppose that, it always presents the same luxuriant appearance, for not many months before this period my persevering, friend Mr. Eyre, on a journey up its banks, could hardly find grass sufficient for his horses. There was not a blade of vegetation on the flats, but little water, in the river, and the whole scenery wore a barren appearance. Countries, however, the summer heat of which is so excessive, as in Australia, always subject to such changes, nor is it any argument against their soil, that it should at one season of the year look bare and herbless. That part of the Darling between Laidley's Ponds and its junction, with the Murray, a distance of about 100 miles in a direct line, had not been previously explored, nor, had I time to lay it regularly down. I should say from the appearance of its channel that it is seldom very deep, frequently dry at intervals, and that its floods are uncertain, sudden, and very temporary. That they rise rapidly may be implied from the fact that in two days the floods we witnessed rose more than nine feet, and that they come from the higher branches of the river there can be no doubt, since the Darling has no tributary between Laidley's Ponds and Fort Bourke. I have no doubt but the whole line of the river will sooner or later be occupied, and that both its soil and climate will be found to suit the purpose both of the grazier and the agriculturist. Be that as it may, I regretted abandoning it, for I felt assured that in doing so our difficulties and trials would commence.

Our camp at Cawndilla was on the right bank of the Williorara, about half a mile above where it enters the lake. Without intending it, we dispossessed the natives of the ground which they had occupied before our arrival, but they were not offended. Our tents stood on a sand bank close to the creek, and was shaded by gum-trees and banksias; behind us to the S.W. there were extensive open plains, and along the edge of the basin of Cawndilla, as well as to some distance in its bed, there was an abundance of feed for our cattle: the locality would be of great value as a station if it were near the located districts of South Australia.

The term Boocolo is I believe generally given to the chief or elder of the tribe, and thus was applied by the natives to me, as chief of the party. The boocolo of the Cawndilla tribe was an old man with grey hairs and rather sharp features, below the ordinary stature, but well made and active. Of all the race with whom I have communicated, his manners were the most pleasing. There was a polish in them, a freedom and grace that would have befitted a drawing-room. It was his wont to visit my tent every day at noon, and to sleep during the heat; but he invariably asked permission to do this before he composed himself to rest, and generally laid down at my feet. Differing from the majority of the natives, he never asked for anything, and although present during our meals kept away from the table. If offered anything he received it with,' becoming dignity, and partook of it without displaying that greedy voracity which the natives generally exhibit over their meals. He was a man, I should, say, in intellect and feeling greatly in advance of his fellows. We all became exceedingly partial to this old man, and placed every confidence in him; although, as he did not understand the language of the Murray natives, we gained little information from him as to the remote country.

The boocolo of Cawndilla had two sons; but as the circumstances under which they were more particularly brought forward occurred on the return of the expedition from the interior, I shall not mention them here; but will conclude these remarks by describing an event that took place the day after our removal from the Darling. The men who had been out chaining left the flags standing after their work, and came to the camp. When Mr. Poole went out the next morning, he found that one of them had been taken away. The natives, when charged with the theft, stoutly denied it, and said that it had been stolen by one of the Darling tribe in returning to the river. I therefore directed him, as he generally superintended the issue of presents and provisions to the natives, to stop all further supplies. The old boocolo failed in his endeavours to recover the flag, and the natives who visited the camp were evidently under restraint. On the following day the boocolo came to my tent, and I spoke angrily to him. "Why," I asked, "has the black fellow taken that which did not belong to him? I do not take anything from you. I do not kill your kangaroos or take your fish." The old man was certainly much annoyed, and went out of the tent to our fire, at which there were several natives with whom he had an earnest conversation; this terminated by two of them starting for the Darling, from whence, on the following day, they brought back the flag and staff, which they said had been taken by three of the Darling natives as they had stated already.

Probably such was the case, and we admitted the excuse. The base line was completed on the 19th, and measured six miles. I was anxious to have made it of greater length, but the ground would not admit of it. The angles were necessarily very acute; but the bearings were frequently repeated, and found to agree. I was the less anxious on the point because my intention was to check any error by another line as soon as I could.

The position we had taken up was a very favourable one, since being on the right or northern bank of the creek, we were, by the flooding of the lake, cut off from the Darling natives. I now therefore determined on making an excursion into the interior to the N.W., to examine the ranges seen by Mr. Poole, and to ascertain if, as he supposed, there was a body of water to the westward of them. With this view I engaged Topar to accompany us, and on the 21st left the camp, with Mr. Browne, Flood, and Morgan, taking the light cart with our provisions, and some water-casks. During the recent rains the weather had been very cold, but excessive heat succeeded it. The day before we started the thermometer rose as high as 112° during a violent hot wind; and certainly if the following day had been equally warm we could not have proceeded on our journey. Fortunately for us, however, the wind shifted to the S.W. during the night, and the morning was cool and refreshing. I should have commenced this trip two or three days earlier, but on the 20th we were surprised by the reappearance of old Nadbuck, who had turned back with some natives he met on the way to our camp, with letters from Moorundi. The old man was really overjoyed to see us again. He said he had left Camboli well advanced on his journey, and that he would have reached Lake Victoria before he (Nadbuck) had reached us. Some of the letters he brought requiring answers, I was unable to arrange for my intended departure on the 19th. The 20th being a day of excessive heat, we could not have ventured abroad; but as I have stated, on the 21st we commenced the journey under more favourable circumstances than we had anticipated. The old boocolo took leave of Mr. Browne and myself, according, I suppose, to the custom of his people, by placing his hands on our shoulders and bending his head so as to touch our breasts; in doing which he shed tears. Topar, seated on the cart, was followed by his mother who never expected to see him again. I had given Topar a blanket, which he now gave to his parent, and thus set off with us as naked as he was born. I mention this the more readily because I have much to detail to his discredit, and therefore in justice, I think, I am bound to record anything to his advantage. At a quarter of a mile from the camp we crossed the little sand hill which separates the two basins of Cawndilla and Minandichi, from which we descended into the flats of the latter, but at a mile rose, after crossing a small creek, to the level of the great plains extending between us and the ranges. Our first course over these plains was on a bearing of 157° to the west of south, or N.N.W. nearly. They were partly covered by brush and partly open; the soil was a mixture of clay and sand, and in many places they resembled, not only in that but in their productions, the plains of Adelaide. A good deal of grass was growing on them in widely distributed tufts, but mixed with salsolaceous plants. The trees consisted of a new species of casuarina, a new caparis, with some hakea, and several species of very pretty and fragrant flowering shrubs. At twelve miles we changed our course to 135° to the west of south, or N.W.W. and kept upon it for the remainder of the day, direct for a prominent hill in the ranges before us.* The hills Mr. Poole had visited then bore a few degrees to the east of north, distant from twelve to fourteen miles, and were much lower than those towards which we were going, continuing north-wards. The country as we advanced became more open and barren. We traversed plains covered with atriplex and rhagodia, in the midst of which there were large bare patches of red clay. In these rain water lodges, but being exceedingly shallow they soon dry up and their surfaces become cracked and blistered. From the point at which we changed our course the ground gradually rose, and at 26 miles we ascended a small sand hill with a little grass growing upon it. From this hill we descended into and crossed a broad dry creek with a gravelly bed, and as its course lay directly parallel to our own, we kept in the shade of the gum-trees that were growing along its banks. At about four miles beyond this point

*Coonbaralba Station, No. 2.

Topar called out to us to stop near a native well he then shewed us, for which we might in vain have hunted. From this we got a scanty supply of bad water, after some trouble in cleaning and clearing it, insomuch that we were obliged to bale it out frequently during the night to obtain water for our horses. This creek, like others, was marked by a line of gum-trees on either side; and from the pure and clean gravel in its bed, I was led to infer that it was subject to sudden floods. We could trace the line of trees upon it running upwards to the N.W. close up to the foot of the ranges, and down southwards, until the channel seemed to be lost in the extensive flats of that depressed region.

Topar called this spot "Murnco Murnco." As the horses had fared indifferently during our stay, and he assured us there was a finer well higher up the creek, we pushed on at an early hour the next morning, keeping on the proper right bank of the creek, and having an open barren country to the south, with an apparent dip to the south-west; to our left, some undulations already noticed by us, assumed more the shape of hills. The surface was in many places covered with small fragments of white quartz, which together with a conglomerate rock cropped out of the ground where it was more elevated. There was nothing green to meet the eye, except the little grass in the bed of the creek itself, and a small quantity on the plains.

At two miles on our former bearing Topar stopped close to another well, but it was dry and worthless; we therefore pushed on to the next, and after removing a quantity of rubbish, found a sufficiency of water both for ourselves and the horses, but it was bitter to the taste, and when boiled was as black as ink from the decoction of gum leaves; the water being evidently the partial and surface drainage from the hills. We stopped here however to breakfast. Whilst so employed, Topar's quick and watchful eye caught sight of some smoke rising from the bed of the creek about a mile above us. He was now all impatience to be off, to overtake the party who had kindled it. Nothing could exceed his vehement impetuosity and impatience, but this was of no avail, as the natives who had probably seen our approach, kept in front of us and avoided a meeting. We rode for five miles on our original bearing of 135° to the west of north, or N.W. the direct bearing of the hill for which we were making, Coonbaralba. At five miles Topar assisted on crossing the creek, and led us over the plains on a bearing of 157° to the west of north, thus, changing his purpose altogether. He assigned as a reason that there was no water in the creek higher up, and that we must go to another place where there was some. I was somewhat reluctant to consent to this, but at length gave way to him; we had not however gone more than two and a half miles, when he again caught sight of smoke due west of us, and was as earnest in his desire to return to the creek as he had been to leave it. Being myself anxious to communicate with the natives I now the more readily yielded to his entreaties. Where we came upon it there was a quantity of grass in its bed, but although we saw the fire at which they had been, the natives again escaped us. Mr. Browne and Topar ran their track up the creek, and soon reached a hut opposite to which there was a well. On ascending a little from its bed they discovered a small pool of water in the centre of a watercourse joining the main branch hereabouts from the hills. Round this little pool there was an unusual verdure. From this point we continued to trace the creek upwards, keeping it in sight; but the ground was so stony and rough and the brush approached so close to the banks that I descended into its bed, and halted at sunset after a fatiguing day's journey without water, about which we did not much care; the horses having had a good drink not long before and their feed being good, the want of water was not much felt by them. Topar wished to go on to some other water at which he expected to find the natives, and did not hesitate for a moment in thus contradicting his former assertion. This however I would not allow him to do alone, but Mr. Browne good-naturedly walked with him up the creek, and at less than a mile came up on a long and beautiful pond He informed me that it was serpentine in shape and more than eighty yards long, but as there was no grass in its neigbourhhood I did not move to it. It was evident that Topar had intended leading us past this water, and it was owing to his anxiety to see the natives that we had now discovered it.

On the following morning I determined to take the direction of our movements on myself, and after we had breakfasted at the long water-hole, struck across the plains, and took up a course of 142° to the west of south for a round hill which I proposed ascending. Topar seeing us determined, got into a state of alarm almost bordering on frenzy; he kept shouting out "kerno, kerno," "rocks, rocks," and insisted that we should all be killed. This however had no effect on us, and we continued to move towards a spur, the ascent of which appeared to be less difficult than any other point of the hills. We reached its base at 10 A.M. and had little trouble in taking the cart up. On gaining the top of the first rise, we descended into and crossed a valley, and ascending the opposite side found ourselves on the summit of the range, the surface being much less broken than might have been anticipated, insomuch that we had every hope that our progress amongst the hills would be comparatively easy; but in pushing for the one I wished to ascend, our advance was checked by a deep ravine, and I was obliged to turn toward another hill of nearly equal height on our left. We descended without much difficulty into a contiguous valley, but the ascent on the opposite side was too rough for the cart. We had pressed up it along a rocky watercourse, in which I was obliged to leave Morgan and Topar. Mr. Browne, myself, and Flood, with our horses reached the top of the hill at half-past twelve. Although the position commanded a considerable portion of the horizon there was nothing cheering in the view. Everything below us was dark and dreary, nor was there any indication of a creek to take us on to the north-west. We could see no gum-trees in that direction, nor indeed could we at an elevation of 1600 feet above the plains distinctly make out the covering of the ground below. It appeared to be an elevated table land surrounded by hills, some of which were evidently higher than that on which we stood.

The descent to the westward was still more precipitous than the side we had ascended. The pass through which the creek issued from the hills was on our left, Coonbaralba being between us and it, but that hill was perfectly inaccessible; I thought it better therefore to return to sleep at the water where we had breakfasted, with a view to running the creek up into the ranges on the following morning. After taking bearings of the principal objects visible from our station, we rejoined

Morgan and descended to the plains. There was a little water in the creek leading from the hill I had at first intended to ascend, to the S.W., which was no doubt a branch of the main creek. On our return we saw that beautiful flower the Clianthus formosa, in splendid blossom on the plains. It was growing amidst barrenness and decay, but its long runners were covered with flowers that gave a crimson tint to the ground.

The principal object I had in view during the excursion I was then employed upon, was if possible to find a proper position to which the party might move; for I foresaw that my absence would be frequent and uncertain, and although my men were very well disposed towards the natives, I was anxious to prevent the chance of collision or misunderstanding. I had now found such a position, for on examining the water-hole I felt satisfied that it might be depended upon for ten days or a fortnight, whilst the grass in its neighbourhood although dry was abundant. Wishing, however, to penetrate the ranges by the gap through which the creek issued from them, I still thought it advisable to prosecute my intended journey up it. Accordingly on the 24th we mounted our horses and rode towards the hills. A little above where we had slept we passed a small junction to the westward, and at 7 miles entered the gap, Coonbaralba, on the bearing of which we had run across the plains, being on our right. We had already passed several small water-holes, but at the entrance of the gap passed some larger ones in which the water was brackish, and these had the appearance of being permanent. Topar had shewn much inclination at our going on, and constantly remonstrated with us as we were riding along; however, we saw two young native dogs about a third grown, after which he bounded with incredible swiftness, but, when they saw him they started off also. It was, soon evident, that both were doomed to destruction, his speed being greater that that of the young brutes, for he rapidly gained upon them. The moment he got within reach of the hindmost he threw a stick which he had seized while running, with unerring precision, and striking it full in the ribs stretched it on the ground. As he passed the animal, he gave it a blow on the head with another stick, and bounding on after the other was soon out of our sight. All we knew further of the chase, was, that before we reached the spot where his first prize lay, he was returning to us with its companion as soon as he had secured his prey he sat down to take out their entrails, a point in which the natives are very particular. He was careful in securing the little fat they had about the kidneys, with which he rubbed his body all over, and having finished this operation he filled their insides with grass and secured them with skewers. This done he put them on the cart, and we proceeded up the pass, at the head of which we arrived sooner than I expected. We then found ourselves at the commencement of a large plain. The hills we had ascended the day before trended to the north, and there was a small detached range running perpendicular to them on our right. To the south there were different points, apparently the terminations of parallel ranges, and westward an unbroken line of hills. The creek seemed to trend to the S.W., and in that direction I determined to follow it, but Topar earnestly entreated us not to do so. He was in great consternation; said here was no water, and promised that if we would follow him he would shew us water in which we could swim. On this condition I turned as he desired, and keeping along the western base of the main or front range, took up a course somewhat obtuse to that by which I had crossed the plains of Cawndilla. The productions on the ground were of a salsolaceous kind, although it was so much elevated above the plains, but amongst them there was not any mesembryanthemum. At about three miles we passed a very remarkable and perfectly isolated hill, of about 150 feet in height. It ran longitudinally from south to north for about 350 yards, and was bare of trees or shrubs, with the exception of one or two casuarinas. The basis of this hill was a slaty ferruginous rock and protruding above the ground along the spine of the hill there was a line of the finest hepatic iron ore I ever saw; it laid in blocks of various sizes, an many tons weight piled one upon the other, without a particle of earth either on their faces or between them. Nothing indeed could exceed the clean appearance of these huge masses. On ascending this hill and seating myself on the top of one of them to take bearings, I found that the compass deviated 37° from the north point, nor could I place any dependance on the angles I here took.

At about nine miles the main range turned to N.N.E., and Topar accordingly keeping near its base changed his course, and at five miles more led us into a pass in some respects similar to that by which we had entered the range. It was however less confined and more open. Steep hills, with rocks in slabs protruding from many parts, flanked it to the south, whilst on its northern side perpendicular rocks, varying in height from 15 to 20 feet, over which the hills rose almost as perpendicularly more than 200 feet higher, were to be seen. Close under these was the stony bed of a mountain torrent, but it was also evident that the whole pass, about 160 yards broad, was sometimes covered by floods. Down this gully Topar now led us, and at a short distance, crossing over to its northern side, he stopped at a little green puddle of water that was not more than three inches deep. Its surface was covered with slime and filth, and our horses altogether rejected it. Some natives had recently been at the place, but none were there when we arrived. I was exceedingly provoked at Topar's treachery, and have

always been at a loss to account for it. At the time, both Mr. Browne and myself attributed it to the machinations of our friend Nadbuck; but his alarm at invading the hilly country was too genuine to have been counterfeited. It might have been that Nadbuck and Toonda expected that they would benefit more by our presents and provisions than if we left them for the interior, and therefore tried by every means to deter us from going: they certainly had long conversations with Topar before he left the camp to accompany us. Still I may do injustice to them in this respect. However, whether this was the case or not, we had to suffer from Topar's misconduct. I turned out of the pass and stopped a little beyond it, in a more sheltered situation. Here Topar coolly cooked his dogs, and wholly demolished one of them and part of the other. In wandering about the gorge of the glen, Mr. Browne found a native well, but there was no water in it.

Our camp at Cawndilla now bore S.S.E. from us, distant 70 odd miles, and having determined on moving the party, I resolved to make the best of my way back to it. On the following morning therefore, we again entered the pass, but as it trended too much to the eastward, I crossed a small range and descended at once upon the plain leading to the camp. At about 17 miles from the hills, Topar led us to a broad sheet of water that must have been left by the recent rains. It was still tolerably full, and water may perhaps be found here when there is none in more likely places in the hills. This spot Topar called Wancookaroo; it was unfortunately in a hollow from whence we could take no bearings to fix its precise position.

We halted at sunset on the top of a small eminence, from which the hills Mr. Poole had ascended bore E.N.E., and the hill at the pass N.W. We were suddenly roused from our slumbers a little before daylight by a squall of wind that carried every light thing about us, hats, caps, etc. all went together, and bushes of atriplex also went bounding along like so many foot-balls. The wind became piercing cold, and all comfort was gone. As morning dawned the wind increased, and as the sun rose it settled into a steady gale. We were here about forty miles from Cawndilla, nor do I remember having ever suffered so severely from cold even in Canada. The wind fairly blew through and through us, and Topar shivered so under it that Morgan gave him a coat to put on. As we seldom put our horses out of a walk, we did not reach the tents until late in the afternoon, but I never was more rejoiced to creep under shelter than on this occasion.

Every thing had gone on well during our absence, and Mr. Poole had kept on the most friendly terms with the natives.

I should have mentioned, that, as we descended from the hills, the quick eye of Topar saw a native at a great distance to our left, and just at the outskirt of a few trees. We should have passed him unperceived, but I requested Mr. Browne to ride up to and communicate with him. The poor fellow had dug a pit, for a Talperos,* big enough to hide himself in, and as he continued to work at it, did not see Mr. Browne approach, who stood mounted right over the hole before he called to him. Dire was the alarm of the poor native when he looked up and saw himself so immediately in contact with such a being as my companion must have appeared to him; but Mr. Browne considerately retired until he had recovered from his astonishment, and Topar, whom I sent to join them, coming up, he soon recovered his composure and approached the cart. As we had prevented the old man from securing his game, I desired Topar to give him the remains of the dog; but this he refused to do. I therefore ordered Morgan to take it from him, and told Topar I would give him an equivalent when we reach the camp. This native did not seem to be aware that the Darling was up, a piece of news that seemed to give him much joy and satisfaction. I kept my promise with Mr. Topar, but he deserved neither my generosity nor consideration.

Mr. Poole informed me that the fluctuations of temperature had been as great at Cawndilla with us; that the day before, the heat likewise had been excessive, the thermometer having risen, to 110° on the day of our return it was down to 38°. The natives appeared really glad to see us again, for I believe they had given us up for lost. My old friend shed tears when he embraced us, and Nadbuck, who still remained with

*A native animal about the size of a rabbit, but longer in shape.

Toonda shewed the most unequivocal signs of joy. Cawndilla bears about W.S.W. from the junction of the Williorara with the Darling, at a distance of from six to seven miles. We broke up our camp there on the 28th of October 1844, but, however easily Mr. Browne and I had crossed the plains to the north-west, it was a journey that I felt assured would try the bullocks exceedingly. The weather had again changed, and become oppressively hot, so that it behoved me to use every precaution, in thus abandoning the Darling river.

At early dawn Mr. Browne started with Flood, Cowley, and Kirby, in the light cart, to enlarge the wells at Curnapaga to enable the cattle to drink out of them. Naturally humane and partial to the natives, he had been particularly kind to Toonda, who in his way was I believe really attached to Mr. Browne. This singular man had made up his mind to remain with his tribe, but when he saw the cart, and Mr. Browne's horse brought up, his feelings evidently overpowered him, and he stood with the most dejected aspect close to the animal, nor could he repress his emotion when Mr. Browne issued from the tents; if our route had been up the Darling, I have no doubt Toonda would still have accompanied us, but all the natives dreaded the country into which we were going, and fully expected that we should perish. It was not therefore surprising that he wavered, more especially as he had been a long-time absent from his people, and there might be objections to his leaving them a second time. The real cause, however, was, I think, the overflowing of the Darling, and the usual harvest of fish, and incessant feasting the natives would have in consequence. Their god certainly is their belly, we must not therefore be surprised that Toonda wished to partake of the general abundance that would soon be at the command of his tribe, and probably that his assistance was required. However his heart failed him when he saw Mr. Browne mount his horse to depart, and he expressed his readiness to accompany us to the hills, but no farther. The Boocolo's son had also volunteered to go so far with his friend the cook: when therefore at 8 A.M. I followed Mr. Browne with the remainder of the party, he and Toonda got on the drays. We took a kind leave of the Boocolo, who put his two hands on my head, and said something which I did not understand. It was however the expression of some kind wish at parting. The cattle got on very well during the early part of the day, and at noon we halted for two hours. After noon our progress was slow, and night closed in upon us, whilst we were yet some distance from the creek. We reached the little sand hill near it, to which we were guided by a large fire Flood had kindled at midnight, for it appeared that the horses had given in, and that Mr. Browne had been obliged to halt there. On leaving Cawndilla I sent Mr. Poole to Scrope's Range, to verify his bearings, and to enable Mr. Stuart to sketch in the hills, but he had not at this time rejoined me. At early dawn on the 29th, I accompanied Mr. Browne to the wells, leaving Mr. Piesse with the horse-cart and drays. We arrived there at nine, and by twelve, the time when the oxen came up, had dug a large pit under a rock on the left bank of the creek, which filled rapidly with water. The horses however were still in the rear, and I was ultimately obliged to send assistance to them. At 1 P.M. Mr. Poole and Mr. Stuart rejoined us. Two of our kangaroo dogs had followed them from Cawndilla, but one only returned, the other fell exhausted on the plains. Mr. Poole informed me that he had seen, but lost sight of Flood's signal fire, and had therefore slept higher up on the creek. The animals, but the cart horses in particular, were still very weak when we left Curnapaga, on the 30th, nor is it probable we should have got them to the long water-hole if we had not fortunately stumbled on another little pool of water in a lateral creek about half way. After breakfasting here, we moved leisurely on, and reached our destination at half-past five, P.M. Sullivan shot a beautiful and new hawk (*Elanus scriptus,* Gould), which does not appear to extend farther south than where we here met it, although it wanders over the whole of the north-west interior as far as we went. There were some beautiful plants also growing in the bed of the creek; but we had previously met with so few things that we might here be said to have commenced our collection.

At this water-hole, "Parnari," we surprised the natives who were strangers. They did not betray any fear, but slept at the tents and left us the following day, as they said to bring more natives to visit us, but we

Parnari.

never saw any thing more of them. They were hill natives. and shorter in stature than the river tribes.

The day succeeding that of our arrival at Parnari was very peculiar, the thermometer did not rise higher than 81°, but the barometer fell to 28.730°, and the atmosphere was so light that we could hardly breathe. I had hoped that this would have been a prelude to rain, but it came not.

The period from the 1st to the 5th of November was employed in taking bearings from the loftiest points of the range, both to the northward and southward of us; in examining the creek to the south-west, and preparing for a second excursion from the camp.

The rock formation of Curnapaga was of three different kinds. A mixture of lime and clay, a tufaceous deposit, and an apparently recent deposit of soapstone, containing a variety of substances, as alumina, silica, lime, soda, magnesia, and iron. The ranges on either side of the glen were generally varieties of gneiss and granite, in many of which feldspar predominated, coarse ferruginous sandstone, and a siliceous rock with mammillary hematite and hornblende. These, and a great mixture of iron ores, composed the first or eastern line of Stanley's Barrier Range.

It will be remembered that in tracing up the creek on the occasion of our first excursion from Cawndilla, that Topar had persuaded me, on gaining the bead of the glen to go to the north, on the faith of a promise that he would take us to a place where there was an abundance of water, and that in requital he took us to a shallow, slimy pool, the water of which was unfit to drink. Mr. Browne and I now went in the direction we should have gone if we had been uninfluenced by this young cub, and at less than a hundred yards came upon a pretty little clear pool of water, that had been hid from our view by a turn of the creek. What motive Topar could have had in thus deceiving us, and punishing himself, is difficult to say. On our further examination of the creek, however, there was no more water to be found, and from the gravelly and perfectly even nature of its bed, I should think it all runs off as fast as the channel filled. Whilst I was thus employed, Mr. Poole and Mr. Stuart were on the ranges, and both, as well as the men generally, continued in good health; but I was exceedingly anxious about Mr. Browne, who had a low fever on him, and was just then incapable of much fatigue; nevertheless he begged so hard to be permitted to accompany me on my contemplated journey, that I was obliged to yield.

I had been satisfied from the appearance of the Williorara, that it was nothing more than a channel of communication between the lakes Cawndilla, and Minandechi and the Darling, as the Rufus and Hawker respectively connect Lakes Victoria and Bonney with the Murray, and I felt assured that as soon as we should leave the former river, our difficulties as regards the supply of water for our cattle would commence, and that although we were going amongst hills of 1500 or 2000 feet elevation, we should still suffer from the want of that indispensable element. Many of my readers judging from their knowledge of an English climate, and living perhaps under hills of less elevation than those I have mentioned, from which a rippling stream may pass their very door, will hardly understand this; but the mountains of south-east Australia bear no resemblance to the moss-covered mountains of Europe. There that spongy vegetation retains the water to give it out by degrees, but the rain that falls on the Australian hills runs off at once, and hence the terrific floods to which their creeks are subject. In the barren and stony ranges, through which I had now to force my way, no spring was to be found. During heavy rains, indeed, the torrents are fierce, and the waters must spread over the plains into which they descend for many miles; but such effects disappear with their cause; a few detached pools only remain, that are fed for a time by under drainage, which soon failing, the thirsty sun completes his work, and leaves that proscribed region—a desert.

Fully satisfied then that the greatest obstacle to the progress of the Expedition would be the want of water, and that it would only be by long and laborious search that we should succeed in gaining the interior, I determined on taking as much as I could on my proposed journey, and with a view to gaining more time for examining the country, I had a tank constructed, which I purposed to send a day or two in advance.

The little pond of which I have spoken at the head of the pass, had near it a beautiful clump of acacias of a species entirely new to us. It was a pretty graceful tree, and threw a deep shade on the ground; but with the exception of these and a few gum-trees the vicinity was clear and open. Our position in the creek on the contrary was close and confined. Heavy gusts of wind were constantly sweeping the valley, and filling the air with sand, and the flies were so numerous and troublesome that they were a preventative to all work. I determined, therefore, before Mr. Browne and I should start for the interior, to remove the camp to the upper part of the glen. On the 4th we struck our tents and again pitched them close to the acacias. Early on the morning of the 5th, I sent Flood with Lewis and Sullivan, having the cart full of water, to preserve a certain course until I should overtake them, being myself detained in camp with Mr. Browne, in consequence of the arrival of several natives from whom we hoped to glean some information; but in this we were disappointed. Toonda had continued with us as far as "Parnari;" but on our moving up higher into the hills, his heart failed him, and he returned to Cawndilla.

At eleven, Mr. Browne and I took leave of Mr. Poole, and pursuing a course of 140° to the west of south, rode on to overtake the cart. At about four miles from the camp we crossed a small ironstone range, from which we saw Flood and his party nearly at the foot of the hill on which I had directed him to move, and at which I intended to cross the ranges if the place was favourable. In this, however, we were disappointed, for the hills were too rugged, although of no great breadth or height. We were consequently obliged to turn to the south, and in going over the rough uneven ground, had the misfortune to burst our tank. I therefore desired Lewis to stop, and gave the horses as much water as they would drink, still leaving a considerable quantity in the tank, of which I hoped we might yet avail ourselves. Although we had found it impracticable to cross the ranges at the proposed point, Mr. Browne and I had managed to scramble up the most elevated part of them. We appeared still to be amidst broken stony hills, from which there was no visible outlet. There was a line of gum-trees, however, in a valley to the south-west of us, as if growing on the side of a creek that would in such case be tributary to the main creek on which our tents were pitched, and we hoped, by running along the base of the hills to the south and turning into the valley, to force our way onwards. At about three and a half miles our anticipations were verified by our arriving opposite to an opening leading northwards into the hills. This proved to be the valley we had noticed. A line of gum-trees marked the course of a small creek, which passing behind a little hill at the entrance of the valley, reappeared on the other side, and then trended to the N.W. Entering the valley and pursuing our way up it, at two miles we crossed another small creek, tributary to the first, and at a mile beyond halted for the night, without having found water. Although there was a little grass on the plains between the camp and the ranges, there was none in the valley in which we stopped. Low bushes of rhagodia and atriplex were alone to be seen, growing on a red, tenacious, yet somewhat sandy soil, whilst the ranges themselves were covered with low brush.

The water had almost all leaked out of the tank when we examined it, so that it was no longer of any service to us. On the morning of the 7th, therefore, I sent Lewis and Sullivan with the cart back to the camp, retaining Flood and Morgan to attend on Mr. Browne and myself.

When we started I directed them to follow up the creek, which did not appear to continue much further, and on arriving at the head of it to cross the range, where it was low, in the hope that they would strike the opposite fall of waters in descending on the other side, whilst I went with Mr. Browne to a hill from which I was anxious to take bearings, although Lewis, who had already been on the top of it, assured me that there was nothing new to be seen. However, we found the view to be extensive enough to enable us to judge better of the character of the country than from any other point on which we had yet been. It was traversed by numerous rocky ridges, that extended both to the north and south beyond the range of vision. Many peaks shewed themselves in the distance, and I was enabled to connect this point with "Coonbaralba," the hill above the camp. The ridge I had directed Flood to cross was connected with this hill, and appeared to create a division of the waters thereabouts. All however to the north or northwest was as yet confused. There was no visible termination of the ranges in any direction, nor could we see any feature to guide us in our movements.

The rock formation of this hill was a fine grained granite, and was in appearance a round and prominent feature. Although its sides were covered with low dark brush, there was a considerable quantity of oat-grass in its deep and sheltered valleys. We soon struck on Flood's track after leaving this hill, which, as Lewis had been the first to ascend, I called "Lewis's Hill;" and riding up the valley along which the men had already passed, at six miles crossed the ridge, which (as we had been led to hope) proved to be the range dividing the eastern and western waters. On our descent from this ridge we proceeded to the north-west, but changed our course to north in following the cart tracks, and at four miles overtook Flood and Morgan on the banks of a creek, the channel of which, and the broad and better grassed valley through which it runs, we ourselves had several times crossed on our way down, and from the first had hoped to find it the main creek on the west side of the ranges.

At the point where we overtook Flood it had increased greatly in size, but we searched its hopeless bed in vain for water, and as it there turned too much to the eastward, for which reason Flood had stopped until we should come up, we left it and crossed the low part of a range to our left; but as we were going too much to the south-west, I turned shortly afterwards into a valley that led me more in the direction in which I was anxious to proceed. The country had been gradually improving from the time we crossed the little dividing range, not so much in soil as in appearance, and in the quality of its herbage. There was a good deal of grass in the valleys, and up the sides of the hills, which were clear and open on the slopes but stony on their summits. After proceeding about two and a half miles, we got into a scrubby part of the hills, through which we found it difficult to push our way, the scrub being eucalyptus dumosa, an unusual tree to find in those hills. After forcing through the scrub for about half a mile, we were suddenly stopped by a succession of precipitous sandstone gullies, and were turned to the eastward of north down a valley the fall of which was to that point. This valley led us to that in which we had rejoined Flood, but lower down; in crossing it we again struck on the creek we had then left, much increased in size, and with a row of gum-trees on either side of it, but its even broad bed composed of the cleanest gravel and sand, precluded the hope of our finding water. At about a mile, however, it entered a narrow defile in the range, and the hills closed rapidly in upon it. Pursuing our way down the defile it gradually narrowed, the bed of the creek occupied its whole breadth, and the rocks rose perpendicularly on either side. We searched this place for water with the utmost care and anxiety, and I was at length fortunate enough to discover a small clear basin not a yard in circumference, under a rock on the left side of the glen. Suspecting that this was supplied by surface drainage, we enlarged the pool, and obtained from it an abundance of the most delicious water we had tasted during our wanderings. Mr. Browne will I am sure bear the Rocky Glen in his most grateful remembrance. Relieved from further anxiety with regard to our animals, he hastened with me to ascend one of the hills that towered above us to the height of 600 feet, before the sun should set, but this was no trifling task, as the ascent was exceedingly steep. The view from the summit of this hill presented the same broken country to our scrutiny which I have before described, at every point excepting to the westward, in which direction the ranges appeared to cease at about six miles, and the distant horizon from S.W. to N.W. presented an unbroken level. The dark and deep ravine through which the creek ran was visible below us, and apparently broke through the ranges at about four miles to the W.N.W. but we could not see any water in its bed. It was sufficiently cheering to us however to know that we were near the termination of the ranges to the westward, and that the country we should next traverse was of open appearance.

I had hoped from what we saw of it from the top of the hill above us, on the previous afternoon, that should have had but little difficulty in following down the creek, but in this we were disappointed.

We started at eight to pursue our journey, and kept for some time in its bed. The rock formation near and at our camp was trap, but at about a mile below it changed to a coarse grey granite, huge blocks of which, traversed by quartz, were scattered about. The defile had opened out a little below where we had slept, but it soon again narrowed, and the hills closed in upon it nearer than before. The bed of the creek at the same time became rocky, and blocked up with immense fragments of granite. We passed two or three pools of water, one of which was of tolerable size, and near it there were the remains of a large encampment of natives. Near to it also there was a well, a sure sign that however deep the water-holes in the glen might now be, there are times when they are destitute of any. There can be no doubt, indeed, but that we owed our present supply of water both at this place and at the Coonbaralba pass, to the rains that fell in the hills during the week we remained at Williorara.

Soon after passing the native camp, our further progress was completely stopped by large blocks of granite, which, resting on each other, prevented the possibility of making a passage for the cart or even of advancing on horseback. In this predicament I sent Flood to climb one of the hills to our left, to see if there was a leading spur by which we could descend to the plains; but on his return to us he said that the country was wholly impracticable, but that he thought we should see more of it from a hill he had noticed about three miles to the north-east. We accordingly left Morgan with the horses and walked to it. We reached the summit after a fatiguing walk of an hour, but neither were we repaid for our trouble, nor was there anything in the view to lead us to hope for any change for the better. The character of the country had completely changed, and in barrenness it far exceeded that through which we had already passed. The line of hills extended from S.E. by S. to the opposite point of the compass, and formed a steep wall to shut out the level country below them.

One might have imagined that an ocean washed their base, and I would that it really had been so, but a very different hue spread between them and the distant horizon than the deep blue of the sea. The nearer plains appeared of a lighter shade than the rest of the landscape, but there were patches of trees or shrubs upon them, which in the distance were blended together in universal scrub. A hill, which I had at first sight taken to be Mount Lyell of Sir Thomas Mitchell, bore 7° to the east of north, distant 18 miles, but as our observations placed us in 31°32'00"S. only, it could not have been that hill. To the south and east our view was limited, as the distant horizon was hid from our sight by higher ground near us, but there was a confused succession of hills and valleys in those directions, the sides of both being covered with low brush and huge masses of granite, and a dark brown sombre hue pervaded the whole scene. We could not trace the windings of the creek, but thought we saw gum-trees in the plains below us, to the N.E., indicating the course of a creek over them. Some of the same trees were also visible to our left (looking westward), and the ranges appeared less precipitous and lower in the same direction. We cast our eyes therefore to that point to break through them, and returned to Morgan with at least the hope of success. In the view I had just then been contemplating, however, I saw all realized of what I had imagined of the interior, and felt assured that I had a work of extreme difficulty before me in the task of penetrating towards the centre.

On our return to the cart, I determined on again taking up my quarters at the little rocky water-hole, and sending Mr. Browne and Flood to the west-ward to find a practicable descent to the plains, before I again moved from the glen.

In the evening, Mr. Browne went with Flood down the creek, but the road was perfectly impracticable even for led horses, so that the only hope of progressing rested on the success that might attend his endeavours on the following day. He accordingly started with Flood at an early hour proposing to return by the way of the creek, if he should succeed in finding a descent to the plains. I and Morgan remained in the glen. My observations placed this well-remembered spot in lat. 31°32'17"S.

I had plenty of occupation during my officer's absence, whilst Morgan was engaged looking over the harness and filling up the water-casks. At four, Mr. Browne returned, having succeeded beyond our most sanguine expectations, not only in finding an uninterrupted descent to the plains, but an abundance of water in the creek at the gorge of the glen; yet he was of opinion that we should not find any water below that point, as the creek there had a broad and even bed of sand and gravel. He said that the aspect of the plains was better than he had expected to find them, and he distinctly saw from the ranges, as he descended, the hills of whose existence we had had some doubt the day before, bearing N.N.W. Thus, then, fortune once more befriended our movements, by enabling us to push on another day in advance, without being dependent on our own resources. Morgan was too glad to empty the casks again, and to lighten the cart-load, with which, on the morning of the 9th, we left the glen, and gradually turned to the westward, until the hill we had walked to on the 7th, and which bore west by north from the place where we had left Morgan with the cart, now bore W.N.W.

Pushing up a narrow valley, we found little difficulty in our way, and leaving the above hill somewhat to our right, we gradually descended by a long and leading spur to the Cis-Darling interior.

We could now look back on the ranges from the depressed region into which we had fallen, nor could the eye follow their outline and glance over the apparently boundless plain beyond them, without feeling a conviction that they had once looked over the waters of the ocean as they then overlooked a sea of scrub.

As soon as we had got well into the plains, we pursued a course of half a point to the eastward of north, nearly parallel to the ranges, until we reached the glen from which the creek issues, and formed our little camp on its banks. The water however was not good, so that we were obliged to send for some from a pool a little above us. In the bed of this creek we found beautiful specimens of Solani, and a few new plants.

I halted at this place in consequence of the resolution I had taken to push into the interior on the following morning. I was therefore anxious that the horses should start as fresh as possible, as we could not say where we should again find water.

The direction of the hills was nearly north and south, extending at either hand to a distance beyond the range of vision or telescope. Our observations here placed us in latitude 31°23'20"S., so that we were still nearly half a degree to the south of Mount Lyell, and a degree to the south of Mount Serle. I had little prospect of success, however, in pursuing a direct westerly course, as it would have led me into the visible scrub there; on the other hand I did not wish to move exactly parallel to the ranges, but, in endeavouring to gain a knowledge of the more remote interior, to keep such a course as would not take me too far from the hills in the event of my being obliged to fall back upon them. We started on the 11th, therefore, on a N.N.W. course, and on the bearing of the low hills we had seen to the westward, and which were now distinctly visible. For the first five miles we travelled over firm and open plains of clay and sand, similar to the soil of the plains of the Murray. At length the ground became covered with fragments of quartz rock, ironstone, and granite. It appeared as if M'Adam had emptied every stone he ever broke to be strewed over this metalled region. The edges of the stones were not, however, rounded by attrition, or mixed together, but laid on the plains in distinct patches, as if large masses of the different rocks had been placed at certain distances from each other, and then shivered into pieces. The plains were in themselves of undulating surface, and appeared to extend to some low elevations on our left, connecting them with the main range as outer features; although in the distance they only shewed as a small and isolated line of hills detached about eleven miles from the principal groups, from which we were gradually increasing our distance. This outer feature prevented our seeing the north-west horizon until we gained an elevated part of it, whence it appeared that we should soon have to descend to lower ground than that on which we had been travelling. There was a small eminence that just shewed itself above the horizon to the N.N.W., and was directly in our course, enabling us to keep up our bearings with the loftier and still visible peaks on the ranges. We found the lower ground much less stony and more even than the higher ground, and our horses got well over it. At 4 P.M. we observed a line of gum-trees before us, evidently marking the line of a creek, the upper branch of which we had already noticed as issuing from a deep recess in the range. At the distance we were from the hills, we had little hope of finding water; on approaching it, however, we alarmed some cockatoos and other birds, and observed the recent tracks of emus in the bed of the creek. Flood, who had ridden ahead, went up it in search for water. Mr. Browne and I went downwards, and from appearances had great hopes that at a particular spot we should succeed by digging, more especially as on scraping away a little of the surface gravel with our hands, there were sufficient indications to induce us to set Morgan to work with a spade, who in less than an hour dug a hole from which we were enabled to supply both our own wants and those of our animals; and as there was good grass in the creek, we tethered them out in comfort. This discovery was the more fortunate, as Flood returned unsuccessful from his search.

The gum-trees on this creek were of considerable size; and many of the shrubs we had found in a creek, at the glen, were in beautiful flower in its broad and gravelly bed, along which the Clyanthus was running with its magnificent blossoms; a situation where I certainly did not expect to find that splendid creeper growing. It was exceedingly curious to observe the instinct which brought the smaller birds to our well. Even whilst Morgan was digging, and Mr. Browne and I sitting close to him, some Diamond birds (Amandina) were bold enough to perch on his spade; we had, in the course of the day, whilst passing over the little stony range, been attracted to a low Banksia, by seeing a number of nests of these little birds in its branches, and of which there were no less than fourteen. In some of them were eggs and in others young birds; so that it appeared they lived in communities, or congregated together to breed. But we had numberless opportunities of observing the habits of this interesting little bird whose note cheered us for months, and was ever the forerunner of good, as indicating the existence of water.

We placed the cart under a gum-tree, in which the cockatoos we had alarmed when descending into the creek bad a nest. These noisy birds (*Plyctolophus Leadbeaterii*) kept incessantly screeching to their young, which answered them in notes that resembled the croaking of frogs, more than anything else.

On the 11th we left the creek, well satisfied with our night's occupation of it, as also, I believe, to the still greater satisfaction of our noisy friends. For about two and a half or three miles there was every appearance of an improving country. It was open, and in many places well covered with grass; and although at three miles it fell off a little, still the aspect on the northern side of the creek was, to a considerable distance, preferable to that on the south side. At 11 A.M. we gained the crest of the little stony hill we had seen the day before to the N.N.W., and from it were enabled not only to take back bearings, but to carry others forward. We were fast losing sight of the hills, whose loftier summits alone were visible, yet we now saw fresh peaks to the north, which satisfied me that they continued in that direction far beyond the most distant one we had seen. From this circumstance I was led to hope that we might fall on another creek, and so gradually, but surely, work our way to the N.W.

On descending from the little hill, however, we traversed an inferior country, and at two miles saw a few scattered Pine-trees. Shortly afterwards, on breaking through a low scrub, we crossed a ridge of sand, on which numerous Pine-trees were growing. These ridges then occurred in rapid succession, separated by narrow flats only; the soil being of a bright red clay covered with Rhagodiæ, and having bare patches on them. The draught over this kind of country became a serious hindrance to our movements, as it was very heavy, and the day excessively hot, the horses in the team suffered much. I therefore desired Morgan to halt, and, with Mr. Browne, rode forward in the hope of finding water, for he had shot a new and beautiful pigeon, on the bill of which some moist clay was adhering; wherefore we concluded that he had just been drinking at some shallow, but still unexhausted, puddle of water near us: we were, however, ever unsuccessful in our search; but crossed pine ridge after pine ridge, until at length I thought it better to turn back to the cart, and, as we had already travelled some 25 miles, to halt until the morning; more especially as there was no deficiency of grass on the sand ridges, and I did not apprehend that our horses would suffer much from the want of water.

Whatever idea I might have had of the character of the country into which we had penetrated, I certainly was not prepared for any so singular that we encountered. The sand ridges, some partially, some thickly, covered with Pine-trees were from thirty to fifty feet high, and about eight yards at their base, running nearly longitudinally from north to south. They were generally well covered with grass, which appeared to have been the produce of recent rains; and several very beautiful leguminous plants were also growing on them. I did not imagine that these ridges would continue much longer, and I therefore determined, the following morning to push on. Our position was in lat. 30°40'S. and in longitude 140°51'E. nearly.

On the morning of the 12th we commenced our day's journey on a N.W. course, as I had proposed to Mr. Browne. Flood had been about half a mile to the eastward, in the hope of finding water before we rose, but was disappointed; the horses did not, however, appear to have suffered from the want of it during the night. On starting I requested Mr. Browne to make a circuit to the N.E. for the same purpose, as we had observed many birds fly past us in that direction; and I sent Flood to the westward, but both returned unsuccessful. Nevertheless, although we could not find any water, the country improved.

The soil was still clay and sand, but we crossed some very fine flats, and only wanted water to enjoy comparative luxury. Both the flats and the ridges were well clothed with grass, and the former had box-trees and hakeas scattered over them; but these favourable indications soon ceased. The pine ridges closed upon each other once more, and the flats became covered with salsolaceous plants. The day was exceedingly hot, and still more oppressive in the brushes, so that the horses began to flag. At 2 P.M. no favourable change had taken place. Our view was limited to the succeeding sand hill; nor, by ascending the highest trees could we see any elevated land at that hour; therefore I stopped, as the cart got on so slowly, and as the horses would now, under any circumstances, be three days without water, I determined on retracing my steps to the creek in which we had dug the well. I directed Mr. Browne, with Flood, however, to push on, till sunset, in the hope that he might see a change. At sunset I commenced my retreat, feeling satisfied that I had no hope of success in finding water so far from the hills. Turning back at so late an hour in the afternoon, it was past midnight when we reached the sand ridge from which we had started in the morning; where we again stopped until dawn, when proceeding onwards, and passing a shallow puddle of surface water, that was so thick with mud and animalculæ as to be unfit to drink, we gained the creek at half-past 4 P.M. Mr. Browne and Flood joined us some little time after sunset, having ridden about 18 miles beyond the point at which we had parted, but had not noticed any change. The sandy ridges, Mr. Browne informed me, continued as far as he went; and, to all appearance, for miles beyond. The day we returned to the creek was one of most overpowering heat, the thermometer at noon being 117° in the shade. I had promised to wait for Mr. Browne at the shallow puddle, but the sun's rays fell with such intense effect on so exposed a spot that

I was obliged to seek shelter at the creek. It blew furiously during the night of the 13th, in heated gusts from the north-east, and on the morning of the 14th the gale continued with unabated violence, and eventually became a hot wind. We were, therefore, unable to stir. The flies being in such myriads around us, so that we could do nothing. It is, indeed, impossible for me to describe the intolerable plague they were during the whole of that day from early dawn to sunset.

On the night of the 14th it rained a little. About 3 A.M. the wind blew round to the north-west, and at dawn we had a smart shower which cooled the air, reducing the temperature to something bearable. The sun rose amidst heavy clouds, by which his fiery beams were intercepted in their passage to the earth's surface. Before we quitted our ground I sent Flood up the creek, to trace it into the hills, an intention I was myself obliged to forego, being anxious to remain with the cart. The distance between the two creeks is about 26 miles, but, as I have already described the intervening country, it may not be necessary to notice it further. I was unable to take many back bearings, as the higher portions of the ranges were enveloped in mist. We reached the glen at half-past 5 P.M., and took up our old berth just at the gorge, preparatory to ascending the hills on the following day. Flood had already arrived there, and informed me that he had not followed the creek to where it issued from the ranges, but had approached very nearly, and could see the point from which it broke through them. That he had not found any surface water, but had tried the ground, in many places, and always found water at two or three inches depth, and that where the water was the most abundant the feed was also the most plentiful.

As I had anticipated, we had heavy rain all night, and in the morning continual flying thunder-storms. We started, however, at eight, and, leaving the cart to push on for the rocky gully, Mr. Browne and I proceeded to ascend some of the higher peaks, which we had not had time to do in our advance. We accordingly turned into a narrow valley, in the middle of which was the bed of a rocky water-course, and on either side of it were large clusters of the Clematis in full flower, that, mixed with low bushes of Jasmine, sent forth a most delicious perfume.

After winding up this valley for about a mile and a half, we were stopped by a wall of rock right across it, and obliged to turn back. We were, however, more fortunate in our next attempt, and succeeded in gaining the summit of one of the loftiest hills on the range, on the very top of which we found large boulders of rocks, imbedded in the soil. They varied in size, from a foot in diameter to less, and were rounded by attrition, just like the rounded stones in the bed of a river, or on the sea shore. The hill itself was of schistose formation, the boulders of different kinds of rocks, and very sparingly scattered through the soil. We had scarcely reached the summit of this hill, when it was enveloped in thick clouds, from which the lightning flashed, and the thunder pealed close to us, and crack after crack reverberated along the valleys. It soon passed away, however, and left us well drenched, but the western horizon was still black with clouds. From this hill we proceeded to another, which at first sight I had thought was of volcanic origin, but proved to be like the first, of schistose formation, and was covered with low scrub. About 2 P.M. we had finished our work, and the sun shone out. On looking back towards the plains we now saw them flashing in the light of waters, and I regretted that we had been forced to retreat before the rains set in. However, seeing that the country was now in a fitter state to travel over, I determined on returning with all speed, to give Mr. Poole an opportunity to pass to the point where I had been, whilst I should move the party over the hills. We struck across the ranges, direct for the rocky gully, from the last hill we ascended, and rode past some very romantic scenery, but I had not time to make any sketch of it. Flood and Morgan had already arrived in the glen, and tethered out the horses in some long grass. At this place we were about 38 miles distant from the camp; but, as the cart could not travel so far in one day, I directed the men to bring it up, and on the morning of the 18th left them for the camp, with Mr. Browne, where we arrived at sunset. But little rain had fallen during the day, still it was easy to foretell that it had not ceased. The wind, for the last three days, had been blowing from the N.W., but on the 19th flew round to the S.E., and although no rain fell during the day, heavy clouds surrounded us.

Considering, however, the rapidity of evaporation in such a climate, and the certainty that the rains would be followed by extreme heat, I was anxious that Mr. Poole should proceed on his journey without delay, he accordingly prepared to leave us on the 20th.

The reader will have inferred, from what I have said on the subject, that my object at this particular time was to attain the meridian of Mount Arden, as soon as circumstances should enable me. Had not this intention influenced me, on my recent journey, I should have kept nearer to the ranges, but, I hoped, by taking a westerly course, that I should strike the N.E. angle of Lake Torrens, or find that I had altogether cleared it; added to this Mr. Eyre had informed me that he could not see the northern shore of that lake; I therefore thought that it might be connected with some more central body of water, the early discovery of which, in my progress to the N.W., would facilitate my future operations. This was a point whereon I was most anxious to obtain information; but, as my horses were knocked up it appeared to me, that Mr. Poole, with fresh horses, would find no difficulty in gaining a distance sufficiently great to enable me to act on the knowledge he might acquire of the distant interior.

In my instructions to that officer therefore, I directed him to pursue a general N.W. course, as the one most likely to determine the questions on the several points to which I called his attention. "Should you,"I said, "reach the shores of Lake Torrens, or any body of water of unknown extent, you will endeavour to gain every information on that head; but if you should not strike any basin of either description, you will do your uttermost to ascertain if a westerly course is open to us, after you shall have reached lat. 30° to enable me to gain the 138° meridian, as soon as circumstances will permit. Should the supply of water which the recent rains will ensure for a time, be likely to fail, or if the rains should not have extended so far as you would desire to go, and your advance be thus rendered hazardous, it will be discretionary with you to return direct to the camp, or turn to the eastward, and proceed along the western flanks of the ranges, but you are on no account to endanger either yourself or party by an attempt to push into the interior, to a distance beyond that which prudence might reasonably justify. Should you return along the ranges you will examine any creek or water-course you may intersect, and bring me the fullest information as to the supply of water and feed. Should you, on the other hand, discover any very extensive sheet of water, you will, after ascertaining its extent and direction, as far as your means will allow, return immediately to the camp; as, in the event of our requiring the boat, many necessary preparations will have to be made, that will take a considerable length of time to complete, during which the examination of the country to the north can be carried on with advantage.

"You will select the men you would wish to accompany you and will provide as well for your comfort as safety; for although these regions do not seem to be inhabited at the present moment, at least in that part from whence I have just returned, it will be necessary for you to be always on your guard, even although no apparent danger may be near."

Mr. Browne had greatly recovered from his late indisposition, and as Mr. Poole intimated to me that he had expressed his willingness to accompany him, I had several reasons for giving my assent to this arrangement.

On the morning of the 20th it still continued to rain, insomuch that I was anxious Mr. Poole should postpone his departure, but clearing up at noon, he left me and proceeded on his journey. In the evening, however, we had heavy and violent showers; all night it poured in torrents with thunder and lightning, but the morning of the 21st was clear and fine. A vast quantity of rain however had fallen. The creek was overflowing its banks, and the ground in such a state that it would have been impossible to have moved the drays. The temperature was exceedingly cold, although the thermometer did not fall below 66° at half-past 2 P.M. the hottest part of the day. Such a temperature I am aware would be considered agreeable in England, but in a climate like that of Australia, where the changes are so sudden, they are more severely felt. Only a few days before the thermometer had ranged from 108° to 117° in the shade, thus at once causing a difference of 42° and 51°, and I am free to say that it was by no means agreeable. On the

22nd I commenced my advance over the ranges, although the ground was hardly then in a condition to bear the weight of the drays. We were indeed obliged to keep on the banks of the creek as they were higher and firmer than the plains, but after all we only made seven miles and halted, I had almost said without water, for notwithstanding the recent rains, there was not a drop in the bed of the creek, nor could we get any other than a scanty supply by digging; Jones, however, one of the bullock drivers, found a shallow pool upon the plains to which the cattle were driven.

On the way I ascended a small hill composed of Mica slate, and on its summit found two or three specimens of tourmaline. The boiling point of water on this hill was 210°, the thermometer stood at 70°.

On the 25th we crossed the little dividing range connected with Lewis's Hill, which last I ascended to verify my bearings, as we had erected three pyramids on the Coonbaralla range that were visible from it. I also availed myself of the slow progress of the drays, to ascend a hill at some little distance from our line, which was considerably higher than any of those near it, and was amply rewarded for my trouble by the extensive view it afforded.

Our specimens and collections were at this period exceedingly limited, nor did there appear to be any immediate chance of increasing them. The most numerous of the feathered race were the owls, *(Strix flameus.)* These birds flew about in broad daylight, and kept the camp awake all night by their screeching, it being at that time the breeding season. The young birds generally sat on a branch near the hole in which they had been hatched, and set up a most discordant noise about every quarter of an hour, when the old ones returned to them with food.

On trying the thermometers, one on Lewis's Hill, and the other on the Black Hill, I found that they boiled at 209° and 208° respectively. On the 26th Jones was unfortunate enough to snap the pole of his dray, and I was consequently detained on the 27th repairing it. I was the more vexed at the accident, being anxious to push over the ranges and gain the plains, in order to prevent Mr. Poole the necessity of re-ascending them. I felt satisfied that I should find a sufficiency both of water and feed at the gorge of the Rocky Glen, to enable me to rest until more thorough knowledge of the country could be gained, whilst by encamping at that place I should save Mr. Poole a journey of 63 miles.

Lower part of the Rocky Glen.

As we descended from the ranges I observed that all the water I had seen glittering on the plains had disappeared; I found too that the larger water-hole in the glen had rather fallen than increased during the rains. The fact however was, that the under drainage had not yet reached the lower part of the gully.

We were now about 24 miles from the second creek Mr. Browne and I had crossed on our recent excursion, and from Flood's examination of it afterwards, I felt assured that unless a party was sent forward to dig a large hole for the cattle I could not prudently advance any farther for the present; but being anxious to push on, and hoping that the late rains increased the supply of water in the creek, I sent Flood on the 28th with two of the men (Joseph and Sullivan) to dig a tank in the most favourable spot he could select, and followed him with the drays on the 29th. Wishing however to examine the country a little to the westward, I desired the men to keep on the plains about two miles from the foot of the ranges, until they should strike the creek or Flood should join them, and did not reach the encampment before eight o'clock.

Flood then told me that he had been to the place where he had before found most surface water; but that, notwithstanding the rains, it was all gone. He had tried the creek downwards, and had at length sunk a tank opposite to a little gully, thinking that it might influence the drainage. The tank was quite full, and continued so for two or three days after, when, without any great call upon it from the cattle, it sensibly diminished, and at length dried up, and we should have been obliged to fall back, if in tracing up the little gully we had not found a pond that enabled us to keep our ground. It often happened that we thus procured water in detached localities when there was not a drop in the main channels the creeks. At this place the boiling point of the thermometer was 212°; thus bringing us again pretty nearly on a level with the ocean, although we were at the time distant from it more than 480 miles.

At this period we had frequent heavy winds, with a heated temperature: yet our animals, if I except the dogs, did not suffer much. The sheep, it is true, would sometimes refuse to stir, and assemble in the shade, when on the march, whilst the dogs too shelter in wambut holes, and poking their heads out, would bark at their charge to very little purpose. It was evident, indeed, that the heat was fast increasing, and what we had already experienced was only an earnest of that which was to follow.

Mr. Poole had now been absent thirteen days, and I began to be anxious for his return. Our march to the second creek had again shortened his homeward journey 70 miles, and as I felt assured he would cross the creek at the point where we had dug the well, I stuck a pole up in it, with instructions, and on the 2nd December he rode into the camp with Mr. Browne, both much fatigued, as well as their horses. I had been engaged the greater part of the day fixing the points for another base line, as I was fearful that the angles of our first were too acute, and found that the party had got back on my return to the camp. Mr. Poole informed me that as soon as the weather cleared, after leaving me on the range, he had pushed on. That on the 24th he left my cart tracks as they turned to the N.W., and continued the N.N.W. course as I had directed. On that day he encamped early at a good water-hole, as the horses had travelled fast; the country thereabouts had become more open, but water was exceedingly scarce. On this day he ascended a small sandstone hill, from which some high peaks on the range bore S.S.E.

On the 26th he had not advanced 10 miles, when the pack-horse fell exhausted by heat. Mr. Poole then consulted with Mr. Browne, and it was thought better by both to travel at night, and they accordingly did so. The country by moonlight appeared more open, and the water seemed to be in greater abundance, as if much more rain had fallen thereabouts than to the south. They continued a N.N.W. course until daylight, when they halted, and Mr. Browne ascended a sand hill, from whence he saw peaks on the range bearing to the north of east, and the Mount Serle range, bearing due west, distant 50 miles. The latter circumstance induced Mr. Poole, when he again resumed his journey, to change his course to west, in the hope that as he had passed the 30th parallel he should find Lake Torrens between himself and the ranges. Accordingly, on starting at 4 P.M. they went on that course, and halted at dawn on a swampy flat, under a gum-tree. Mr. Poole subsequently ascertained that the swamp was the head of a little creek falling into the Sandy Lake, where he afterwards terminated his journey.

The country had now assumed a very barren appearance. At sunrise Mr. Poole and Mr. Browne ascended another sand hill, from whence they again saw the hills to the westward, seemingly very high and steep; but there was no sign of an intermediate basin, the country towards the ranges bearing a most Sterile aspect. Here Mr. Browne saw a new pigeon, which had a very singular flight.

On the afternoon of the 28th the party moved on a course of 10° to the south of west, down a leading valley, the country becoming still more barren, the sand ridges quite bare, and only an occasional hakea on the flats. At eight miles on the above course, and from the top of a sandy ridge at the distance of two miles, they saw a sheet of water about a mile and a half in length, in a sandy bed extending to the north, without any visible termination. There was another sheet of water to the south of this in the same kind of bed, connected with the larger one by a dry channel. It appeared from the lay of the country that these sheets of water were formed by drainage from the barren ranges from which Mr. Poole calculated he was 15 to 18 miles distant. The lakes were about three miles in length, taking the two together, the water was slightly brackish, and in Mr. Poole's opinion they might during the summer season be dry. He again ascended the sandy ridge and observed that he was immediately opposite to three remarkable peaks, similar to those marked down by Mr. Eyre. The party then turned homewards, and encamped on the creek at the head of which they had slept the night before, where they could hardly rest for the swarms of mosquitos. Pursuing their journey towards the camp on the following morning, keeping some few miles to the westward of their former line, they passed through a similar country. At noon, on the 1st of December they were still amongst the pine ridges; after noon the country began to improve, and they rode across large plains well grassed and covered with acacia trees of fine growth, but totally destitute of water; they were in consequence obliged to tether the horses all night. They reached the creek in which I had erected the pole, early on the following morning, and there found the paper of instructions informing them of the removal of the camp to within a mile of where they then were.

It was evident from the result of this excursion and from the high northerly point Mr. Poole had gained, that he had either struck the lower part of the basin of Lake Torrens or some similar feature. It was at the same time, however, clear that the country was not favourable for any attempt to penetrate, since there was no surface water. I felt indeed that it would be imprudent to venture with heavily loaded drays into such a country; but although I found a westerly course as yet closed upon me, I still hoped that we should find larger waters in the north-west interior, from the fact of the immense number of bitterns, cranes, and other aquatic birds, the party flushed in the neighbourhood of the lakes. Whence could these birds (more numerous at this point than we ever afterwards saw them) have come from? To what quarter do they go? They do not frequent the Murray or the Darling in such numbers, neither do they frequent the southern portion of the coast. If then they are not to be found in those localities, what waters do they inhabit in the interior?

On the 4th I sent Flood to the north in search of water, directing him to keep at a certain distance from the ranges, with especial instructions not to proceed beyond 60 or 70 miles, but in the event of his finding water within that distance to return immediately to the camp. During his absence I was abundantly occupied, and anxious that Mr. Poole and Mr. Browne should have a little rest after their late journey. Both those gentlemen were however too interested in the service in which they were engaged to remain idle when they could be usefully employed. Mr. Poole went out with me on the 5th and 6th to assist in the measurement of the new base line I had deemed it prudent to run, for the purpose, as I have said, of correcting any previous error. Mr. Piesse examined the pork, and according to my instructions made out a list of the stores on hand, when I found it necessary to make a reduction in the allowance of tea and sugar, in consequence of the loss of weight. The former from 4 oz. to 3 oz. per week, the latter from 2 lb. to 1½ lb.

The heat had now become excessive, the thermometer seldom falling under 96°, and rising to 112° and 125° in the shade. The surface of the ground never cooled, and it was with difficulty that we retained any stones in our hands that had been exposed to the sun; still we had not as yet experienced a hot wind. The existing heat was caused by its radiation from the earth's surface and the intensity of the solar rays.

The horses Mr. Poole had out with him, had suffered a good deal, and considering that if the country should continue as heretofore, and we should be obliged to hunt incessantly for water, we should require relays, I thought it advisable to do away with the horse-team, as the consumption of provision now enabled me to divide the load the horses had drawn equally amongst the bullocks. We finished the base line on the 7th, and I was glad to find that it was of sufficient length to ensure a favourable result, it being rather more than 10 miles.

All drainage in the creek had now ceased, and we were therefore dependent on the water in the gully, which, although invaluable as a present supply, would soon have been exhausted, where our total consumption could not have been less than 1000 to 1100 gallons a day, for the horses and bullocks drank a fearful quantity. Had Flood been unsuccessful in the object of his journey, therefore, I should in the course of a few days have been obliged to fall back, but he returned on the 7th, bringing news that he had found a beautiful little creek, in which there were long deep water-holes shaded by gum-trees, with an abundance of grass in its neighbourhood. This creek he said was about 40 miles in advance, but there was no water between us and it. He also confirmed an impression I had had on my mind from our first crossing the Barrier Range, that it would not continue to any great distance northwards; Flood said that from what he could observe the hills appeared to be gradually declining, as if they would soon terminate. He saw three native women at the creek, but did not approach them, thinking it better not to excite their alarm. These were the first natives we had seen on the western side of the hills.

On the 9th we again moved forward, on a course a little to the eastward of north, over the barren, stony, and undulating ground that lies between the main and outer ranges. The discovery of this creek by Flood, so much finer than any we had hitherto crossed, led me to hope that if the mountains should cease I might fall in with other ranges beyond them coming from the north-east, as forming the northwest slope of the valley of the. Darling. I was anxious, therefore, to examine the ranges as we advanced, and leaving the party in Mr. Poole's charge, rode away to ascend some of the hills and to take bearings from them to some particular peaks, the bearing of which had already been taken from different elevations; but from no hill to which I went could a view of the south-west horizon be obtained, so much lower had the hills become, and from their general aspect I was fully satisfied that we should soon arrive at their termination. From the last point I ascended, as from others, there was a large mountain bearing N.E. by N. from me, distant 50 or 60 miles, which I rightly judged to be Mount Lyell. It was a bold, round hill, without any particular feature, but evidently the loftiest connected with the Barrier Range. Mount Babbage bore N. by E. and was only just visible above the dark scrubs between me and it. The teams were keeping rather nearer the hills than Flood had gone, and were moving directly for a line of trees apparently marking the course of a creek. On my way to overtake the party, I met Mr. Browne and Flood on the plains, with whom I rode back. As we crossed these plains we flushed numerous pigeons—a pair, indeed, from under almost every bush of rhagodia that we passed. This bird was similar to one Mr. Browne had shot in the pine forest, and this was clearly the breeding season; there were no young birds, and in most of the nests only one egg. We should not, however, have encumbered ourselves with any of the young at that time, but looked to a later period for the chance of being able to take some of that beautiful description of pigeon home with us. The old birds rose like grouse, and would afford splendid shooting if found in such a situation at any other period than that of incubation; at other times however, as I shall have to inform the reader, they congregate in vast flocks, and are migratory.

Fortunately, at that part of the creek where the party struck it, there was a small pool of water, at which we gladly halted for the night, having travelled about 28 miles; our journey to Flood's Creek on the following day was comparatively short. Flood had not at all exaggerated his account of this creek, which, as an encouragement, I named after him. It was certainly a most desirable spot to us at that time; with plenty of water, it had an abundance of feed along its banks; but our tents were pitched on the rough stony ground flanking it, under cover of some small rocky hills. To the north-west there was a very pretty detached range, and westward large flooded flats, through which the creek runs, and where there was also an abundance of feed for the stock.

Although, as I have observed, the heat was now very great, the cereal grasses had not yet ripened their seed, and several kinds had not even developed the flower. Everything in the neighbourhood of the creek looked fresh, vigorous, and green, and on its banks (not, I would observe, on the plains, because on them there was a grass peculiar to such localities) the animals were up to their knees in luxuriant vegetation. We there found a native wheat, a beautiful oat, and a rye, as well as a variety of grasses; and in hollows on the plains a blue or purple vetch not unusual on the sand ridges, of which the cattle were very fond. In crossing the stony plains to this creek we picked up a number of round balls, of all sizes, from that of a marble to that of a cannon ball; they were perfect spheres, and hollow like shells, being formed of clay and sand cemented by oxide of iron. Some of these singular balls were in clusters like grape-shot, others had rings round them like Saturn's ring; and as I have observed, the plains were covered with them in places. There can be no doubt, I think, but that they were formed by the action of water, and that constant rolling, when they were in a softer state, gave them their present form.

The day succeeding that of our arrival at Flood's Creek was one of tremendous heat; but in the afternoon the wind flew round to the S.W. from the opposite point of the compass, and it became cooler. On the 11th, I detached Mr. Poole and Mr. Browne, with a fortnight's provisions, to the N. E. in search of water. It may appear that I had given these officers but a short respite from their late labours; but the truth is that a camp life is a monotonous one, and both enjoyed such excursions, and when there was no necessity for other arrangements, as they evinced a great interest in the expedition, I was glad to contribute to their pleasures, and should have rejoiced if it had fallen to their lot to make any new and important discovery.

My instructions to Mr. Poole on these occasions were general. To keep a course somewhat to the eastward of north, but to be guided by circumstances. I thought it better to give him that discretionary power, since I could not know what changes might take place in the country. I sent Flood at the same time to ride along the base of the ranges; but desired him not to be absent more than three or four days, as I myself contemplated an excursion to the eastward, to examine the country on that side as I passed up it.

The reader will observe, that although slowly, we were gradually, and, I think, steadily working our way into the interior. At that time I hoped with God's blessing we should have raised the veil that had so long hung over it, more effectually than we did. Up to that period we had been exceedingly fortunate; nothing had occurred to disturb the tranquillity of our proceedings; no natives to interrupt our movements; no want either of water or grass for our cattle, however scarce the parties scouring the country might have found it; no neglect on the part of the men, and a consequent efficient state of the whole party. But time brings round events to produce a change in all things; the book of fate being closed to our inspection, it is only from the past that we discover what its pages before concealed from us.

CHAPTER V.

NATIVE WOMEN—SUDDEN SQUALL—JOURNEY TO THE EASTWARD—VIEW FROM MOUNT LYELL—INCREASED TEMPERATURE—MR. POOLE'S RETURN—HIS REPORT—LEAVE FLOOD'S CREEK—ENTANGLED IN THE PINE FOREST—DRIVE THE CATTLE TO WATER—STATE OF THE MEN—MR. POOLE AND MR. BROWNE LEAVE THE CAMP—PROCEED NORTHWARDS—CAPT. STURT LEAVES FOR THE NORTH—RAPID DISAPPEARANCE OF WATER—MUDDY CREEK—GEOLOGICAL FORMATION—GYPSUM—MOVE ON TO THE RANGES—RETURN TO THE CREEK—AGAIN ASCEND THE RANGE—FIND WATER BEYOND THEM—PROCEED TO THE W.N.W.—RETURN TO THE RANGES—ANTS AND FLIES—TURN TO THE EASTWARD—NO WATER—RETURN TO THE CAMP—MR. POOLE FINDS WATER—MACK'S ADVENTURE WITH THE NATIVES—MOVE THE CAMP.

I WAS much surprised that the country was not better inhabited than it appeared to be; for however unfit for civilized man, it seemed a most desirable one for the savage, for there was no want of game of the larger kind, as emus and kangaroos, whilst in every tree and bush there was a nest of some kind or other, and a variety of vegetable productions of which these rude people are fond. Yet we saw not more than six or seven natives during our stay in the neighbourhood of Flood's Creek.

One morning some of the men had been to the eastward after the cattle, and on their return informed me that they had seen four natives at a distance. On hearing this I ordered my horse to be saddled, with the intention of going after them; but just at that moment Tampawang called out that there were three blacks crossing from the flats, to the eastward, I therefore told him to follow me, and started after them on foot. The ground was very stony, so that the poor creatures, though dreadfully alarmed, could not get over it, and we rapidly gained upon them. At last, seeing there was no escape, one of them stopped, who proved to be an old woman with two younger companions. I explained to her when she got calm, for at first she was greatly frightened, that my camp was on the creek, and I wanted the blackfellows to come and see me; and taking Tampawang's knife, which hung by a string round his neck, I shewed the old lady the use of it, and putting the string over her head, patted her on the back and allowed her to depart. To my surprise, in about an hour and a half after, seven natives were seen approaching the camp, with the slowness of a funeral procession. They kept their eyes on the ground, and appeared as if marching to execution. However, I made them sit under a tree; a group of seven of the most miserable human beings I ever saw. Poor emaciated creatures all of them, who no doubt thought the mandate they had received to visit the camp was from a superior being, and had obeyed it in fear and trembling. I made them sit down, gave them a good breakfast and some presents, but could obtain no information from them; when at length they slunk off and we never saw anything more of them. The men were circumcised, but not disfigured by the loss of the front teeth, perfectly naked, rather low in stature, and anything but good looking.

On the 12th, about midnight, we had a violent squall that at once levelled every tent in the camp to the ground. It lasted for about half an hour with terrific fury, but gradually subsided as the cloud from behind which it burst passed over us. A few drops of rain then fell and cooled the air, when I called all hands to replace the tents. I was up writing at the time, and of a sudden found myself sitting without anything above me save the blue vault of heaven. My papers, etc. were carried away, and the men could scarcely hear one another, so furiously did the wind howl in the trees.

On the 13th I left the camp in charge of Mr. Piesse my store-keeper, and with Mr. Stuart and Flood crossed the ranges to the eastward, intending to examine the country between us and the Darling. Immediately on the other side of the range there was a plain of great width, and beyond, at a distance of between 50 and 60 miles, was a range of hills running parallel to those near the camp. They terminated however at a bold hill, bearing E.N.E. from me, it was evidently of great height; beyond this hill there was another still higher to the north-east, which I believe was Mount Lyell. The first portions of the plain were open, and we could trace several creeks winding along them, but the distant parts were apparently covered with dense and black scrub. Descending to the eastward towards the plains we rode down a little valley, in which we found a small pool of water; at this we stopped for a short time, but as the valley turned too much to the north I left it, and pursuing an easterly course over the plains halted at seven miles, and slept upon them, under some low bushes. The early part of the day had been warm, with the wind at N.E., but in the evening it changed to the south, and the night was bitterly cold. On the morning of the 14th we were obliged to wrap ourselves up as well as we could, the wind still blowing keenly from the south. We travelled for more than five miles over grassy plains, and crossed the dry beds of several lagoons, in which not very long before there might have been water. At nine miles we entered a dense brush of pine trees, acacia and other shrubs growing on pure sand. Through this we rode for more than 15 miles, to the great labour of our animals, as the soil was loose, and we had constantly to turn suddenly to avoid the matted and fallen timber. In this forest the temperature was quite different from that on the plains, and as we advanced it became perfectly oppressive. At about 15 miles we ascended a small clear sandy knoll, from whence we had a full view of Mount Lyell. I had expected that we should have found some creek near it, but the moment my eye fell on that naked and desolate mountain, my hopes vanished. We had now approached it within five miles, and could discover its barren character. Although of great height (2000 feet), there did not appear to be a blade of vegetation, excepting on the summit, where there were a few casuarinas, but the pines grew high up in its rugged ravines, and the brush continued even to its base. I still however hoped that from the top we should see some creek or other, but in this expectation we were also disappointed. The same kind of dark and gloomy brush extended for miles all round, nor could we either with the eye or the telescope discover any change. Again to the eastward there were distant ranges, but no prominent hill or mountain to be seen. One dense forest lay between us and them, within which I could not hope to find water, and as we had been without from the time we left the little creek in the ranges near the camp, I determined on retracing my steps, my object in this journey having been fully gratified by the results. The country through which we had passed was barren enough, but that towards the Darling was still worse. I should, however, have pushed on to Mount Babbage, which loomed large and bore a little to the eastward of north; but I did not see that I should gain anything by prolonging my journey. We were now about 56 miles from the camp, and there was little likelihood of our finding any water on our way back; when we descended from the hill, therefore, I pressed into the pine forest, as far as could, and then halted. On the following morning we crossed the plains more to the north than we had before done. About 11 A.M. we struck a creek, and startled a native dog in its bed, which ran along the bank. In following this animal we stumbled on a pool of water, and stopped to breakfast. Wishing to examine the country there as far to the north as possible on my way back, I passed over the northern extremity of the ranges. They there appeared gradually to terminate, and a broad belt of pine scrub from the westward stretched across the country, below me, to the east, until it joined the forest, through a lower part of which we had penetrated to Mount Lyell; but beyond this scrub nothing was to be seen. On my return to the camp I examined the drays, and found that the hot weather had had a tremendous effect on the wheels; the felloes had shrunk greatly, and the tyres of all were loose. I therefore had them wedged and put into serviceable condition.

The heat at this period was every day increasing, and it blew violently from whatever point of the compass the wind came.

On the 17th I examined the stock, and was glad to find they were all in good condition, the horses fast recovering from their late fatigues, the cattle in excellent order, and the sheep really fat.

Mr. Stuart was generally employed over the chart, which now embraced more than 80 miles of a hilly country, and I was happy to find that our angles agreed.

As I have already observed, there were a great variety of the cereal grasses about Flood's Creek, but they merely occupied a small belt on either side of it. All the grasses were exceedingly green, and there was a surprising appearance of verdure along the creek. Beyond it, on both sides, were barren stony plains, on which salsolaceous plants alone grew. About 13 miles to the westward the pine ridges commenced, and between us and these were large flats of grassy land, over which the waters of the creek spread in times of flood.

The white owl here appeared, like other birds, at noon-day; but there were also numerous other night birds. Here too the black-shouldered hawk collected in flights of thirty or forty constantly on the wing, but we never saw them take any prey; nor, (although we invariably examined their gizzards), could we discover upon what they lived.

Our lunars placed us in long. 141°18'2"E. and lat. 30°49'29"S. Up to this point we had traversed nothing but a desert, which, as far as our examinations had extended, was worse on either side than the line on which we were moving; how much further that gloomy region extended, or rather how far we were destined to wander into it, was then a mystery.

The heat now became so great that it was almost unbearable, the thermometer every day rose to 112° or 116° in the shade, whilst in the direct rays of the sun from 140° to 150°. I really felt much anxiety on account of Mr. Poole and Mr. Browne, who did not return to the camp until the 25th. So great was the heat, that the bullocks never quitted the shade of the trees during the day, and the horses perspired from their exertions to get rid of the mosquitos. On the 22nd the natives fired the hills to the north of us, and thus added to the heat of the atmosphere, and filled the air with smoke.

At 7 A.M. on the morning of that day the thermometer stood at 97°; at noon it had risen 10° and at 3 P.M., the hottest period of the day, it rose to 118° in the shade. The wind was generally from the E.S.E., but it drew round with the sun, and blew fresh from the north at mid-day, moderating to a dead calm at sunset, or with light airs from the west. A deep purple hue was on the horizon every morning and evening, opposite to the rising and setting sun, and was a sure indication of excessive heat.

On the 23rd I sent Flood and Lewis to the N.E., with instructions to return on Christmas-day. At this time the men generally complained of disordered bowels and sore eyes, but I attributed both to the weather, and to the annoyance of the flies and mosquitos. The seeds were ripening fast along the banks of the creek, and we collected as many varieties as we could; but they matured so rapidly, and the seed-vessels burst so suddenly that we had to watch them.

The comet, which we had first noticed on the 17th of the month, now appeared much higher and brighter than at first. Its tail had a slight curve, and it seemed to be rather approaching the earth than receding from it.

On the morning of the 24th, about 5 A.M., I was roused from sleep by an alarm in the camp, and heard a roaring noise as of a heavy wind in that direction. Hastily throwing on my clothes, I rushed out, and was surprised to see Jones's dray on fire; the tarpaulin was in a blaze, and caused the noise I have mentioned. As this dray was apart from the others, and at a distance from any fire, I was at a loss to account for the accident; but it appeared that Jones had placed a piece of lighted cow dung under the dray the evening before, to drive off the mosquitos, which must have lodged in the tarpaulin and set it on fire. Two bags of flour were damaged, and the outside of the medicine chest was a good deal scorched, but no other injury done. The tarpaulin was wholly consumed, and Jones lost the greater part of his clothes, a circumstance I should not have regretted if he had been in a situation to replace them.

Flood returned on the 25th, at 2 P.M., having found water in several places, but none of a permanent kind like that in the creek. He had fallen on a small and shallow lagoon, and had seen a tribe of natives, who ran away at his approach, although he tried to invite them to remain.

About an hour before sunset Mr. Poole and Mr. Browne returned, to the great relief of my mind; for, with every confidence in their prudence, I could not help being anxious in such a situation as that in which I was placed, my only companions having then been many days absent. They had nearly reached the 28th parallel, and had discovered an abundance of water, but Mr. Poole was more sanguine than Mr. Browne of its permanency.

The first water they found at the commencement of their journey, was at a distance of 40 miles and upwards, and as I felt assured we should have great difficulty in taking the cattle so far without any, I sent Flood, on the 26th, to try if he could find some intermediate pool at which I could stop. Mr. Poole informed me that the ranges still continued to the north, but that they were changed in character, and he thought they would altogether terminate ere long.

He also reported to me that the day he left the camp he pursued a N.N.E. course, skirting an acacia scrub, and that arriving at a small puddle of water at 12 miles, he halted. That on the 12th he started at six, and after travelling about three miles first got a view of distant ranges to the north; he soon afterwards entered an acacia scrub, and at 15 miles crossed a creek, the course of which was to the S.W., but there was no water in it. At five the party reached the hills, the acacia scrub continuing to within a mile of them; and as the day had been exceedingly warm, Mr. Poole encamped in a little gully. He then walked with Mr. Browne to the top of the nearest hill, and from it observed two lines of gum-trees in the plains below them to the north, which gave them hopes of finding water in the morning, as they were without any. Saw two detached ranges bearing 320° and 329° respectively, and a distant flat-topped hill, bearing 112° from them, the country appearing to be open to the north.

On the 13th, the party pushed on at an early hour for the gum-trees, but found no water. Observed numerous flights of pigeons going to the N. W. Traced the creek down for two miles, when they arrived at a place where the natives had been digging for water; here Mr. Poole left Mr. Browne and went further down the creek, when he succeeded in his search; but finding on his return, that Mr. Browne and Mack had cleared out the well and got a small supply of water, with which they had relieved the horses and prepared breakfast, he did not return to the water he had discovered, but proceeded to the next line of gum-trees where there was another creek, but without water in it; coming on a small quantity in its bed at two miles, however, they encamped. A meridian altitude of Aldebaran here gave their latitude 30°10'0"S. On the following morning Mr. Poole started on a W. N. W. course for a large hill, from whence he was anxious to take bearings, and which he reached and ascended after a journey of 22 miles. From this hill. which he called the Magnetic Hill (Mount Arrowsmith), because on it the north point of the compass deviated to within 3° of the south point, he saw high ranges to the north and north-east; a hill they had already ascended bore 157°30', and the flat-topped hill 118°30'. From the Magnetic Hill, Mr. Poole went to the latter, and ascended the highest part of it. The range was rugged, and composed of indurated quartz, and there was a quantity of gypsum in round flat pieces scattered over the slopes of the hills. The country to the W. and W.N.W. appeared to be very barren. The range on which they were was perfectly flat at the top, and covered with the same vegetation as the plains below. From this point Mr. Poole went to the north, but at 12 miles changed his course to the N.E. for three miles, when he intersected a creek with gum-trees, and shortly afterwards found a large supply of permanent water. Their latitude at this point was 29°47'S., and up to it no change for the better had taken place in the appearance of the country. On Monday, the 15th, Mr. Poole ascended several hills to take bearings before he moved on; he then proceeded up the creek to the north-west, and passed from fifteen to twenty large water-holes. At about three miles, Mr. Poole found himself on an open table land, on which the creek turned to the west. He, therefore, left it, and at two miles crossed a branch creek with water and grass. At 7½ miles farther to the north crossed another creek, followed it for a mile, when it joined a larger one, the course of which was to the north-east. In this creek there were numerous large pools of water. Crossing it, Mr. Poole ascended a hill to take bearings, from which he descended to a third creek, where he stopped for the night. On the following morning he continued his journey to the north, being anxious to report to me the character of the ranges. At 12 miles over open plains he intersected a creek trending to the eastward, in which there was an abundant supply of water; but this creek differed from the others in having muddy water, and but little vegetation in its neighbourhood. Passed some native huts, and saw twenty wild turkeys. At 10 miles from this creek Mr. Poole struck another, the ranges being still 12 miles distant. The horses having travelled for the last 10 miles over barren stony plains, had lost their shoes, and were suffering greatly. Mr. Poole, therefore, stopped at this place, and on consulting with Mr. Browne, determined to return to the camp without delay. Accordingly on the following morning he rode to the hills with Mr. Browne, leaving Mack with the other horses to await his return, and at 10 A.M. ascended the range. The view from it was not at all encouraging. The hills appeared to trend to the N.E., and were all of them flat-topped and treeless. The country to the west and north-west was dark with scrub, and the whole region barren and desolate. After taking bearings, Mr. Poole descended, returned to the creek on which he had left Mack, and as I have already stated, reached the camp on the evening of the 25th.

It will be obvious to the reader that the great danger I had to apprehend was that of having my retreat cut off from the failure of water in my rear; or if I advanced without first of all exploring the country, of losing the greater number of my cattle. It may be said that my officers had now removed every difficulty; but notwithstanding that Mr. Poole was sanguine in his report of the probable permanency of the water he had found, I hesitated whether to advance or not; but considering that under all circumstances the water they had found would still be available for a considerable time, and that it would enable me to push still further to the north, I decided on moving forward at once; but the weather was at this time so terrifically hot, that I hardly dared move whilst it continued, more especially as we had so great a distance to travel without water. I kept the party in readiness, however, to move at a moment's notice. On the 27th we had thunder, but no rain fell, and the heat seemed rather to increase than to decrease. On the 28th, at 2 P.M., the wind suddenly flew round to the south, and it became cooler. In hopes that it would continue, I ordered the tents to be struck, and we left Flood's Creek at half-past 4. As soon as I had determined on moving, I directed Mr. Poole to lead on the party in the direction he thought it would be best to take, and mounting my horse, rode with Mr. Browne and Mr. Stuart towards the ranges, to take bearings from a hill I had intended to visit, but had been prevented from doing in consequence of the extreme heat of the weather. I did not, indeed, like leaving the neighbourhood without going to this hill. The distance, however, was greater than it appeared to be, and it was consequently late before we reached it; but once on the top we stood on the highest and last point of the Barrier Range; for although, as we shall learn, other ranges existed to the north, there was a broad interval of plain between us and them, nor were they visible from our position. We stood, as it were, in the centre of barrenness. I feel it impossible, indeed, to describe the scene, familiar as it was to me. The dark and broken line of the Barrier Range lay behind us to the south; eastward the horizon was bounded by the hills I had lately visited, and the only break in the otherwise monotonous colour of the landscape was caused by the plains we had crossed before entering the pine forest. From the south-west round to the east northwards, the whole face of the country was covered with a gloomy scrub that extended like a sea to the very horizon. To the north-west, at a great distance, we saw a long line of dust, and knowing it to be raised by the party, after having taken bearings and tried the point of boiling water, we descended to overtake it. In doing this we crossed several spurs, and found tolerably wide and grassy flats between them. Following one of these down we soon got on the open plains, and about half-past seven met Mr. Poole, who had left the party to go to a fire he had noticed to the eastward, which he thought was a signal from us that we had found water; but such had not been our good fortune.

I now halted the party until the moon should rise, and we threw ourselves on the ground to take a temporary repose, the evening being cool and agreeable. At 11 we again moved on, keeping a north course, under Mr. Poole's guidance, partly over stony plains, and partly over plains of better quality, having some little grass upon them, until 8 A.M. of the morning of the 29th, when we stopped for an hour. As day dawned, Mr. Poole had caught sight of the hill, as he thought, to the base of which he wished to lead the party, and under this impression

Chaining over the Sand Hills to Lake Torrens.

Mus conditor. Gould.

Native Village in the Northern Interior.

Note. — Mr. Arrowsmith has prepared a detailed Map of Captn. Sturt's late

Routes into the Interior, in 2 Sheets, which may be had in a cover — Price 7s.

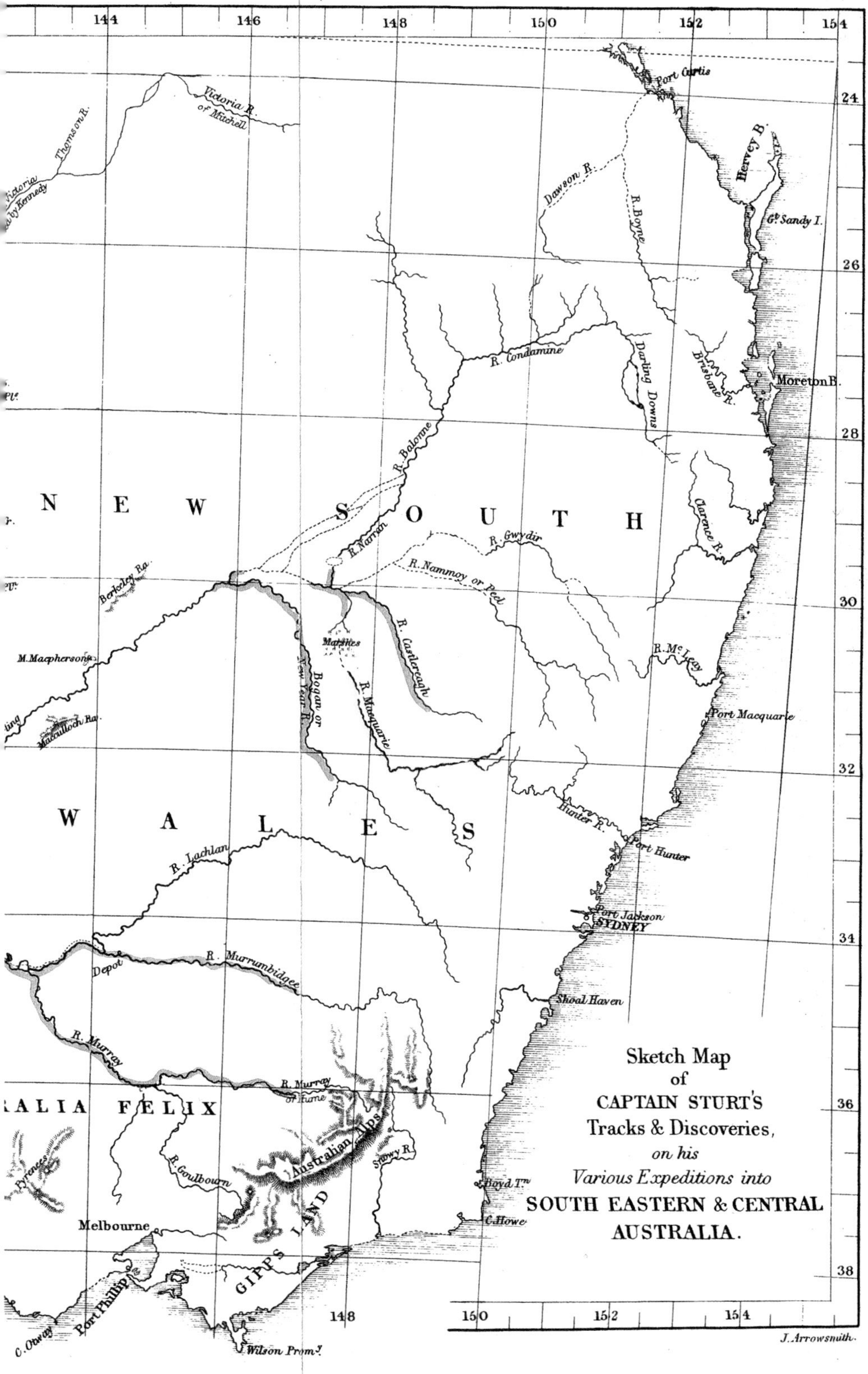

Sketch Map
of
CAPTAIN STURT'S
Tracks & Discoveries,
on his
Various Expeditions into
SOUTH EASTERN & CENTRAL
AUSTRALIA.

J. Arrowsmith.

Milvus affivis Gould.

1. Geophaps plumifera Gould.
2. Peristera histrionica Gould.

Cinclosoma cinnamomeus. Gould.

we continued our northerly course at 9, until by degrees we entered a low brush, and from it got into a pine forest and amongst ridges of sand. Mr. Poole had crossed a similar country; but the sandy ridges had soon ceased, and in the hope that such would now be the case he pushed forward until it was too late to retreat, for the exertion had already been very great to the animals in so heated and inhospitable a desert. In vain did the men urge their bullocks over successive ridges of deep loose sand, the moment they had topped one there was another before them to ascend. Seeing that they were suffering from the heat, I desired the men to halt, and sending Mr. Poole and Mr. Stuart forward with the spare horses and sheep to relieve them as soon as possible, I remained with the drays, keeping Mr. Browne with me. We had not travelled more than half a mile, on resuming our journey, when we arrived at a dry salt lagoon, at which the sheep had stopped. I here determined on leaving two of the drays, in the hope that by putting an additional team into each of the others we should get on, although before this we had discovered that Mr. Poole had mistaken his object, and had inadvertently led us into the thickest of the pinery. The drivers, however, advanced but slowly with the additional strength I had given them, and it was clear they would never get out of their difficulties, unless some other plan were adopted. I therefore again stopped the teams, and sent Mr. Browne to the eastward to ascertain how far the ridges extended in that direction, since Mr. Poole's track appeared to be leading deeper into them. On his return he informed me that the ridges ceased at about a mile and a quarter; in consequence of which I turned to the north-east, but the bullocks were now completely worn out and refused to pull. To save them, therefore, it became necessary to unyoke and to drive them to water, and as Mr. Browne felt satisfied he could lead the way to the creek, I adopted that plan, and telling the men with the sheep to follow on our tracks, we left the drays, at 6 P. m., taking two of the men only with us, and clearing the sand ridges at dusk, entered upon and traversed open plains. We then stopped to rest the cattle until the moon should rise, and laid down close to them; but although we kept watch, they had well nigh escaped us in search for water. At half-past ten we again moved on, and at midnight reached a low brush, in which one of the bullocks fell, and I was obliged to leave him. About two hours afterwards another fell, but these were the total of our casualties. We reached the creek at 3 in the morning of the 30th, and rode to a fire on its banks, where we found Davenport and Joseph with the cart; they had separated from Mr. Poole, who was then encamped about a quarter of a mile to the westward of them, although Davenport did not know where he was, nor had he found water. Our situation would have been exceedingly perplexing, if Mr. Browne, who had led me with great precision to this point, had not assured me that he recognised the ground, and that as soon as day dawned he would take me to the water. Just at this moment we saw another fire to the eastward, to which I sent Morgan on horseback, who returned with Mr. Poole, when we were enabled to give the poor animals the relief they so much required.

Having thus secured the horses and bullocks, I turned my attention to the men in the forest, with regard to whom I had no occasion to feel any alarm, as I had left ten gallons of water for their use, and strictly cautioned them not to be improvident with it. However, as soon as he had had a little rest, I sent Morgan with a spare horse for their empty casks to replenish them. At 2 o'clock I sent Flood with four gallons of water to the nearest bullock that had fallen. About 11 Brock came up with the sheep all safe and well. Flood returned at 7, with information that the bullock was dead, but night closed in without our seeing anything of Morgan, and having nothing to eat we looked out rather anxiously for him. The water on which we rested was at some little distance from the creek, in a long narrow lagoon, but we had scarcely any shade from the intense heat of the sun, the water being muddy, thick, and full of frogs and crabs. I have observed upon the extreme and increasing heat that prevailed at this time. Notwithstanding this, however, the night was so bitterly cold that we were glad to put on anything to keep us warm. Our situation may in some measure account for this extreme variation of temperature, as we were in the bed of the creek which might yet have been damp, as its surface had only just dried up; perhaps also from exposure to such heat during the day we were more susceptible of the least change. Be that as it may, certain it is that as morning dawned on this occasion, when the thermometer stood at 67°, we crept nearer to our fires for warmth, and in less than six hours afterwards were in a temperature of 104°.

As we passed through the acacia scrub, we observed that the natives had lately been engaged collecting the seed. The boughs of the trees were all broken down, and there were numerous places where they had thrashed out the seed, and heaped up the pods. These poor people must indeed be driven to extremity if forced to subsist on such food, as its taste is so disagreeable that one would hardly think their palates could ever be reconciled to it. Natives had evidently been in our neighbourhood very lately, but we saw none.

At this time I was exceedingly anxious both about Mr. Poole and Mr. Browne, who were neither of them well. The former particularly complained of great pain, and I regretted to observe that he was by no means strong.

About 10 o'clock on the morning of the last day of the year 1844, I was with Tampawang at the head of the lagoon, trying to capture one of the building rats, a nest of which we had found under a polygonum bush. We had fired the fabric, and were waiting for the rats to bolt, when we saw Morgan riding up to us. He stopped when he got to the water, and throwing himself on the ground drank long at it. Seeing that he came without anything for which he had been sent, I began to apprehend some misfortune; but on questioning him I learnt that he had been at the drays, and was on his return, when, stopping on the plains to let his horses feed, he fell fast asleep, during which time they strayed, and he was obliged to leave everything and walk until he overtook his horse near the creek. He said the men had consumed all the water I had left with them, and were in great alarm lest they should die of thirst; I was exceedingly provoked at Morgan's neglect, more particularly as the comfort of the other men was involved in the delay, although they deserved to suffer for the prodigal waste of their previous supply. But it is impossible to trust to men in their sphere of life under such circumstances, as they are seldom gifted with that moral courage which ensures calmness in critical situations. I made every allowance too for their being in so hot a place, and it only remained for me to relieve them as soon as I could. I sent the ever ready Flood for the casks and provisions Morgan had left behind him, but it was necessarily late before he returned; I then directed him to get up two teams of the strongest bullocks, and with him and another of the men left Mr. Poole and Mr. Browne to go myself to the pine forest for two of the drays. About seven miles from the creek we met Lewis, who was on our tracks. He said he apprehended that Morgan had lost himself and that he came on to ensure relief to the other men, who he said were suffering greatly from the want of water. At 9 P.M. we rounded up the cattle until the moon should rise, and made fires to prevent their escape. At 11 she rose, but it was behind clouds, so that it was 12 before we could move on. About two miles from the drays we saw Kirby wandering away from the track and called to him. This man would infallibly have been lost if we had not thus accidentally seen him. On reaching the party I found that Lewis had somewhat exaggerated the state of affairs, still the men were bad enough, although they had not then been 36 hours without water.

Notwithstanding that the moon had risen behind clouds, the first sun of the new year (1845) rose upon us in all his brightness, and the temperature increased as he advanced to the meridian. As Jones was with the hindmost drays, I sent Sullivan on my horse with some water for him, and ordered Flood to precede me with two of the drays along a flat I had noticed as I rode along, by which they would avoid a good many of the ridges. Sullivan returned with Jones about half-past ten, who, he told me, so far from wanting water had given all I had sent him to the dogs. As there were twelve bullocks to each dray I was obliged to give the drivers assistance, and consequently had to leave Jones by himself in the forest. I allowed him however to keep two of the dogs, and gave him four gallons of water, promising to send for him in two days. I then mounted my horse to overtake the teams, which by the time I came up with them had got on better than I expected. But the heat was then so intense that I feared the bullocks would drop. I therefore ordered the men to come slowly and steadily on, and as I foresaw that they would want more water ere long, I rode ahead to send them some. On my arrival at the creek I was sorry to find both Mr. Poole and Mr. Browne complaining, and very much indisposed. During the short time we had been at this spot, the water in the lagoon had rapidly diminished, and was now not more than a foot deep and very muddy. Fearing that the quality of the water was disagreeing with my officers, I ordered a well to be dug in the bed of the creek, from which we soon got a small quantity both clearer and better. Having despatched Joseph with a fresh supply for the party with the drays, I sat down to break my own fast which I had not done for many hours. In speaking to Mr. Browne of the intense heat to which we had been exposed in the pine forest, he informed me that the day had not been very hot with them, the thermometer not having risen above 94° at 2 P.M.

The drays reached the creek at 3 A.M. on the morning of the 2nd, both men and cattle fairly worn out. I had hoped they would have arrived earlier, but the men assured me that shortly after I left them the heat was so great they could hardly move onwards. The ground became so heated that the bullocks pawed it to get to a cool bottom, every time they stopped to rest. The upper leathers of Mack's shoes were burnt as if by fire, and Lewis's back was sadly blistered. The dogs lost the skin off the soles of their feet. and poor Fingall, one of our best, perished on the road.

Amidst all the sufferings of the other animals the sheep thrived exceedingly well under Tampawang's charge who was a capital shepherd. Their fleeces were as white as snow, and some of them were exceedingly fat. On the 3rd I sent Mr. Stuart to the Magnetic hill, Mount Arrowsmith, to verify Mr. Poole's bearings, in consequence of the great deviation of the compass from its true point, and also to sketch in that isolated group of hills; but as he found the same irregularity in his compass, I did not trust to the bearings either he or Mr. Poole had taken. The rock of which that hill was composed is a compact sandstone, with blocks of specular iron ore scattered over it, highly magnetic.

In the hope that a ride would do both my officers good, I sent them on the 4th to trace the creek up, and to fix on our next halting place. I also despatched Flood to the pine forest for the remaining drays, sending an empty one to lighten the other loads; a precaution that proved of great advantage, as the bullocks got on much easier than on the former occasion, but the day also was much cooler.

Mr. Poole and Mr. Browne returned at 11 on the 5th, but I was sorry to observe that Mr. Browne looked very unwell, and Mr. Poole continued to complain. They had however succeeded in their mission, and as I was very anxious to get them to better water, our lagoon being all but dry, I determined on moving northward on the 7th.

Flood re-crossed the creek on the morning of the 6th, when the bullocks completed a task of about 170 miles in eight days.

As I had determined on moving on the 7th, it became necessary to examine the drays, and I was vexed to find that they wanted as much repair as they had done at Flood's Creek. The men were occupied wedging them up, and greasing them on the 6th, and finished all but that of Lewis, the repair of which threw it late in the day on the 7th, before we proceeded on our journey. Independently, however, of my anxiety on account of my officers, several of the men were indisposed, and I was glad to break up our camp and fix it in a healthier spot than this appeared to be.

We started at 5 P.M., but as we had only about eight miles to go, it was not a matter of much consequence. We arrived at our destination at 10 P.M., but had some difficulty in finding the water, nor do I think we should have done so if we had not been guided to it by the hoarse and discordant notes of a bull-frog.

I had sent Mr. Stuart in the morning to some hills on our left, and Mr. Browne had ridden in the same direction to collect some seeds of a purple Hibiscus, and neither had joined the party when it reached the creek, as soon therefore as the cattle were unyoked, I fired a shot which they fortunately heard. Our collection of natural history still continued scanty. A very pretty tree, a new species of Grevillia, out of flower, however, and which I only concluded to be a Grevillia from its habit, and the appearance of its bark, had taken the place of the gum-trees on the creeks, and the jasmine was everywhere common, but, with the exception of a few solani and some papilionaceous plants, we had seen nothing either new or rare.

Of birds the most numerous were the new pigeon and the black-shouldered hawk; but there was a shrike that frequented the creeks which I should have noticed before. This bird was about the size of a thrush, but had the large head and straight hooked bill of its species; in colour it was a dirty brownish black, with a white bar across the wings. Whilst we were staying at Flood's Creek, one of these birds frequented the camp every morning, intimating his presence by a shrill whistle, and would remain for an hour trying to catch the tunes the men whistled to him. His notes were clear, loud, metallic and yet soft; their variety was astonishing, and his powers of imitation wonderful; there was not a bird of the forest that he did not imitate so exactly as to deceive. I would on no account allow this songster to be disturbed, and the consequence was that his rich note was the first thing heard at dawn of day, during the greater part of our residence in that neighbourhood.

We passed several native huts shortly after leaving the creek that were differently constructed from any we had seen. They were all arched elliptically by bending the bough of a tree at a certain height from the ground, and resting the other end on a forked stick at the opposite side of the arch. A thick layer of boughs was then put over the roof and back, on which there was also a thick coating of red clay, so that the hut was impervious to wind or heat. These huts were of considerable size, and close to each there was a smaller one equally well made as the larger. Both were left in perfect repair, and had apparently been swept prior to the departure of their inmates.

On the 8th we started at 5 A.M., and reached our destination (a place to which Mr. Poole had already been) at 11. We crossed barren stony plains, having some undulating ground to our left, and the magnetic hill as well as another to the south of it shewed as thunder clouds above the horizon. On our arrival at the creek we found about 30 fires of natives still burning, whom we must have frightened away.

We did not see any of them, nor did I attempt to follow on their tracks which led up the creek.

As I have already stated the fall of Flood's Creek was to the west. The creek from which we had just removed, as well as the one on which we then were, fell in the opposite direction or to the eastward, terminating after short courses either in grassy plains or in shallow lagoons.

On the 9th I remained stationary, and thus gave Mr. Piesse an opportunity to examine a part of our stores. He reported to me that the flour had lost weight nearly 10 per cent, some of the bags not weighing their original quantity by upwards of sixteen pounds. As the men had their full allowance of meat, I thought it advisable, in consequence of this, to reduce the ration of flour to 7 lb. per week, and I should be doing an injustice to them if I did not give them credit for the readiness with which they acquiesced in this arrangement.

The 10th of the month completed the fifth of our wanderings. We left our position rather late in the day, and halted a little after sunset at the outskirt of a brush, into which I was afraid to enter by that uncertain light, and as the animals had been watered at a small creek we crossed not long before, I had no apprehension as to their suffering. We started at 4 A.M. on the morning of the 11th, and soon passed the scrub; we then traversed open plains thickly covered in many places with quartz, having crossed barren sandy plains on the other side of the scrub. We now found the country very open, and entirely denuded of timber, excepting on the creeks, the courses of which were consequently most distinctly marked. Keeping a little to the eastward to avoid the gullies connected with some barren stony hills to our left, we descended to the ground Mr. Poole had fixed upon as our next temporary resting place. To the eye of an inexperienced bushman its appearance was in every respect inviting; there was a good deal of grass in its neighbourhood; the spot looked cheerful and picturesque, with a broad sheet of water in the creek, which when Mr. Poole first saw it must have been much larger and deeper; but in the interval between his first and second visit, it had been greatly reduced, and now presented a broad and shallow surface, and I felt assured that it would too soon dry up. Convinced therefore of the necessity of exertion, to secure to us if possible a supply of water, on which we could more confidently rely, I determined on undertaking myself the task of looking for it without delay. Both Mr. Poole and Mr. Browne were better, and the men generally complained less than they had done. On Sunday, the 12th, we had thunder with oppressive heat, but no rain. On Monday the wind, which had kept with the regularity of a monsoon to the E.S.E., flew round to the N.W., the thermometer at noon standing at 108° in the shade.

From the period at which we left Flood's Creek we had not seen any hills to the eastward, the ranges having terminated on that side. The hills we had passed were detached from each other, and to the westward of our course. The fall of the creek on which we were at this time encamped was consequently to the eastward, but there was a small hill about five miles to the E.N.E., under which it ran; that hill was the southern extremity of the ranges Mr. Poole and Mr. Browne had lately visited.

I left the camp on the 14th of the month, in the anxious hope that I should succeed in finding some place of more permanent safety than the one we then occupied, for we could almost see the water decrease, so powerful was the evaporation that was going on. I was accompanied by Mr. Browne and Mr. Poole, with Flood, Joseph, and Mack; but Mr. Poole only attended me with a view to his returning the next day with Mack, in the event of our finding water, to which he might be able to remove during my absence. We traced the creek upwards to the north-west, and at about four miles came to another, joining it from the westward. There was no water, but a good deal of grass about its banks, and it was evidently a tributary of no mean consequence. Crossing this we traced up the main creek on a more northerly course, having the Red Hill, subsequently called Mount Poole, on our left. We were obliged to keep the banks of the creek to avoid the rough and stony plains on either side. A little above the junction of the creek I have noticed, we passed a long water-hole, at which Mr. Poole and Mr. Browne had stopped on their excursion to the north; but it was so much diminished that they could hardly recognise it. The fact however shewed how uncertain our prospects were at this period. The bed of the creek was grassy, but broad, level, and gravelly. At almost every turn to which we came Mr. Poole assured me there had been, when he passed, a large sheet of water; but not a drop now remained, nor could we by scratching find the least appearance of moisture. Yet it was evident that this creek was at times highly flooded, there being a great accumulation of rubbish at the butts of the trees on the flats over which its waters must sweep, and the trunks of trees were lodged at a considerable height in the branches of those growing in its bed. Following its general course for 14 miles, we were led somewhat to the eastward of north, towards some hills in that direction, from which the creek appeared to issue, and then halted for the night, after a vain search for water. The Red Hill bore S.47°W., and some hills of less elevation were seen more to the westward of it, but beyond the last towards the north there were vast open and stony plains, destitute of timber and with very little vegetation upon them. On the morning of the 15th, at 5 P.M., we traversed these plains on a north course, and at 11 miles struck the creek of which Mr. Poole had spoken as containing muddy water, and found it precisely as he described. There were long water-holes about twenty-five feet broad, and three or four deep; but the water was exceedingly muddy. The banks were of a stiff, light-coloured clay, without any vegetation either on them or the contiguous flats, except a few bushes of polygonum growing under box-trees.

We here stopped to breakfast, although there was but little for the horses to eat. We then proceeded on a south-east course down the creek, keeping close upon its banks to avoid the macadamised plains on either side. To our left there were some undulating hills, and beyond them the summits of some remarkable flat-topped hills were visible. After leaving the place where we had breakfasted, we did not find any more water in the bed of the creek, but halted late in the afternoon at a small lagoon, not far from it. This lagoon was surrounded by trees; but like those of the creek its waters were muddy and not more than 18 inches deep. Our latitude at this point was 29°14'S., and our longitude 141°42'E.; the variation being 5°5'E.

Not wishing to keep Mr. Poole any longer away from the party, I sent him back to the camp on the 16th, with Mack, directing him to examine the creek we had crossed on his way homewards; as it appeared to me to break through some hills about three miles from its junction with the main creek, and I thought it probable he might there find water. I also directed him during my absence to trace the creek on which the camp was established downwards, to ascertain if there was water in it below us.

In the mean time Mr. Browne and I pushed on for the ranges, which presented a very singular appearance as we surveyed them from the lagoon.

The geological formation of these hills was perfectly new, for they were now composed almost exclusively of indurated or compact quartz. The hills themselves no longer presented the character of ranges, properly so called, but were a group of flat-topped hills, similar to those figured by Flinders, King, and other navigators. Some were altogether detached from the main group, not more than two thirds of a mile in length, with less than a third of that breadth, and an elevation of between three and four hundred feet. These detached hills were perfectly level at the top, and their sides declined at an angle of 54°. The main group as we now saw it appeared to consist of a number of projecting points, connected by semicircular sweeps of greater or less depth. There was no vegetation on the sides either of the detached hills or of the projecting points, but they consisted of a compact white quartz, that had been split by solar heat into innumerable fragments in the form of parallelograms. Vast heaps of these laid at the base of the hills, and resembled the ruins of a town, the edifices of which had been shaken to Pieces by an earthquake, and on a closer examination it appeared to me that a portion of the rock thus scaled off periodically. We approached these hills by a gradual ascent, over ground exceedingly stony in places; but as we neared them it became less so, the soil being a decomposition of the geological structure of the hills. It was covered with a long kind of grass in tufts, but growing closer together than usual. There were bare patches of fine blistered soil, that had as it were been raised into small hillocks, and on these, rounded particles, or stools, if I may so call them, of gypsum rested, oval or round, but varying in diameter from three to ten inches or more. These stools were perfectly flat and transparent, the upper surface smooth, but in the centre of the under surface a pointed projection, like that in a bull's eye in window glass was buried in the ground, as if the gypsum was in process of formation.

On leaving the lagoon, we crossed the creek, riding on a north-east course over stony plains, and at five miles struck another creek in which we found a good supply of water, coming direct from the hills, and continuing to the S.S.E., became tributary to the one we had just left. I had taken bearings of two of the most prominent points on the ranges from the lagoon, and directing Flood to go to one of them with Joseph, and wait for me at the base, I rode away with Mr. Browne to ascend the other; but finding it was much farther than we had imagined, that it would take us out of our way, and oblige us to return, we checked our horses and made for the other hill, at the foot of which Flood had already arrived. The ascent was steep and difficult, nor did the view from its summit reward our toil. If there was anything interesting about it, it was the remarkable geological formation of the ranges. The reader will understand their character and structure from the accompanying cut, better than from any description I can give. They were, in fact,

wholly different in formation from hills in general. To the westward there was a low, depressed tract, with an unbroken horizon and a gloomy scrub. Southwards the country was exceedingly broken, hilly, and confused; but there was a line of hills bounding this rugged region to the eastward, and immediately beyond that range were the plains I had crossed in going to Mount Lyell. From the point on which we stood there were numerous other projecting points, similar to those of the headlands in the channel, falling outwards at an angle of 55°, as if they had crumbled down from perpendicular precipices. The faces of these points were of a dirty white, without any vegetation growing on them; they fell back in semicircular sweeps, and the ground behind sloped abruptly down to the plains. The ranges were all flat-topped and devoid of timber, but the vegetation resembled that of the country at their base, and the fragments of rock scattered over them were similar: that is to say, milky quartz, wood opal, granite, and other rocks (none of which occurred in the stratification of these ranges), were to be found on their summits as on the plains, and in equal proportion, as if the whole country had once been perfectly level, and that the hills had been forced up. Such indeed was the impression upon Mr. Poole's mind, when he returned to me from having visited these ranges. "They appear," he remarked, "to have been raised from the plains, so similar in every respect are their tops to the district below." Our eyes wandered over an immense expanse of country to the south, and we were enabled to take bearings of many of the hills near the camp, although there was some uncertainty in our recognition of them at the distance of 40 miles. The Red Hill, however, close to the camp bore south, and was full that distance from us. We could also see the course of the creeks we had been tracing, ultimately breaking through the range to the eastward and passing into the plains beyond. Behind us to the north there were many projecting points appearing above the level of the range. These seemed to be the northern termination of these hills, and beyond them the country was very low. The outline of the projecting points was hilly, and they were so exactly alike that it would have been impossible to have recognised any to which we might have taken bearings; but there were two little cones in a small range to the north upon which I felt I could rely with greater certainty. They respectively bore 302 and 306 from me; and as they were the only advanced points on which I could now keep up bearings, although in the midst of hills, I determined as soon as I should have examined the neighbourhood a little more, to proceed to them. From our first position we went to the next, a hill of about 450 feet in height, perfectly flat-topped, and detached from the main group.

In crossing over to this point the ground was stony, but there was a good deal of grass growing in tufts upon it, and bare patches of blistered earth on which flat stools of gypsum were apparently in process of formation. Immediately to the left there were five remarkable conical hills. These we successively passed, and then entered a narrow, short valley, between the last of these cones and the hill we were about to ascend. The ground was covered with fragments of indurated quartz (of which the whole group was composed), in parallelograms of different dimensions. The scene was like that of a city whose structures had been shaken to pieces by an earthquake—one of ruin and desolation. The faces of the hills, both here and in other parts of the group, were cracked by solar heat, and thus the rock was scaling off.

Part of the Northern Range.

We were here obliged to dismount and walk. The day being insufferably hot, it was no pleasant task to climb under such exposure to an elevation of nearly 500 feet. We had frequently to take breath during our ascent, and reached the summit of the hill somewhat exhausted. The view was precisely similar to that we had overlooked from the opposite point, which bore W. by N. from us. Again the two little peaks were visible to the N.N.W., and after taking bearings of several distant points, we descended, as I had determined on returning for the night to the creek we had passed in the morning, and tracing it into the hills on my way to the westward. Accordingly, on the following morning we commenced our journey up it at an early hour, not knowing where we should next find the water. At about six miles we had entered a valley, with high land on either side, and at a mile beyond reached the head of the creek, and had the steep brow of a hill to ascend, which I thought it most prudent first to attempt on foot. Mr. Browne and I, therefore, climbed it, and on looking back to the north-east, saw there was a declining plain in that direction. Over the level outline the tops of the projections of this range were to be seen all exactly alike; but there was an open space to the north-east, as if the fall of waters was to that point. There were also some low scattered trees upon the plain, seeming to mark the course of a creek. Anxious to ascertain if we had been so fortunate, I looked for a practicable line for the horses to ascend. and having got them up the hill, we pushed forward. On arriving at the first trees, there was a little channel, or rather gutter, and a greener verdure marked its course along the plain to the next trees. Gradually it became larger, and at last was fully developed as a creek. After tracing it down for some miles, having stony barren plains on both sides, we turned to look for the hill we had so lately left, and only for a red tint it had peculiar to itself, should we again have recognised it. We now pushed on in eager anticipation that sooner or later water would appear, and this hope was at last gratified by our arrival at a fine pool, into which we drove a brood of very young ducks, and might, if we had pleased, shot the mother; but although a roast duck would have been very acceptable, we spared her for her children's

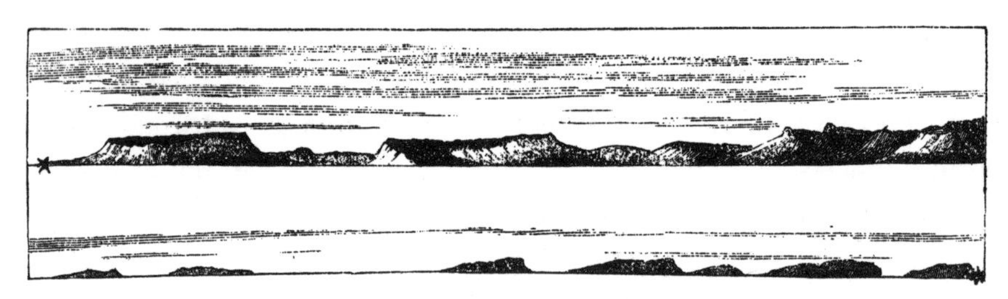

General appearance of the Northern Ranges at their Termination.

sake. This was a nice pond, but small. It was shaded by gum trees, and there was a cavernous clay bank on the west side of it, in which gravel stones were embedded. Here we staid but for a short time, as it was early in the day. We had flushed numerous pigeons as we rode along, and flights came to the water while we stopped, but were not treated with the same forbearance as the duck. We shot two or three, and capital eating they were. About 3, we had left the creek, as it apparently turned to the eastward, and was lost on the plain, and crossing some stony ground, passed between two little ranges. We then found ourselves on the brow of a deep valley that separated us from the little cones we purposed ascending. The side of it which trended to the north-west was very abrupt and stony, and it was with some difficulty we descended into it; but that done, we left Morgan and Flood with the cart, and ascended the nearer peak.

From the summit of the highest of the cones we had a clear view round more than one half of the horizon. Immediately at the base of the ranges northwards, there was a long strip of plain, and beyond it a dark and gloomy scrub, that swept round from S.W. to E., keeping equidistant from the hills, excepting at the latter point where it closed in upon them. On the N.W. horizon there was a small low undulating range, apparently unconnected with any other, and distant about 40 miles. No change had taken place in the geological formations of the main range. The same abrupt points, and detached flat-topped hills, characterised their northern as well as the southern extremity. We had now however reached their termination northwards, but they continued in an easterly direction until they were totally lost in the dark mass of scrub that covered and surrounded them, not one being of sufficient height to break the line of the horizon. To the S.W. a column of smoke was rising in the midst of the scrub, otherwise that desolate region appeared to be uninhabited. On descending from the peak, we turned to the N.W. along the line of a watercourse at the bottom of the valley, tracing it for about four miles with every hope of finding the element we were in search of in its green bed, but we gained the point where the valley opened out upon the plains, and halted under disappointment, yet with good grass for the horses. Our little bivouac was in lat. 29°2'14"S. The above outline will enable the reader to judge of the character of the hills, that still existed to the eastward of us, and the probability of their continuance or cessation. I must confess that they looked to me as if they had been so many small islands, off the point of a larger one. They rose in detached groups from the midst of the plains, as such islands from the midst of the sea, and their aspect altogether bore such a striking resemblance to many of the flat-topped islands round the Australian continent described by other travellers, that I could not but think they had once been similarly situated.

On the 18th I passed into the plains until we had cleared the hills, when we rode along their base on a course somewhat to the east of north. We kept about half a mile from the foot of the ranges, with the brush about three miles to our left, and a clear space between us and them. I had been induced to take this direction in the hope that if there were any creeks falling from the hills into the plains we should intersect them, and accordingly after a ride of about seven miles we observed some gum-trees, about two miles ahead. On a nearer approach we saw flights of pigeons, cockatoos, and parrots winging round about them, and making the air resound with their shrill notes. The anticipations these indications of our approach to water raised, were soon verified by our arrival on the banks of a small creek coming from the hills. Under the trees there were two little puddles, rather than pools of water. The one had been reduced to its last dregs, and smelt offensively, the other was very muddy but drinkable, and such as it was we were most grateful for it. The horses requiring rest here, I halted for the night, more especially as the day was unusually hot, and as we could see the creek line of trees extending to the N.W., towards the low range we had noticed in that direction from the little peak, I determined therefore to run it down in the morning, and to make for them, in the hope that something new would develop itself.

On the other side of the creek from that on which we remained, there was a new but unfinished hut. Round about it were the fresh impressions of feet of all sizes, so that it was clear a family of natives must have been engaged in erecting this simple edifice when we were approaching, and that we must have frightened them away. Under this idea Mr. Browne and I tried to find them, perhaps hid in some low brush near us, but we could not. The plains were exceedingly open on both sides, so that they must have seen us at a great distance, and thus had time for flight.

On the 19th we started at daylight, as I proposed if possible to gain the hills before sunset, that being as much as the horses would do. Running the creek down at three and a half miles we were again attracted by a number of birds, pigeons, the rose cockatoo, the crested paroquet, and a variety of others flying round a clump of trees at no great distance from us, but they were exceedingly wild and watchful. We found a pool under, or rather shaded by the trees, of tolerable size, and much better than the water nearer to the hills. Close to it also, on a sloping bank, there was another more than half finished hut from which the natives could only just have retreated, for they had left all their worldly goods behind them; thus it appeared we had scared these poor people a second time from their work. I was really sorry for the trouble we had unintentionally given them, and in order to make up for it, I fastened my own knife with a glittering blade, to the top of a spear that stood upright in front of the hut; not without hopes that the owner of the weapon seeing we intended them no harm, would come to us on our return from the hills.

Below this water-hole the creek sensibly diminished. Crossing and abandoning it we struck away to the N.W. At about half a mile we entered the scrub, which had indeed commenced from the water, but which at that distance became thick. We were then in a perfect desert, from the scrub we got on barren sandy flats, bounded at first by sandy ridges at some little distance from each other, but the formation soon changed, and the sand ridges succeeded each other like waves of the sea. We had no sooner descended one than we were ascending another, and the excessive heat of so confined a place oppressed us greatly. We had on our journey to the westward found an abundance of grass on the sand ridges as well as the flats; but in this desert there was not a blade to be seen. The ridges were covered with spinifex, through which we found it difficult to force a way, and the flats with salsolaceous productions alone. There were no pine trees, but the brush consisted of several kinds of acacia, casuarina, cassia, and hakeae, and these were more bushes than shrubs, for they seldom exceeded our own height, and had leaves only at the termination of their upper branches, all the under leaves having dropped off, withered by the intensity of the reflected surface heat. At one we stopped to rest the horses, but mounted again at half-past one, and reached the hills at 5 P.M. The same dreary desert extended to their base, only that as we approached the hills the flats were broader, and the fall of waters apparently to the east. The surface of the flats was furrowed by water, and there were large bare patches of red soil, but with the exception of a flossy grass that grew sparingly on some of them, nothing but rhagodia and atriplex flourished.

I had tried the temperature of boiling water at the spot where we stopped in the Rocky Glen, and found it to be 211° and a small fraction; and as we descended a little after leaving the creek, we could not have been much above the sea level at one period of the day, although now more than 450 miles from the coast. Our ascent to the top of the little range was very gradual; its sides destitute alike of trees and vegetation, being profusely covered with fragments of indurated quartz, thinly coated with oxide of iron: when on the summit we could not have risen more than 120 feet. It extended for some miles to the N.E., apparently parallel to the ranges from which we had come, whose higher points were visible from it, but to the north and west the horizon was as level as that of the ocean. A dark gloomy sea of scrub without a break in its monotonous surface met our gaze, nor was there a new object of any kind to be seen indicative of a probable change of country. Had other hills appeared to the north I should have made for them, but to have descended into such a district as that below me, seemed to be too hazardous an experiment at this stage of our journey. I determined therefore to return to the main range, and examine it to the north-east. I could not but think, however, from the appearance of the country as far as we had gone, that we could not be very far from the outskirts of an inland sea, it so precisely resembled a low and barren sea coast. This idea I may say haunted me, and was the cause of my making a second journey to the same locality; but on the present occasion, as the sun had set, I retraced my steps to a small flat where we had noticed a little grass, and tethering our horses out laid down to rest.

The desert ridden through the day before, seemed doubly desolate as we returned. The heat was intolerable, in consequence of a hot wind that blew upon us like a sirocco from the N.W., and the air so rarefied that we could hardly breathe, and were greatly distressed. To our infinite relief we got back to the creek at half-past two, after a ride of about 37 miles.

The first thing we did on arriving, was to visit the hut of the natives to see if they had been there during our absence, but as my knife still dangled on the spear, we were led to conclude they had not. On examining the edifice, however, we missed several things that had been left untouched by us, and from the fresh footsteps, of natives over our own of the day before, it was clear they had been back. The knife which was intended as a peace-offering, seems to have scared them away in almost as much haste as if we had been at their heels. There can be no doubt but that they took it for an evil spirit, at which they were, perhaps, more alarmed than at our uncouth appearance. Be that as it may, we departed from the creek without seeing anything of these poor people.

At a little distance from the creek to the N.W., upon a rising piece of ground, and certainly above the reach of floods, there were seven or eight huts, very different in shape and substance from any we had seen.

They were made of strong boughs fixed in a circle in the ground, so as to meet in a common centre; on these there was, as in some other huts I have had occasion to describe, a thick seam of grass and leaves, and over this again a compact coating of clay. They were from eight to ten feet in diameter, and about four and a half feet high, the opening into them not being larger than to allow a man to creep in. These huts also faced the north-west, and each had a smaller one attached to it as shewn in the sketch. Like those before seen they had been left in the neatest order by their occupants, and were evidently used during the rainy season, as they were at some little distance from the creek, and near one of those bare patches in which water must lodge at such times. At whatever season of the year the natives occupy these huts they must be a great comfort to them, for in winter they must be particularly warm, and in summer cooler than the outer air; but the greatest benefit they can confer on these poor people must be that of keeping them from ants, flies, and mosquitos: it is impossible to describe to the reader the annoyance we experienced from the flies during the day, and the ants at night. The latter in truth swarmed in myriads, worked under our covering, and creeping all over us, prevented our sleeping. The flies on the other hand began their attacks at early dawn, and whether we were in dense brush, on the open plain, or the herbless mountain top, they were equally numerous and equally troublesome. On the present occasion Mr. Browne and I regretted we had not taken possession of the deserted huts, as, if we had, we should have got rid of our tormentors, for there were not any to be seen near them. From the fact of these huts facing the north-west I conclude that their more inclement weather is from the opposite point of the compass. It was also evident from the circumstance of their being unoccupied at that time (January), that they were winter habitations, at which season the natives, no doubt, suffer greatly from cold and damp, the country being there much under water, at least from appearances. I had remarked that as we proceeded northwards the huts were more compactly built, and the opening or entrance into them smaller, as if the inhabitants of the more northern interior felt the winter's cold in proportion to the summer heat.

Our position at this point was in latitude 29°43'S., and in longitude 141°14'E., the variation being 5°21' East. I had intended pushing on immediately to the ranges, and examining the country to the northeast; but I thought it prudent ere I did this to ascertain the farther course of this creek, as it appeared from observations we had just made that the fall of waters was to the eastward. We accordingly started at daylight on the 20th, but after tracing it for a few miles, found that it turned sharp round to the westward and spread over a flat, beyond which its channel was nowhere to be found. I therefore turned towards the ranges, and arriving at the upper water-hole at half-past two, determined to stop until the temperature should cool down in the afternoon before I proceeded along the line of hills to the N.E., for the day had been terrifically hot, and both ourselves and our horses were overpowered with extreme lassitude. At a quarter past 3 P.M. on the 21st of January, the thermometer had risen to 131° in the shade, and to 154° in the direct rays of the sun. In the evening however we pushed on for about ten miles, and halted on a plain about a mile from the base of the hills, without water.

On the 22nd we continued our journey to the north-east, through a country that was anything but promising. Although we were traversing plains, our view was limited by acacias and other trees growing upon them. Notwithstanding that we kept close in to the ranges, the watercourses we crossed could hardly be recognised as such, as they scarcely reached to a greater distance than a mile and a half on the plains, before they spread out and terminated. As we advanced the brush became thicker, nor was there anything to cheer us onwards. In the afternoon therefore I turned towards the hills, and ascended one of them, to ascertain if there was any new object in sight, but here again disappointment awaited us.

The hills were more detached than in other places, and much lower. The brush swept over them, and we could see it stretching to the horizon on the distant plains between them. Excepting where the nearer hills rose above it, that horizon was unbroken; nor were the hills, although detached groups still existed to the north-east, distinguishable from the dark plains round them, as the brush extended over all, and the same sombre hue pervaded everything. I should still, however, have persevered in exploring that hopeless region; but my mind had for the last day or two been anxiously drawn to the state of the camp, and the straits to which I felt assured it would have been put, if Mr. Poole had not succeeded in finding water in greater quantity than that on which the people depended when Mr. Browne and I left them. Having been twelve days absent, I felt convinced that the water in the creek had dried up, and thought it more than probable that Mr. Poole had been forced to move from his position. Under such circumstances, I abandoned, for the time, any further examination of the north-east interior, and turning round to the south-west, passed up a flat rather than a valley between the hills, and halted on it at half-past 6 P.M. On the 23rd, we continued on a south-west course, and gradually ascended the more elevated part of the range; at 2 P.M. reached the water-hole we discovered the day we crossed the hills to the little peaks. Our journey back to the camp was only remarkable for the heat to which we were exposed. We reached it on the 24th of the month, and were really glad to get under shelter of the tents. All the water in the different creeks we passed in going out, had sunk many inches, and as I had feared, that at the camp had entirely vanished, and Mr. Poole having been obliged to dig a hole in the middle of the creek, was obtaining a precarious supply for the men, the cattle being driven to a neighbouring pond, which they had all but exhausted.

As the reader will naturally conclude, I was far from satisfied with the result of this last excursion. It had indeed determined the cessation of high land to the north and north-east; for although I had not reached the termination of the ranges in the latter direction, no doubt rested on my mind but that they gradually fell to a level with the plains. We had penetrated to lat. 28°43'S., and to long. 141°4'30"; but had found a country worse than that over which we had already passed—a country, in truth, that under existing circumstances was perfectly impracticable. Yet from appearances I could not but think that an inland sea existed not far from the point we had gained. As I have already observed, the fall of all the creeks from Flood's Creek had been to the eastward, and from what we could judge at our extreme north, the dip of the country was also to the eastward. I thought it more than probable, therefore, that we were still in the valley of the Darling, and that if we could have persevered in a northerly course, we should have crossed to the opposite fall of waters, and to a decided change of country.

We had hitherto made but few additions to our collections. A new hawk and a few parrots were all the birds we shot; and if I except another new and beautiful species of Grevillia, we added nothing to our botanical collections. The geological formation was such as I have already described a compact quartz of a dirty white. Of this adamantine rock all the hills were now composed.

A remarkable feature in the geology of the hills we had recently visited was, as I have remarked, that they were covered with the same productions and the same stones as the plains below, of which they seemed to have formed a part. Milky quartz was scattered over them, although no similar formation was visible; of manganese, basalt, and ironstone, with other substances, there were now no indications. None of these fragments had been rounded by attrition, but still retained their sharp edges and seemed to be little changed by time.

Mr. Poole informed me, that the day he returned to the party he proceeded towards the little range I had directed him to examine; in which, I should observe, both he and Mr. Browne thought there might be water, as they had passed to the westward of it, on their last journey towards the hills, and had then noticed it. Mr. Poole stated, that on approaching the range he arrived at a line of gum. trees, under which there was a long deep sheet of water; that crossing at the head of this, he entered a rocky glen, where there were successive pools in stony basins, in which he considered there was an inexhaustible supply of water for us; but that although the water near the camp had dried up, he had been unwilling to move until my return. The reader may well imagine the satisfaction this news gave me; for had my officer not been so fortunate, our retreat upon the Darling would have been inevitable, whatever difficulties might have attended such a movement—for we were in some measure cut off from it, or should only have made the retreat at an irreparable sacrifice of animals. Mr. Poole had also been down the creek whereon the camp was posted, and had found that it overflowed a large plain, but failing to recover the channel, he supposed it had there terminated. He met a large tribe of natives, amounting in all to forty or more, who appeared. to be changing their place of abode. They were very quiet and inoffensive, and seemed rather to avoid than to court any intercourse with the party.

Foulkes, one of the bullock drivers, had had a sharp attack of illness, but was in some degree recovered. In all other respects everything was regular, and the stock at hand in the event of their being wanted.

I was exceedingly glad to find that the natives had not shewn any unfriendly disposition towards Mr. Poole and his men; but I subsequently learnt from him a circumstance that will in some measure account for their friendly demonstrations. It would appear that Sullivan and Turpin when out one day, during my absence, after the cattle, saw a native and his lubra crossing the plains to the eastward, with some stones for grinding their grass seed, it being their harvest time. Sullivan went after them; but they were exceedingly alarmed, and as he approached the woman set fire to the grass; but on seeing him bound over the flaming tussocks, they threw themselves on the ground, and as the lad saw their terror he left them and returned to his companion. No sooner, however, had these poor creatures escaped one dreaded object than they encountered another, in the shape of Mack, who was on horseback. As soon as they saw him they took to their heels; but putting his horse into a canter, he was up with them before they were aware of it; on this they threw down their stones, bags, net, and fire-stick, and scrambled up into a tree. The fire-stick set the grass on fire, and all their valuables would have been consumed, if Mack had not very properly dismounted and extinguished the flames, and put the net and bags in a place of safety. He could not, however, persuade either of the natives to descend, and therefore rode away. Mack happened to be with Mr. Poole at the time he met the tribe, and was recognised by the man and woman, who offered both him and Mr. Poole some of their cakes. Had the behaviour of my men been different, they would most likely have suffered for it; but I was exceedingly pleased at their strict compliance with my orders in this respect, and did not fail to express my satisfaction, and to point out the beneficial consequences of such conduct.

Mr. Poole having thus communicated with the natives, I was anxious to profit by it, and if possible to establish a friendly intercourse; the day after my arrival at the camp, therefore, I went down the creek with Mack in the hope of seeing them. I took a horse loaded with sugar and presents, and had every anticipation of success; but we were disappointed, since the whole tribe had crossed the plains, on the hard surface of which we lost their tracks. On this ride I found a beautiful little kidney bean growing as a runner amongst the grass, on small patches of land subject to flood. It had a yellow blossom, and the seed was very small and difficult to collect, as it appeared to be immediately attacked by insects.

The fact of the natives having crossed the plain confirmed my impression that the creek picked up beyond it, and I determined on the first favourable opportunity to ascertain that fact. It now, however, only remained for me to place the camp in a more convenient position. To do this we moved on the 27th, and whilst Mr. Browne led the party across the plains, I rode on ahead with Mr. Poole to select the ground on which to pitch our tents. At the distance of seven miles we arrived at the entrance of the little rocky glen through which the creek passes, and at once found ourselves on the brink of a fine pond of water, shaded by trees and cliffs. The scenery was so different from any we had hitherto seen, that I was quite delighted, but the ground being sandy was unfit for us, we therefore turned down the creek towards the long sheet of water Mr. Poole had mentioned, and waited there until the drays arrived, when we pitched our tents close to it, little imagining that we were destined to remain at that lonely spot for six weary months. We were not then aware that our advance and our retreat were alike cut off.

CHAPTER VI.

THE DEPÔT—FURTHER PROGRESS CHECKED—CHARACTER OF THE RANGES—JOURNEY TO THE NORTH-EAST—RETURN—JOURNEY TO THE WEST—RETURN—AGAIN PROCEED TO THE NORTH—INTERVIEW WITH NATIVES—ARRIVE AT THE FARTHEST WATER—THE PARTY SEPARATES—PROGRESS NORTHWARDS—CONTINUE TO ADVANCE—SUFFERINGS OF THE HORSE—CROSS THE 28TH PARALLEL—REJOIN MR. STUART—JOURNEY TO THE WESTWARD—CHARACTER OF THE COUNTRY—FIND TWO PONDS OF WATER—THE GRASSY PARK—RETURN TO THE RANGE—EXCESSIVE HEAT—A SINGULAR GEOLOGICAL FEATURE—REGAIN THE DEPÔT.

AS the reader will have learnt from what I have stated at the conclusion of the last chapter, we pitched our tents at the place to which I have led him, and which I shall henceforth call the "Depôt," on the 27th of January, 1845. They were not struck again until the 17th of July following.

This ruinous detention paralyzed the efforts and enervated the strength of the expedition, by constitutionally affecting both the men and animals, and depriving them of the elasticity and energy with which they commenced their labours. It was not however until after we had run down every creek in our neighbourhood, and had traversed the country in every direction, that the truth flashed across my mind, and it became evident to me, that we were locked up in the desolate and heated region, into which we had penetrated, as effectually as if we had wintered at the Pole. It was long indeed ere I could bring myself to believe that so great a misfortune had overtaken us, but so it was. Providence had, in its allwise purposes, guided us to the only spot, in that widespread desert, where our wants could have been permanently supplied, but had there stayed our further progress into a region that almost appears to be forbidden ground. The immediate effect, however, of our arrival at the Depôt, was to relieve my mind from anxiety as to the safety of the party.

There was now no fear of our encountering difficulties, and perhaps perishing from the want of that life-sustaining element, without which our efforts would have been unavailing, for independently of the beautiful sheet of water, on the banks of which the camp was established, there was a small lagoon to the S.E. of us, and around it there was a good deal of feed, besides numerous waterholes in the rocky gully. The creek was marked by a line of gum-trees, from the mouth of the glen to its junction with the main branch, in which, excepting in isolated spots, water was no longer to be found. The Red Hill (afterwards called Mount Poole), bore N.N.W. from us, distant 3½ miles; between us and it there were undulating plains, covered with stones or salsolaceous herbage, excepting in the hollows, wherein there was a little grass. Behind us were level stony plains, with small sandy undulations, bounded by brush, over which the Black Hill, bearing S.S.E. from the Red Hill, was visible, distant 10 miles. To the eastward the country was, as I have described it, hilly. Westward at a quarter of a mile the low range, through which Depôt Creek forces itself, shut out from our view the extensive plains on which it rises. This range extended longitudinally nearly north and south, but was nowhere more than a mile and a half in breadth. The geological formation of the range was slate, traversed by veins of quartz, its interstices being filled with magnesian limestone. Steep precipices and broken rugged gullies alternated on either side of this creek, and in its bed there were large slabs of beautiful slate. The precipices shewed the lateral formation with the rock split into the finest laminæ, terminating in sharp points. But neither on the ranges or on the plains behind the camp was there any feed for the cattle, neither were the banks of the creek or its neighbourhood to be put in comparison with Flood's Creek in this respect, for around it there was an abundance as well as a variety of herbage. Still the vegetation on the Depôt Creek was vigorous, and different kinds of seeds were to be procured. I would dwell on this fact the more forcibly, because I shall, at a future stage of this journey, have to remark on the state of the vegetation at this very spot, that is to say, when the expedition was on its return from the interior at the close of the year.

A few days after we had settled ourselves at the Depôt, Mr. Browne had a serious attack of illness, that might have proved fatal; but it pleased God to restore him to health and reserve him for future usefulness. At this time, too, the men generally complained of rheumatism, and I suspected that I was not myself altogether free from that depressing complaint, since I had violent pains in my hip joints; but I attributed them to my having constantly slept on the hard ground, and frequently in the bed of some creek or other. It eventually proved, however, that I had been attacked by a more fearful malady than rheumatism in its worst stage.

There being no immediate prospect of our removal, I determined to complete the charts up to the point to which we had penetrated. I therefore sent Mr. Stuart, on the 2nd February, to sketch in the ranges to the eastward, and connect them with the hills I had lately crossed over. I directed Lewis, who had been in the survey, to assist Mr. Stuart, and sent Flood with them to trace down the creek I had noticed from several of our stations on the northern ranges, as passing through a gap in the hills to the eastward. They returned to the camp on the 4th, Mr. Stuart having been very diligent in his work. Flood had also obeyed my orders; but could find no water in the lower branches of the creek, although there was so much in it nearer the hills. The party had fallen in with a small tribe of natives, for whom Flood had shot an emu. Mr. Stuart informed me that they were very communicative; but their language was unknown to him. He understood from them that they intended to visit the camp in a couple of days; but as I had some doubts on this head, and was anxious to establish a communication, and induce them to return with me to the camp, I rode on the 5th with Mr. Browne across the plain, at the farther extremity of which they were encamped near a little muddy puddle. Flood and Joseph in the light cart accompanied us.

Great as the heat had been, it appeared rather to increase than diminish. The wind constantly blew from the E.S.E. in the morning, with the deep purple tint to the west I have already had occasion to notice. It then went round with the sun, and blew heavily at noon; but gradually subsided to a calm at sunset, and settled in the west, the same deep tint being then visible above the eastern horizon which in the morning had been seen in the west. The thermometer ranged from 100° to 117° in the shade at 3 P.M.; the barometer from 29.300° to 29.100°. Water boiled at 211° and a fraction; but there was no dew point. I should have stated, that both whilst Mr. Browne and I were in the hills and at the camp, there was thunder and rain on the 23rd, 24th, and 25th, but the showers were too light even to lay the dust, and had no effect whatever on the temperature.

The morning we started to pay a visit to the blacks was more than usually oppressive even at daybreak, and about 9 it blew a hot wind from the N.E. As we rode across the stony plain lying between us and the hills, the heated and parching blasts that came upon us were more than we could bear. We were in the centre of the plain, when Mr. Browne drew my attention to a number of small black specks in the upper air. These spots increasing momentarily in size, were evidently approaching us rapidly. In an incredibly short time we were surrounded by several hundreds of the common kite, stooping down to within a few feet of us, and then turning away, after having eyed us steadily. Several approached us so closely, that they threw themselves back to avoid contact, opening their beaks and spreading out their talons. The long flight of these birds, reaching from the ground into the heavens, put me strongly in mind of one of Martin's beautiful designs, in which he produces the effect of distance by a multitude of objects gradually vanishing from the view. Whatever the reader may think, these birds had a most formidable aspect, and were too numerous for us to have overpowered, if they had really attacked us. That they came down to see what unusual object was wandering across the lonely deserts over which they soar, in the hope of prey, there can be no doubt; but seeing that we were likely to prove formidable antagonists, they wheeled from us in extensive sweeps, and were soon lost to view in the lofty region from whence they had descended.

When we reached the place where the natives had been, we were disappointed in not finding them. They had, however, covered up their fires and left their nets, as if with the intention of returning. Nevertheless we missed them, and reached the tents late in the evening, after a ride of 40 miles.

After my return from this excursion, I was busily employed filling in the charts; but the ink in our pens dried so rapidly, that we were obliged to have an underground room constructed to work in, and it proved of infinite service and comfort, insomuch that the air in it was generally from 7° to 8° cooler than that of the outer air.

Our observations and lunars placed us in latitude 29°40'14"S., and in longitude 141°30'41"E. Mount Hopeless, therefore, bore N.N.W. of us, as we were still 25 miles to the south of it, the difference of longitude being about 171 miles, and our distance from the eastern shore of Lake Torrens about 120. The result of our lunars, however, placed us somewhat to the westward of the longitude I have given; and when I came to try my angles back from the Depôt to Williorara, I found that they terminated considerably to the westward of Sir Thomas Mitchell's position there. My lunars at Williorara, however, had not been satisfactory, and I therefore gave that officer credit for correctness, and in the first chart I transmitted to the Secretary of State assumed his position to be correct. There was a small range, distant about 20 miles to the westward of the stony range connected with the Depôt Creek. It struck me that we might from them obtain a distant view of Mount Serle, or see some change of country favourable to my future views. Under this impression, I left the camp on the 7th of the month, with Mr. Poole and two of the men. The ranges were at a greater distance than I had imagined, but were of trifling elevation, and on arriving at them I found that the horizon to the westward was still closed from my view, by rising ground that intervened. I should have pushed on for it, but Mr. Poole was unfortunately taken ill, and I felt it necessary to give him my own horse, as having easier paces than the one he was riding. It was with difficulty I got him on his way back to the camp as far as the upper waterhole, just outside the Rocky Glen, at which we slept, and by that means reached the tents early on the following morning. I had anticipated rain before we should get back, from the masses of heavy clouds that rose to the westward, after the wind, which had been variable, had settled in that quarter; but they were dispersed during the night, and the morning of the 8th was clear and warm. We had felt it exceedingly hot the day we left the camp—there the men were oppressed with intolerable heat, the thermometer having risen to 112° in the shade. We had not ourselves felt the day so overpowering, probably because we were in motion, and it is likely that a temporary change in the state of the atmosphere, had influenced the temperature, as the eastern horizon was banded by thunder clouds, though not so heavy as those to the westward, and there was a good deal of lightning in that quarter.

I have said that I was not satisfied with the result of my last excursion with Mr. Browne to the north. I could not but think that we had approached to within a tangible distance of an inland sea, from the extreme depression and peculiar character of the country we traversed. I determined, therefore, to make another attempt to penetrate beyond the point already gained, and to ascertain the nature of the interior there; making up my mind at the same time to examine the country both to the eastward and westward of the northern ranges before I should return to the camp. Mr. Poole and Mr. Browne being too weak to venture on a protracted excursion of such a kind, I took Mr. Stuart, my draftsman, with me. I should have delayed this excursion for a few days, however, only that I feared the total failure of the creeks in the distant interior; I proposed, in the first place, to make for the last and most distant water-hole in the little creek beyond the ranges. Thence to take the light cart with one horse, carrying as much water as he could draw, and with one man, on foot, to pursue a due north course into the brush. I hoped by this arrangement to gain the 27th parallel, and in so doing to satisfy myself as to the point on which I was so anxious. I selected a fine young lad to accompany me, named Joseph Cowley, because I felt some confidence in his moral courage in the event of any disaster befalling us. On this occasion I had the tank reconstructed, and took all the barrels I could, to enable me to go as far as possible, and the day after I returned to the camp with Mr. Poole, again left it with

Mr. Stuart, Joseph, and Flood, in whose charge I intended to leave my horse during my absence—during which I also proposed that Mr. Stuart should employ his time tracing in the hills.

We reached the muddy creek at the foot of the hills at 2 P.M., after a ride of 25 miles, over the stony and barren plains I have described, and as the distance to the next water was too great for us to attempt reaching it until late, we stopped here for the night. Some natives had been on the creek in the early part of the day, and had apparently moved down it to the eastward. The water had diminished fearfully since the time we passed on our return from the north.

The day was cool and pleasant, as the wind blew from the south, and the thermometer did not rise above 95°.

We had not ridden four miles on the following morning, when we observed several natives on the plain at a little distance to the south, to whom we called out, and who immediately came to us. We stopped with these people for more than two hours, in the hope that we should gain some information from them, either as to when we might expect rain, or of the character of the distant interior, but they spoke a language totally different from the river tribes, although they had some few words in common, so that I could not rely on my interpretation of what they said. They were all of them circumcised, and all but one wanted the right front tooth of the upper jaw. When we left these people I gave them a note for Mr. Poole, in the faint hope that they would deliver it, and I explained to them that he would give them a tomahawk and blankets, but, as I afterwards learnt, they never went to the camp.

When Mr. Browne and I were in this neighbourhood before, he had some tolerable sport shooting the new pigeon, the flesh of which was most delicious. At that time they were feeding upon the seed of the rice grass, and were scattered about, but we now found them, as well as many other birds, congregated in vast numbers preparing to migrate to the north-east, apparently their direct line of migration; they were comparatively wild, so that our only chance of procuring any was when they came to water.

On the 9th we slept at the water in the creek at the top of the ranges; but, on the 10th, instead of going through the pass, and by the valley, under the two little peaks, through which we had entered the plains on the first journey, we now turned to the westward in order to avoid that rugged line, and discovered that the creek, instead of losing itself in the flat to the eastward, continued on a westerly course to our left; for being attracted by a flight of pigeons, wheeling round some gum-trees, we might otherwise have overlooked it; I sent Flood to examine the ground, who returned with the pleasing information that the creek had reformed, and that there was a pool of water under the trees, nearly as large as the one we had just left.

I was exceedingly pleased at this discovery and determined to send Mr. Stuart back to it, as it would place him nearer his work. We reached the farthest water, from which we had the second time driven the poor native, late in the afternoon, and on examining the hut, found he had ventured back to it and taken away his traps; but the water in the creek was almost dried up; thick, muddy, and putrid, we could hardly swallow it, and I regretted that we had not brought water with us from the hills, but I had been influenced by a desire to spare my poor horse, as I knew the task that was before him, although the poor brute was little aware of it. About sunset an unfortunate emu came to water, and unconsciously approached us so near that Flood shot it with his fusee. This was a solitary wanderer, for we had seen very few either of these birds or kangaroos in these trackless solitudes.

On the morning of the 10th we were up early, and had loaded the cart with 69 gallons of water before breakfast, when Joseph and I took our departure, and Mr. Stuart with Flood returned to the hills. I had selected one of our best horses for this journey, an animal I had purchased from Mr. Frew, of Adelaide. He was strong, powerful, and in good condition, therefore well qualified for the journey. I had determined on keeping a general north course, but in the kind of country in which I soon found myself it was impossible to preserve a direct line. At about four miles from the creek the brush became thick, and the country sandy, and at six miles the sand ridges commenced.

Wishing to ease the horse as much as possible, Joseph endeavoured to round them by keeping on the intervening flats, but this necessarily lengthened the day's journey, and threw me more to the eastward than I had intended. At noon I halted for two hours, and then pushed on, the day being cool, with the wind as it had been for the last three or four days from the south. Had the country continued as it was, we might have got on tolerably, but as we advanced it changed greatly for the worse. We lost the flats, on a general coating of sand thickly matted with spinifex, through which it was equally painful to ourselves and poor Punch to tread. We crossed small sandy basins or hollows, and were unable to see to any distance. The only trees growing in this terrible place were a few acacias in the hollows, and some straggling melaleuca, with hakeæ, and one or two other common shrubs, all of low growth; there was no grass, neither were the few herbs that grew on the hollows such as the horse would eat. We stopped a little after sunset having journeyed about 22 miles, on a small flat on which there were a few acacias, and some low silky grass as dry as a chip, so that if we had not been provident in bringing some oats poor Punch would have gone without his supper. A meridian altitude of Capella placed us in lat. 28°41'00". Our longitude by account being 141°15'E. When I rose at daylight on the following morning, I observed that the horse had eaten but little of the dry and withered food on which he had been tethered; however, in consequence of our tank leaking, I was enabled to give him a good drink, when he seemed to revive, but no sooner commenced pulling than he perspired most profusely. We kept a more regular course than on the previous day, over a country that underwent no change. Before we started I left a nine gallon cask of water in a small flat to ease the horse, and as the water in the tank had almost all leaked out, his load was comparatively light. Still it was a laborious task to draw the cart over such a country. Fortunately for us the weather was cool, as the wind continued south, for I do not know what we should have done if we had been exposed to the same heat Mr. Browne and myself had experienced on our return from the little stony ranges now about 10 miles to the westward of us. A little before noon the wind shifted to the N.E.; I had at this time stopped to rest the horse, but we immediately experienced a change of temperature, and the thermometer which stood at 81° rose before we again started to 93°, and at half-past three had attained 119°. We were then in one of the most gloomy regions that man ever traversed. The stillness of death reigned around us, no living creature was to be heard; nothing visible inhabited that dreary desert but the ant, even the fly shunned it, and yet its yielding surface was marked all over with the tracks of native dogs.

We started shortly after noon, and passed a pointed sand hill, from whence we could not only see the stony range but also the main range of hills. The little peak on which Mr. Browne and I took bearings on our last journey bore 150°, the pass through which we had descended into the plains 170°, when I turned however to take bearings of the stony range it had disappeared, having been elevated by refraction above its true position. It bore about N.W.1/2W., distant from eight to nine miles. It was again some time after sunset before we halted, on a small flat that might contain two or at the most three acres. There was some silky grass upon it, but this I knew the horse would not eat, neither had I more than a pint of oats to give him. Our latitude here was 28°22'00".

On the morning of the 13th we still pushed on, leaving, as before, a cask of water to pick up on our return. I had been obliged to limit the horse to six gallons a day, but where he had been in the habit of drinking from 25 to 30, so small a quantity would not suffice. We had not gone many miles when he shewed symptoms of exhaustion, and rather tottered than walked. He took no pains to avoid anything, but threw Joseph into every bush he passed. The country still continued unchanged, sand and spinifex were the universal covering of the land, and only round the edges of the little flats were a few stunted shrubs to be seen. It was marvellous to me that such a country should extend to so great a distance without any change. I could at no time see beyond a mile in any direction. Several flights of parrots flew over our heads to the north-west, at such an elevation as led me to suppose they would not pitch near us; but not a bird of any kind did we see in the desert itself. The day being exceedingly hot I stopped at one, rather from necessity than inclination, having travelled 12 or 14 miles. Both Joseph and myself had walked the whole way, and our legs were full of the sharp ends of the spinifex, but it was more in mercy to poor Punch than to ourselves that I pulled up, and held a consultation with Joseph as to the prudence of taking the cart any further, when it was decided that our doing so would infallibly lead to Punch's destruction. According to my calculation we were now in latitude 28°9'00" or thereabouts. I had hoped to have advanced some 60 miles beyond this point, but now found that it would be impossible to do so. There was no indication of a change of country from any rising ground near us, and as it was still early in the day I resolved on pushing forward until I should feel satisfied that I had passed into the 27th parallel; my reason for this being a desire to know what the character of the country, so far in the interior from, and in the same parallel as Moreton Bay, would be. I had intended tethering Punch out, and walking with Joseph, but as he remonstrated with me, and it did not appear that my riding him would do the horse any harm, I mounted, though without a saddle, and taking our guns, with a quart of water, we commenced our journey. We moved rapidly on, as I was anxious to return to the cart whilst there was yet daylight, to enable us to keep our tracks, but no material change took place in the aspect of the country. We crossed sandridge after sandridge only to meet disappointment, and I had just decided on turning, when we saw at the distance of about a quarter of a mile from us, a little rounded hill some few feet higher than any we had ascended. It was to little purpose however that we extended our ramble to it. At about a mile from where we left the cart, we had crossed two or three small plains, if pieces of ground not a quarter of a mile long might be so termed, on which rhagodia bushes were growing, and I had hoped that this trifling change would have led to a greater, but as I have stated such did not prove to be the case. From the top of the little hill to which we walked (and from which we could see to a distance of six or eight miles, but it was difficult to judge how far the distant horizon was from us), there was no apparent change, but the brush in the distance was darker than that nearer to us, as if plains succeeded the sandy desert we had passed over. The whole landscape however was one of the most gloomy character, and I found myself obliged to turn from it in disappointment. As far as I could judge we passed about a mile beyond the 28th parallel. Our longitude by account only being 141°18'E. The boiling point of water was 211°75/100. The evening had closed in before we got back to the cart, but our course was fortunately true, and having given poor Punch as liberal a draught as reason would justify we laid down to rest.

It was with great difficulty that we got our exhausted animal on, the following morning, although I again gave him as much water as I could spare. His docility under urgent want of food was astonishing. He was in fact troublesomely persevering, and walked round and round the cart and over us as we sat drinking our tea, smelling at the casks, and trying to get his nose into the bung holes, and implored for relief as much as an animal could do so by looks. Yet I am satisfied that a horse is not capable of stronger attachment to man, but that he is a selfish brute, for however kindly he may be treated, where is the horse that will stay, like the dog, at the side of his master to the last, although hunger and thirst are upon him, and who, though carnivorous himself, will yet guard the hand that has fed him and expire upon its post? but, turn the horse loose at night, and where will you find him in the morning, though your life depended on his stay?

We reached the creek on the morning of the 14th, about half-past 10, having still a gallon of water remaining, that was literally better than the water in the muddy puddle from which we had originally taken it. I had thought it probable that we might find either Flood or Mr. Stuart awaiting our return, but not seeing any trace of recent feet I concluded they were in the ranges, and as the distance was too great for the horse to travel in a day, in his exhausted state, I pushed on at 4 P.M., and halted on the plains after having ridden about 6 miles. It was well indeed that I did so, for we did not gain the ranges until near sunset on the following day. Our exhausted horse could hardly drag one leg after the other, although he pricked up his ears and for a time quickened his pace as he fell into the track of the cart coming out. Both Mr. Stuart and Flood were astonished at the manner in which he had fallen off, nor did he ever after recover from the effects of that journey.

Mr. Stuart had completed his work with great accuracy, and had filled in the chart so much that he saved me a good deal of trouble. The 16th being Sunday, was a day of rest to us all, but one of excessive heat. Mr. Stuart had stationed himself in the bed of the creek, which sloped down on either side, and was partially shaded by gum-trees. The remains of what must have been a fine pond of water occupied the centre, and although it was thick and muddy it was as nectar to myself and Joseph. I was surprised and delighted to see that the creek had here so large a channel, and Flood, who had ridden down it a few miles, assured me that it promised very well. During my absence he had shot at and wounded one of the new pigeons, which afterwards reached my house alive.

I had intended proceeding to the eastward on my return from the north, but was prevented by the total failure of water. I therefore determined to trace the creek down, in the hope that it would favour my advance with the party into the interior. On the 17th, therefore, leaving Joseph to take care of Punch, I mounted my horse, and with Mr. Stuart and Flood, rode away to the westward. At first the creek held a course between S.W. and W.S.W. occasionally spreading over large flats, but always reforming and increasing in size. It ran through a flat valley, bounded by sand hills, against which it occasionally struck. The soil of the valley was not bad, but there was little or no vegetation upon it. At 15 miles we arrived at the junction of another creek from the south, and running down their united channels, at three miles found a small quantity of water in a deep and shaded hollow. It was but a scanty supply however, yet being cleaner and purer than any we had for some time seen, I stopped and had some tea. There was a native's hut on the bank, from which the owner must have fled at our approach; it was quite new, and afforded me shelter during our short halt. The fugitive had left some few valuables behind him, and amongst them a piece of red ochre. From this point the creek trended more to the north, spreading over numerous flats in times of flood, dividing its channels into many smaller ones, but always uniting into one at the extremity of the flats. At 21 miles the creek changed its course to 20° to the west of north, and the country became more open and level. There were numerous traces of natives along its banks, and the remains of small fires on either side of it as far as we could see. It was, therefore, evident that at certain seasons of the year they resorted to it in some numbers, and I was then led to hope for a favourable change in the aspect of the country.

The gum-trees as we proceeded down the creek increased in size, and their foliage was of a vivid green. The bed of the creek was of pure sand, as well as the plains through which it ran, although there was alluvial soil partially mixed with the sand, and they had an abundance of grass upon them, the seed having been collected by the natives for food. At about 14 miles from the place where we stopped, the creek lost its sandy bed, and got one of tenacious clay. We soon afterwards pulled up for the night, at two pools of water that were still of considerable size, and on which there were several new ducks. They must, indeed, have been large deep ponds not many weeks before, but had now sunk several feet from their highest level, and, however valuable to a passing traveller, were useless in other respects, as our cattle would have drained them in three or four days. From this place also the natives appeared to have suddenly retreated, since there was a quantity of the Grass* spread out on the sloping bank of the creek to dry, or ripen in the sun. We could not, however, make out to what point they had gone. The heat during the day had been terrific, in so much that we were unable to keep our feet in the stirrups, and the horses perspired greatly, although never put out of a walk.

It was singular that we had no moisture on our skin; the reason why, perhaps, we were at that time much distressed by violent headaches. At about a quarter of a mile below the ponds the creek spreads over an immense plain, almost as large as that of Cawndilla. A few trees marked its course to a certain distance, but beyond them all trace of its channel was lost, nor was it possible from the centre of the plain to judge at what point its waters escaped. The plain was surrounded by sand hills of about thirty feet in elevation, covered with low scrub. When we started in the morning we crossed it on a west course, but saw nothing to attract our notice from the tops of the sand hills. We then turned to the northward, and at about two miles entered a pretty, well wooded, but confined valley, in the bottom of which we once more found ourselves on the banks of the creek. Running it down in a north-west direction for seven miles, we were at length stopped by a bank of white saponaceous clay, crossing the valley like a wall. As we rode down the creek we observed large plains of red soil, precisely similar to the plains of the Darling, receding from it to a great distance on either side. These plains had deep water-worn gutters leading into the valley, so that I conclude the lateral floods it receives are as copious as those from the hills. On arriving at the bank running across the channel there were signs of eddying waters, as if those of the creek had been thrown back; but there was a low part in the bank over which it is evident they pour when they

* *"Panicum lœvinode"* of Dr. Lindley.

rise to its level. Mr. Stuart and Flood were the first to ascend the bank, and both simultaneously exclaimed that a change of country was at hand. On ascending the bank myself, I looked to the west and saw a beautiful park-like plain covered with grass, having groups of ornamental trees scattered over it. Whether it was the suddenness of the change, from barrenness and sterility to verdure and richness, I know not; but I thought, when I first gazed on it, that I never saw a more beautiful spot. It was, however, limited in extent, being not more than eight miles in circumference. Descending from the bank we crossed the plain on a south course. It was encircled by a line of gum-trees, between whose trunks the white bank of clay was visible. We crossed the plain amidst luxuriant grass; but the ground was rotten, and the whole area was evidently subject to flood. It was also clear, that the creek exhausted itself in this extensive basin, from which, after the strictest search, we could find no outlet. On reaching the southern extremity of the plain, we crossed a broad bare channel, having a row of gum-trees on either side, and ascending a continuation of the clay bank, at once found ourselves in the scrub and amidst barrenness again; and at less than a mile, on a north-west course, beheld the sand ridges once more rising before us. I continued on this course, however, for eight miles, when I turned to the north-east, in order to cut any watercourse that might be in that direction, and to assure myself of the failure of the creek. After riding for five miles, I turned to the south, with the intention of ascending a sand hill at some distance, that swept the horizon in a semicircular form and was much higher than any others. Mr. Poole had informed me that he noticed a similar bank just before he made Lake Torrens, and I was anxious to see if it hid any similar basin from my view; but it did not. Sand hills of a similar kind succeeded it to the westward, but there was no change of country. Although we had travelled many miles, yet the zigzag course we had taken had been such that at this point we were not more than sixteen miles from the pools we had left in the morning; and as the day had been intolerably hot, and we had found no water, I determined on returning to them; but I was obliged to stop for a time for Flood, who complained of a violent pain in his head, occasioned by the intense heat. There was no shelter, however, for him under the miserable shrubs that surrounded us; but I stopped for half an hour, during which the horses stood oppressed by languor, and without the strength to lift up their heads, whilst their tails shook violently. Being, anxious to get to water without delay, I took a straight line for the waterholes, and reached them at half-past 6 P.M., after an exposure, from morning till night, to as great a heat as man ever endured, but if the heat of this day was excessive, that of the succeeding one on which we returned to Joseph was still more so. We reached our destination at 3 P.M. as we started early, and on looking at the thermometer fixed behind a tree about five feet from the ground, I found the mercury standing at 132°; on removing it into the sun it rose to 157°. Only on one occasion, when Mr. Browne and I were returning from the north, had the heat approached to this; nor did I think that either men or animals could have lived under it.

On the 20th we again crossed the ranges, and after a journey of 32 miles, reached the lateral creek at their southern extremity, where I had rested on my former journey. There was more water in it than I expected to have found; but it was nevertheless much reduced, and in a week afterwards was probably dry. On the 21st we gained the Muddy Creek, but had to search for water where only a few days before there had been a pond of more than a third of a mile in length. Here, on the following day, I was obliged to leave Flood and Joseph, as the wheels of the cart had shrunk so much that we could not take it on. I should have gained the camp early in the day, but turned to the eastward to take bearings from some hills intermediate between Mount Poole and the Northern Range, as the distance between these points was too great. Our ride was over a singularly rugged country, of equally singular geological formation, nor can I doubt but that at one time or other there were currents sweeping over it in every direction. At one place that we passed there was a broad opening in a rocky but earth-covered bank. Through this opening the eye surveyed a long plain, which at about two miles was bounded by low dark hills. Along this plain the channel of a stream was as distinctly marked in all its windings by small fragments of snow-white quartz as if water had been there instead. On either side the landscape was dark; but the effect was exceedingly striking and unusual. From the hills we ascended I obtained bearings to every station of consequence, and was quite glad that I had thus turned from my direct course. It was dark, the night indeed had closed in before we reached the tents; but I had the satisfaction to learn that both Mr. Poole and Mr. Browne were better, though not altogether well, and that every thing had gone on regularly during my absence. On the following morning, I sent Lewis and Jones with a dray to fetch the cart, and for the next three or four days was occupied charting the ground we had travelled over.

The greatest distance I went northwards on this occasion was to the 28th parallel, and about 17 miles to the eastward of the 141st meridian. Our extreme point to the westward being lat. 28°56', and long. 140°54'. From what I have said, the reader will be enabled to judge what prospects of success I had in either quarter; for myself I felt that I had nothing to hope either in the north or the east; for even if I had contemplated crossing eastward to the Darling, which was more than 250 miles from me, the dreadful nature of the country would have deterred me; but such an idea never entered my head—I could not, under existing circumstances, have justified such a measure to myself; having therefore failed in discovering any change of country, or the means of penetrating farther into it, I sat quietly down at my post, determined to abide the result, and to trust to the goodness of Providence to release me from prison when He thought best.

CHAPTER VII.

MIGRATION OF THE BIRDS—JOURNEY TO THE EASTWARD—FLOODED PLAINS—NATIVE FAMILY—PROCEED SOUTH, BUT FIND NO WATER—AGAIN TURN EASTWARD—STERILE COUNTRY—SALT LAGOON—DISTANT HILLS TO THE EAST—RETURN TO THE CAMP—INTENSE HEAT—OFFICERS ATTACKED BY SCURVY—JOURNEY TO THE WEST—NO WATER—FORCED TO RETURN—ILLNESS OF MR. POOLE—VISITED BY A NATIVE—SECOND JOURNEY TO THE EASTWARD—STORY OF THE NATIVE—KITES AND CROWS—ERECT A PYRAMID ON MOUNT POOLE—PREPARATIONS FOR A MOVE—INDICATIONS OF RAIN—INTENSE ANXIETY—HEAVY RAIN—MR. POOLE LEAVES WITH THE HOME RETURNING PARTY—BREAK UP THE DEPÔT—MR POOLE'S SUDDEN DEATH—HIS FUNERAL—PROGRESS WESTWARD—THE JERBOA—ESTABLISHMENT OF SECOND DEPÔT—NATIVE GLUTTONY—DISTANT MOUNTAINS SEEN—REACH LAKE TORRENS—EXAMINATION OF THE COUNTRY N.W. OF IT—RETURN TO THE DEPÔT—VISITED BY NATIVES—PREPARATIONS FOR DEPARTURE AGAIN INTO THE NORTH-WEST INTERIOR.

THE THREE last days of February were cool in comparison to the few preceding ones. The wind was from the south, and blew so heavily that I anticipated rough weather at the commencement of March. But that rough month set in with renewed heat, consequent on the wind returning to its old quarter the E.S.E. There were however some heavy clouds floating about, and from the closeness of the atmosphere I hoped that rain would have fallen, but all these favourable signs vanished, the thermometer ascending to more than 100°.

When we first pitched our tents at the Depôt the neighbourhood of it teemed with animal life. The parrots and paroquets flew up and down the creeks collecting their scattered thousands, and making the air resound with their cries. Pigeons congregated together; bitterns, cockatoos, and other birds; all collected round as preparatory to migrating. In attendance on these were a variety of the Accipitrine class, hawks of different kinds, making sad havoc amongst the smaller birds. About the period of my return from the north they all took their departure, and we were soon wholly deserted. We no longer heard the discordant shriek of the parrots, or the hoarse croaking note of the bittern. They all passed away simultaneously in a single day; the line of migration being directly to the N.W., from which quarter we had small flights of ducks and pelicans.

On the 5th of March I sent Mr. Browne to the S.W., to a small creek similar to that in the Rocky Glen and in the same range, in the hope that as we had seen fires in that direction he might fall in with the natives, but he was unsuccessful.

On the 6th I sent Flood to the eastward to see if he could recover the channel of the main creek on the other side of the plain on which Mr. Poole had lost it; he returned the following day, with information that at 25 miles from the Depôt he had recovered it, and found more water than he could have supposed. The day of Flood's return was exceedingly hot and close, and in the evening we had distant thunder, but no rain.

In consequence of his report, I now determined on a journey to the eastward to ascertain the character of the country between us and the Darling, and left the camp with this intention on the 12th instant. I should have started earlier than that day had not Mr. Poole's illness prevented me, but as he rallied, I proceeded on my excursion, accompanied by Mr. Browne, Flood, and another of the men. We observed several puddles near our old camp on the main creek as we rode away, so that rain must have fallen there though not at the Depôt. After passing the little conical hill of which I have already spoken, we traced the creek down until we saw plains of great extent before us, and as the creek trended to the south, skirting them on that side, we rode across them on a bearing of 322° or N.W.1/2N. They were 7 or 8 miles in breadth, and full 12 miles in length from east to west; their soil was rich and grassed in many places. At the extremity of the plains was a sand hill, close to which we again came on the creek, but without water, that which Flood had found being a little more to the eastward.

Its channel at this place was deep, shaded, and moist, but very narrow. I was quite surprised when we came to the creek where Flood had been to find so much water; there was a serpentine sheet, of more than a quarter of a mile in length, which at first sight appeared to be as permanent as that at the Depôt. The banks were high and composed of light rich alluvial soil, on which there were many new shrubs growing; the whole vegetation seemed to be more forward on this side of the hills than on that where the Depôt was. Just as we halted we saw a small column of smoke rise up due south, and on looking in that direction observed some grassy plains spreading out like a boundless stubble, the grass being of the kind from which the natives collect seed for subsistence at this season of the year.

Early on the morning of the 14th March we again saw smoke in the same direction as before, but somewhat to the eastward, as if the grass or brush had been fired. In hopes that we should come upon some of the natives on the plains, through which the creek appeared to run, I determined on examining them before I proceeded to the eastward. We accordingly crossed its channel when we mounted our horses after breakfast, and rode at some little distance from it on a course of 80° or nearly east, over flooded lands of somewhat sandy soil, covered with different kinds of grass, of which large heaps that had been thrashed out by the natives were piled up like hay cocks. At about two and a half miles we ascended a sandy rise of about fifty feet in elevation, whence we obtained bearings of the little conical hill at the western termination of the plain, and of the hill we had called the Black Hill. These bearings with our latitude made the distance we had travelled 33 miles. From the sand hill we overlooked plains of great extent to the N.E.; partly grassed and partly bare, but to the eastward there was low brush and a country similar to that we had traversed before the commencement of the sandy ridges. There were low sandy undulations to be seen; but of no great height. I now turned for the smoke on a bearing of 187°, or nearly south, traversing a barren sandy level intermediate between the sand hill and the plains now upon our right, at length we entered upon the flooded ground, it was soft and yielding, and marked all over with the tracks of the natives; at 7 miles arrived at a large clump of gum-trees, and under them the channel of the creek which we had lost on the upper part of the plains was again visible. It was here very broad, but quite bare except a belt of polygonum growing on either side, which had been set on fire, and was now in flames. We were fortunate enough soon after to find a long shallow sheet of water, in the bed of the creek, where we rested ourselves. It was singular enough that we should have pulled up close to the camp of some natives, all of whom had hidden themselves in the polygonum, except an old woman who was fast asleep, but who did not faint on seeing Mr. Browne close to her when she awoke. With this old lady we endeavoured to enter into conversation, and in order to allay her fears gave her five or six cockatoos we had shot, on which two other fair ones crept from behind the polygonum. and advanced towards us. Finding that the men were out hunting, and only the women with the children were present, I determined to stop at this place until the following morning, we therefore unloaded the horses and allowed them to go and feed. A little before sunset, the two men returned to their families. They were much astonished at seeing us quietly seated before their huts, and approached us with some caution, but soon got reconciled to our presence. One of them had caught a talpero and a lizard, but the other had not killed any thing, so we gave him a dinner of mutton. The language of these people was a mixture between that of the river and hill tribes; but from what reason I am unable to say, although we understood their answers to general questions, we could not gather any lengthened information from them. I gave the elder native a blanket, and to the other a knife, with both of which they seemed highly delighted, and in return I suppose paid us the compliment of sending their wives to us as soon as it became dusk, but as we did not encourage their advances, they left us after a short visit. The native who had killed the talpero, skinned it the moment he arrived in the camp, and, having first moistened them, stuffed the skin with the leaves of a plant of very astringent properties. All these natives were very poor, particularly the men, nor do I think that at this season of the year they can have much animal food of any kind to subsist on. Their principal food appeared to be seeds of various kinds, as of the box-tree, and grass seeds, which they pound into cakes and bake, together with different kinds of roots.

On the 15th we started at 7 A.M., and crossing at the head of the water, pursued a south course over extensive flooded plains, on which we again lost the channel of the creek, as, after winding round a little contiguous sand hill, it split into numberless branches; but although the plains hereabouts were grassed, the soil was not so good as that on the plains above them. At six miles we ascended a sand hill, from which we could see to the extremity of the plain; but it had no apparent outlet excepting to the E.S.E. I therefore proceeded on that course for three miles, when we lost sight of all gum-trees, and found ourselves amongst scrub. Low bushes bounded the horizon all round, and hid the grassy plains from our view; but they were denser to the south and east than at any other point. Mount Lyell, the large hill south, bore 140° to the east of north, distant between forty and fifty miles. A short time after we left the grassy flats we crossed the dry bed of a large lagoon, which had been seen by Mr. Poole on a bearing of 77° from the Magnetic Hill. In the richer soil, a plant with round, striped fruit upon it, of very bitter taste, a species of cucumber, was growing. We next proceeded to the eastward, and surveying the country from higher ground, observed that the creek had no outlet from the plains, and that it could not but terminate on them.

As I had no object in a prolonged journey to the south, I turned back from this station, and retracing my steps to the water where we had left the natives, reached it at half-past six. All our friends were still there; we had, therefore, the pleasure of passing another afternoon with them, during which they were joined by two other natives, with their families, who had been driven in from the south, like ourselves, by the want of water. They assured us that all the water in that quarter had disappeared, "that the sun had taken it," and that we should not find a drop to the eastward, where I told them I was going. All these men, excepting one, had been circumcised. The single exception had the left fore-tooth of his upper jaw extracted, and I therefore concluded that he belonged to a different tribe. I had hoped to have seen many more natives in this locality; but it struck me, from what I observed, that they were dispersed at the different water-holes, there being no one locality capable of supporting any number.

The low and flooded track I have been describing must be dreadfully cold during the winter season, and the natives, who are wholly unprovided for inclemency of any kind, must suffer greatly from exposure; but at this time the temperature still continued very high, and the constant appearance of the deep purple tint opposite to the rising and setting sun seemed to indicate a continuance of it.

As our horses had had some long journeys for the last three days, we merely returned to our first bivouac on the creek, when we left the natives, with whom we parted on very good terms, and a promise on their part to come and see us. On the 17th started at quarter-past six for the eastward, with as much water as we could carry in the cart, as from the accounts of the natives we scarcely hoped to find any. For the first five miles we kept a course rather to the north of east, nearly E.N.E. indeed, to round some sand-hills we should otherwise have been obliged to cross. There were very extensive plains to our left, on which water must lie during winter; but their soil was not good, or the vegetation thick upon them. We could just see the points of the northern flat-topped ranges beyond them. At five miles we turned due east, and crossed several small plains, separated by sandy undulations, not high enough to be termed ridges; the country, both to the south and east, appearing to be extremely low. At about fifteen miles, just as we were ascending a sand hill, Mr. Browne caught sight of a native stealing through the brush, after whom he rode; but the black observing him, ran away. On this Mr. Browne called out to him, when he stopped; but the horse happening to neigh at the moment, the poor fellow took to his heels, and secreted himself so adroitly, that we could not find him. He must, indeed, have been terribly alarmed at the uncouth sound he heard.

A short time before our adventure with the native we had seen three pelicans coming from the north. They kept very low to the ground, and wheeled along in circles in a very remarkable manner, as if they had just risen from water; but at length they soared upwards, and flew straight for the lagoon where we had left the natives. With the exception of these three birds, no other was to be seen in those dreary regions. Both Mr. Browne and I, however, rode over a snake, but our horses fortunately escaped being bitten; this animal had seized a mouse, which it let go on being disturbed, and crept into a hole; it was very pretty, being of a bright yellow colour with brown specks. Arriving at the termination of the sand hills, we looked down upon an immense shallow basin, extending to the north and south-east further than the range of vision, which must, I should imagine, be wholly impassable during the rainy season. There was scarcely any vegetation, a proof, it struck me, that it retains water on its surface till the summer is so advanced that the sun's rays are too powerful for any plants that may spring up, or that the heat bakes the soil so that nothing can force itself through. There was little, if any grass to be seen; but the mesembryanthemum reappeared upon it, with other salsolaceous plants. The former was of a new variety, with flowers on a long slender stalk, heaps of which had been gathered by the natives for the seed. Of the timber of these regions there was none; a few gum-trees near the creeks, with box-trees on the flats, and a few stunted acacia and hakea on the small hills, constituted almost the whole. Water boiled on this plain at 212°; that is to say at our camp were we slept, about two miles advanced into it, but the plain extended about five miles further to the eastward. After crossing this on the following morning, we traversed a country which Mr. Browne informed me was very similar to that near Lake Torrens. It consisted of sand banks, or *drifts*, with large bare patches at intervals: the whole bearing testimony to the violence of the rains that must sometimes deluge it. We then traversed a succession of flats (I call them so because they did not deserve the name of plains) separated from each other by patches of red sand and clay, that were not more than a foot and a half above the surface of the flats. At nine miles the country became covered with low scrub, and we soon after passed the dry bed of a lagoon, about a mile in circumference, on which there was a coating of salt and gypsum resting on soft black mud.

About a mile from this we passed a new tree, similar to one we had seen on the Cawndilla plain. From this point the land imperceptibly rose, until at length we found ourselves on some sandy elevations thickly covered with scrub of acacia, almost all dead, but there was a good deal of grass around them, and the spot might at another season, and if the trees had been in leaf, have looked pretty. We pushed through this scrub, the soil being a bright red sand for nine miles, when we suddenly found ourselves at the base of a small stony hill, of about fifty feet in height. From the summit we overlooked the region round about. To the eastward, as a medium point, it was covered with a dense scrub, that extended to the base of a range of hills, distant about 33 miles, the extremities of which bore 71° and 152° respectively from us. But although the country under them was covered with brush, the hills appeared to be clear and denuded of brushes of any kind. Our position here was about 138 miles from the Darling, and about 97 from the Depôt. My object in this excursion had been to ascertain the characteristic of the country between us and the Darling, but I did not think it necessary to run any risks with my horses, by pushing on for the hills, as I could not have reached them until late the following day, when in the event of not finding water, their fate would have been sealed; for we could not have returned with them to the creek. They had already been two days without, if I except the little we had spared them from the casks. I had deemed it prudent to send Joseph and Lewis back to the creek for a fresh supply, with orders to return and meet at a certain point, and there to await our arrival, for without this supply I felt satisfied we should have great difficulty as it was in getting our animals back to the creek. We descended from the hill therefore to some green looking trees, of a foliage new to me, to rest for an hour before we turned back again. There were neither flowers nor fruit on the trees, but from their leaf and habit, I took them to be a species of the Juglans. At sunset we mounted our horses and travelled to the edge of the acacia scrub to give our horses some of the grass, and halted in it for the night, but started early on the following morning to meet Joseph. We reached the appointed place, about 10, but not finding him there continued to journey onwards, and at five miles met him. We then stopped and gave the horses 12 gallons of water each, after which we tethered them out, but they were so restless that I determined to mount them, and pushing on reached the creek at half-past 1 A.M. The animals requiring rest, I remained stationary the next day, and was myself glad to keep in the shade, not that the day was particularly hot, but because I began to feel the effects of constant exposure. Having expressed some opinion, however, that there might have been water to the north of us, in the direction whence the pelicans came, Mr. Browne volunteered to ride out, and accordingly with Flood left me about 10, but returned late in the afternoon without having found any. He ascertained that the creek I had sent Flood to trace when Mr. Stuart went to sketch in the ranges, terminated in the barren plain we had crossed, and such, the reader will observe, is the general termination of all the creeks of these singular and depressed regions.

We returned to the camp on the 21st, and from that period to the end of the month I remained stationary, employed in various ways. On the 24th and 29th we took different sets of lunars, which gave our longitude as before, nearly 141°29', the variation of the compass being 5°14' East.

The month of April set in without any indication of a change in the weather. It appeared as if the flood gates of Heaven were closed upon us for ever. We now began to feel the effects of disappointment, and watched the sky with extreme anxiety, insomuch that the least cloud raised all our hopes. The men were employed in various ways to keep them in health. We planted seeds in the bed of the creek, but the sun burnt them to cinders the moment they appeared above the ground. On the evening of the 3rd there was distant thunder, and heavy clouds to the westward. I thought it might have been that some shower had approached sufficiently near for me to benefit by the surface water it would have left to push towards Lake Torrens, and therefore mounted my horse and rode away to the westward on the 4th, but returned on the night of the 7th in disappointment. Time rolled on fast, and still we were unable to stir. Mr. Piesse, who took great delight in strolling out with my gun, occasionally shot a new bird.

On the 4th the wind blew strong from the south; but although the air was cooled, no rain fell, nor indeed was there any likelihood of rain with the wind in that quarter. Still as this was the first decided shift from the points to which it had kept so steadily, we augured good from it. On the 7th a very bright meteor was seen to burst in the south-east quarter of the heavens; crossing the sky with a long train of light, and in exploding seemed to form numerous stars. Whether it was fancy or not we thought the temperature cooled down from this period. On this day also we had a change of moon, but neither produced a variation of wind or weather of any immediate benefit to us. On the 14th we tried to ascertain the dew point, but failed, as in previous instances. The thermometer in our underground room stood at 78° of Farenheit, but we could not reduce the moist bulb below 49°; nor was I surprised at this, considering we had not had rain for nearly four months, and that during our stay at the Depôt we had never experienced a dew. The ground was thoroughly heated to the depth of three or four feet, and the tremendous heat that prevailed had parched vegetation and drawn moisture from everything. In an air so rarefied, and an atmosphere so dry, it was hardly to be expected that any experiment upon it would be attended with its usual results, or that the particles of moisture so far separated, could be condensed by ordinary methods. The mean of the thermometer for the months of December, January, and February, had been 101°, 104°, and 101° respectively in the shade. Under its effects every screw in our boxes had been drawn, and the horn handles of our instruments, as well as our combs, were split into fine laminæ. The lead dropped out of our pencils, our signal rockets were entirely spoiled; our hair, as well as the wool on the sheep, ceased to grow, and our nails had become as brittle as glass. The flour lost more than eight per cent of its original weight, and the other provisions in a still greater proportion. The bran in which our bacon had been packed, was perfectly saturated, and weighed almost as heavy as the meat; we were obliged to bury our wax candles; a bottle of citric acid in Mr. Browne's box became fluid, and escaping, burnt a quantity of his linen; and we found it difficult to write or draw, so rapidly did the fluid dry in our pens and brushes. It was happy for us, therefore, that a cooler season set in, otherwise I do not think that many of us could much longer have survived. But, although it might be said that the intense heat of the summer had passed, there still were intervals of most oppressive weather.

About the beginning of March I had had occasion to speak to Mr. Browne as to certain indications of disease that were upon me. I had violent headaches, unusual pains in my joints, and a coppery taste in my mouth. These symptoms I attributed to having slept so frequently on the hard ground and in the beds of creeks, and it was only when my mouth became sore, and my gums spongy, that I felt it necessary to trouble Mr. Browne, who at once told me that I was labouring under an attack of scurvy, and I regretted to learn from him that both he and Mr. Poole were similarly affected, but they hoped I had hitherto escaped. Mr. Browne was the more surprised at my case, as I was very moderate in my diet, and had taken but little food likely to cause such a malady. Of we three Mr. Poole suffered most, and gradually declined in health. For myself I immediately took double precautions, and although I could not hope soon to shake off such a disease, especially under such unfavourable circumstances as those in which we were placed, I was yet thankful that I did not become worse. For Mr. Browne, as he did not complain, I had every hope that he too had succeeded in arresting the progress of this fearful distemper. It will naturally occur to the reader as singular, that the officers only should have been thus attacked; but the fact is, that they had been constantly absent from the camp, and had therefore been obliged to use bacon, whereas the men were living on fresh mutton; besides, the same men were seldom taken on a second journey, but were allowed time to recover from the exposure to which they had been subjected, but for the officers there was no respite.

On the 18th the wind, which had again settled in the S.E. changed to the N.E., and the sky became generally overcast. Heavy clouds hung over the Mount Serle chain, and I thought that rain would have fallen, but all these favourable indications vanished before sunset. At dawn of the morning of the 19th, dense masses of clouds were seen, and thunder heard to the west; and the wind shifting to that quarter, we hoped that some of the clouds would have been blown over to us, but they kept their place for two days, and then gradually disappeared. These distant indications, however, were sufficient to rouse us to exertion, in the hope of escaping from the fearful captivity in which we had so long been held. I left the camp on the 21st with Mr. Browne and Flood, thinking that rain might have extended to the eastward from Mount Serle, sufficiently near to enable us to push into the N.W. interior, and as it appeared to me that a W. by N. course would take me abreast of Mount Hopeless, I ran upon it. At 16 miles I ascended a low range, but could not observe anything from it to the westward but scrub. Descending from this range we struck the head of a creek, and at six miles came on the last dregs of a pool of water, so thick that it was useless to us. We next crossed barren stony undulations and open plains, some of them apparently subject to floods; and halted at half-past six, after a journey of between thirty and forty miles without water, and with very little grass for our horses to eat. Although the course we kept, had taken us at times to a considerable distance from the creek, we again came on it before sunset, and consequently halted upon its banks; but in tracing it down on the following morning we lost its channel on an extensive plain, and therefore continued our journey to the westward. At seven miles we entered a dense scrub, and at fifteen ascended a sand hill, from which we expected to have had a more than usually extensive view, but it was limited to the next sand hill, nor was there the slightest prospect of a change of country being at hand. At four miles from this position we came upon a second creek seemingly from the N.E., whose appearance raised our hopes of obtaining water; but as its channel became sandy, and turned southwards, I left it, and once more running on our old course, pulled up at sunset under a bank of sand, without anything either for ourselves or our horses to drink. During the latter part of the evening we had observed a good deal of grass on the sand hills, nor was there any deficiency of it round our bivouac; but, notwithstanding that there was more than enough for the few horses we had, a herd of cattle would have discussed the whole in a night. It was evident from the state of the ground that no rain had fallen hereabouts, and I consequently began to doubt whether it had extended beyond the mountains. Comparing the appearance of the country we were in, with that through which Mr. Browne passed for 50 miles before he came upon Lake Torrens, and concluding that some such similar change would have taken place here if we had approached within any reasonable distance of that basin, I could not but apprehend that we were still a long way from it.

The horses having refused the water we had found in the creek, I could hardly expect they would drink it on their return, so that I calculated our distance from water at about 68 miles; and I foresaw that unless we should succeed in finding some early in the day following, it would be necessary for us to make for the Depôt again. Close to where we stopped there was a large burrow of Talperos, an animal, as I have observed, similar to the rabbit in its habits, and one of which the natives are very fond, as food. The sandy ridges appeared to be full of them, and other animals, that must live for many months at a time without water. Whilst we were sitting in the dusk near our fire, two beautiful parrots attracted by it, I suppose, pitched close to us; but immediately took wing again, and flew away to the N.W. They, no doubt, thought that we were near water, but like ourselves were doomed to disappointment. During the evening also some plovers flew over us, and we heard some native dogs howling to the south-west. At daylight, therefore, we rode in that direction, with the hope of finding the element we now so much required. At three miles a large grassy flat opened out to view upon our right, similar to that at the termination of the Depôt creek. It might have contained 1,000 acres, but there was not at the first glance, a tree to be seen upon it. This flat was bounded to the S.W. by a sand bank, lying at right angles to the sand ridges we had been crossing. The latter, therefore, ran down upon this bank in parallel lines, some falling short of, and others striking it; so that, as the drainage was towards the embankment, the collected waters lodged against it. After crossing a portion of the plain we saw some box-trees in a hollow, towards which we rode, and then came upon a deep dry pond, in whose bottom the natives had dug several wells, and had evidently lingered near it as long as a drop of water remained. It was now clear that our farther search for water would be useless. I therefore turned on a course of 12° to the north of east for the muddy water we had passed two days before, and halted there about an hour after sunset, having journeyed 42 miles. We fell into our tracks going out about four miles before we halted, and were surprised to observe that a solitary native had been running them down. On riding a little further however, we noticed several tracks of different sizes, as if a family of natives had been crossing the country to the north-west. It is more than probable that their water having failed in the hills, they were on their way to some other place where they had a well.

Although we had ourselves been without water for two days, the mud in the creek was so thick that I could not swallow it, and was really astonished how Mr. Browne managed to drink a pint of it made into tea. It absolutely fell over the cup of the panakin like thick cream, and stuck to the horses' noses like pipe-clay. They drank sparingly however, and took but little grass during the night. As we pursued our journey homewards on the following day, we passed several flights of dotterel making to the south, this being the first migration we had observed in that direction. These birds were in great numbers on the plains of Adelaide the year preceding, and had afforded good sport to my friend Torrens; we also observed a flight of pelicans, wheeling about close to the ground, as they had before done to the eastward, as well as a flight of the black-shouldered hawks hovering in the air. Our day's ride had been very long and fatiguing, as the horses were tired, but we got relieved by our arrival at the camp a little before sunset on the 25th: and thus terminated another journey in disappointment. We regretted to find that Mr. Poole was seriously indisposed. His muscles were now attacked and he was suffering great pain, but, as the disease appeared inclined to make to the surface, Mr. Browne had some hopes of a favourable change. Both Mr. Browne and myself found that the sameness of our diet began to disagree with us, and were equally anxious for the reappearance of vegetation, in the hope that we should be able to collect sow-thistles or the tender shoots of the rhagodia as a change. We had, whilst it lasted, taken mint tea, in addition to the scanty supply of tea to which we were obliged to limit ourselves, but I do not think it was wholesome.

The moon entered her third quarter on the 27th, but brought no change; on the contrary she chased away the clouds as she rose, and moved through the heavens in unshrouded and dazzling brightness. Sometimes a dark mass of clouds would rise simultaneously with her, in the west, but as the queen of night advanced in her upward course they gradually diminished the velocity with which they at first came up; stopped, and fell back again, below the horizon. Not once, but fifty times have we watched these apparently contending forces, but whether I am right in attributing the cause I will not say.

At this time (the end of April) the weather was very fine, although the thermometer ranged high. The wind being steady at south accounted for the unusual height of the barometrical column, which rose to 30·600. On the night of the 20th we had a heavy dew, the first since our departure from the Darling. On the morning of the 28th it thundered, and a dense cloud passed over to the north, the wind was unsteady, and I hoped that the storm would have worked round, but it did not. At ten the wind sprung up from the south, the sky cleared and all our hopes were blighted.

Notwithstanding that we treated the natives who came to the creek with every kindness, none ever visited us, and I was the more surprised at this, because I could not but think that we were putting them to great inconvenience by our occupation of this spot. Towards the end of the month, it was so cold that we were glad to have fires close to our tents. Mr. Poole had gradually become worse and worse, and was now wholly confined to his bed, unable to stir, a melancholy affliction both to himself and us, rendering our detention in that gloomy region still more painful. My men generally were in good health, but almost all had bleeding at the nose; I was only too thankful that my own health did not give way, though I still felt the scurvy in a mitigated form, but Mr. Browne had more serious symptoms about him.

The 10th of May completed the ninth month of our absence from Adelaide, and still we were locked up without the hope of escape, whilst every day added fresh causes of anxiety to those I had already to bear up against. Mr. Poole became worse, all his skin along the muscles turned black, and large pieces of spongy flesh hung from the roof of his mouth, which was in such a state that he could hardly eat. Instead of looking with eagerness to the moment of our liberation, I now dreaded the consequent necessity of moving him about in so dreadful a condition. Mr. Browne attended him with a constancy and kindness that could not but raise him in my estimation, doing every thing which friendship or sympathy could suggest.

On the 11th about 3 P.M. I was roused by the dogs simultaneously springing up and rushing across the creek, but supposing they had seen a native dog, I did not rise; however, I soon knew by their continued barking that they had something at bay, and Mr. Piesse not long after came to inform me a solitary native was on the top of some rising ground in front of the camp. I sent him therefore with some of the men to call off the dogs, and to bring him down to the tents. The poor fellow had fought manfully with the dogs, and escaped injury, but had broken his waddy over one of them. He was an emaciated and elderly man, rather low in stature, and half dead with hunger and thirst; he drank copiously of the water that was offered to him, and then ate as much as would have served me for four and twenty dinners. The men made him up a screen of boughs close to the cart near the servants, and I gave him a blanket in which he rolled himself up and soon fell fast asleep. Whence this solitary stranger could have come from we could not divine. No other natives approached to look after him, nor did he shew anxiety for any absent companion. His composure and apparent self-possession were very remarkable, for he neither exhibited astonishment or curiosity at the novelties by which he was surrounded. His whole demeanour was that of a calm and courageous man, who finding himself placed in unusual jeopardy, had determined not to be betrayed into the slightest display of fear or timidity.

From the period of our return from the eastward, I had remained quiet in the camp, watching every change in the sky; I was indeed reluctant to absent myself for any indefinite period, in consequence of

Mr. Poole's precarious state of health. He had now used all the medicines we had brought out, and none therefore remained either for him or any one else who might subsequently be taken ill. As however he was better, on the 12th, I determined to make a second excursion to the eastward, to see if there were any more natives in the neighbourhood of the grassy plains than when I was last there. Wishing to get some samples of wood I took the light cart and Tampawang also, in the hope that he would be of use.

Although the water in the creek had sunk fearfully there was still a month's supply remaining, but if it had been used by our stock it would then have been dry. Close to the spot where we had before stopped, there were two huts that had been recently erected. Before these two fires were burning, and some troughs of grass seed were close to them, but no native could we see, neither did any answer to our call. Mr. Browne, however, observing some recent tracks, ran them down, and discovered a native and his lubra who had concealed themselves in the hollow of a tree, from which they crept as soon as they saw they were discovered. The man, we had seen before, and the other proved to be the frail one who exhibited such indignation at our rejecting her addresses on a former occasion; being a talkative damsel, we were glad to renew our acquaintance with her. We learnt from them that the second hut belonged to an absent native who was out hunting, the father of a pretty little girl who now obeyed their signal and came forth. They said the water on the plain had dried up, and that the only water-holes remaining were to the west, viz. at our camp, and to the south, where they said there were two water-holes. As they had informed us, the absent native made his appearance at sunset, but his bag was very light, so we once more gave them all our mutton; he proved to be the man Mr. Browne chased on the sand hills, the strongest native we had seen; he wanted the front tooth, but was not circumcised.

In the evening we had a thunder-storm, but could have counted the drops of rain that fell, notwithstanding the thunder was loud and the lightning vivid. We returned to the Depôt on the 13th, and on crossing the plain Mr. Browne had well nigh captured a jerboa, which sprang from under my horse's legs, but managed to elude him, and popped into a little hole before he could approach sufficiently near to strike at it. On reaching the tents we had the mortification to find Mr. Poole still worse, but I attributed his relapse in some measure to a depression of spirits. The old man who had come to the camp the day before we left it, was still there, and had apparently taken up his quarters between the cart and my tent. During our absence the men had shewn him all the wonders of the camp, and he in his turn had strongly excited their anticipations, by what he had told them.

He appeared to be quite aware of the use of the boat, intimating that it was turned upside down, and pointed to the N.W. as the quarter in which we should use her. He mistook the sheep net for a fishing net, and gave them to understand that there were fish in those waters so large that they would not get through the meshes. Being anxious to hear what he had to say I sent for him to my tent, and with Mr. Browne cross-questioned him.

It appeared quite clear to us that he was aware of the existence of large water somewhere or other to the northward and westward. He pointed from W.N.W. round to the eastward of north, and explained that large waves higher than his head broke on the shore. On my shewing him the fish figured in Sir Thomas Mitchell's work he knew only the cod. Of the fish figured in Cuvier's works he gave specific names to those he recognised, as the hippocampus, the turtle, and several sea fish, as the chetodon, but all the others he included under one generic name, that of "guia," fish.

He put his hands very cautiously on the snakes, and withdrew them suddenly as if he expected they would bite him, and evinced great astonishment when he felt nothing but the soft paper. On being asked, he expressed his readiness to accompany us when there should be water, but said we should not have rain yet. I must confess this old native raised my hopes, and made me again anxious for the moment when we should resume our labours, but when that time was to come God only knew.

It had been to no purpose that we had traversed the country in search for water. None any longer remained on the parched surface of the stony desert, if I except what remained at the Depôt, and the little in the creek to the eastward. There were indeed the ravages of floods and the vestiges of inundations to be seen in the neighbourhood of every creek we had traced, and upon every plain we had crossed, but the element that had left such marks of its fury was no where to be found.

From this period I gave up all hope of success in any future effort I might make to escape from our dreary prison. Day after day, and week after week passed over our heads, without any apparent likelihood of any change in the weather. The consequences of our detention weighed heavily on my mind, and depressed my spirits, for in looking over Mr. Piesse's monthly return of provisions on hand, I found that unless some step was taken to enable me to keep the field, I should on the fall of rain be obliged to retreat. I had by severe exertion gained a most commanding position, the wide field of the interior lay like an open sea before me, and yet every sanguine hope I had ever indulged appeared as if about to be extinguished. The only plan for me to adopt was to send a portion of the men back to Adelaide. I found by calculation that if I divided the party, retaining nine in all, and sending the remainder home, I should secure the means of pushing my researches to the end of December, before which time I hoped, (however much it had pleased Providence to stay my progress hitherto) to have performed my task, or penetrated the heartless desert before me, to such a distance as would leave no doubt as to the question I had been directed to solve. The old man left us on the 17th with the promise of returning, and from the careful manner in which he concealed the different things that had been given to him I thought he would have done so, but we never saw him more, and I cannot but think that he perished from the want of water in endeavouring to return to his kindred.

I have repeatedly remarked that we had been deserted by all the feathered tribes. Not only was this the case, but we had witnessed a second migration of the later broods; after these were gone, there still remained with us about fifty of the common kites and as many crows: these birds continued with us for the offals of the sheep, and had become exceedingly tame; the kites in particular came flying from the trees when a whistle was sounded, to the great amusement of the men, who threw up pieces of meat for them to catch before they fell to the ground. When the old man first came to us, we fed him on mutton, but one of the men happening to shoot a crow, he shewed such a decided preference for it, that he afterwards lived almost exclusively upon them. He was, as I have stated, when he first came to us a thin and emaciated being, but at the expiration of a fortnight when he rose to depart, he threw off his blanket and exhibited a condition that astonished us all. He was absolutely fat, and yet his face did not at all indicate such a change. If he had been fed in the dark like capons, he could not have got into better condition. Mr. Browne was anxious to accompany him, but I thought that if his suspicions were aroused he would not return, and I therefore let him depart as he came. With him all our hopes vanished, for even the presence of that savage was soothing to us, and so long as he remained, we indulged in anticipations as to the future. From the time of his departure a gloomy silence pervaded the camp; we were, indeed, placed under the most trying circumstances; everything combined to depress our spirits and exhaust our patience. We had gradually been deserted by every beast of the field, and every fowl of the air. We had witnessed migration after migration of the feathered tribes, to that point to which we were so anxious to push our way. Flights of cockatoos, of parrots, of pigeons, and of bitterns, birds also whose notes had cheered us in the wilderness, all had taken the same high road to a better and more hospitable region. The vegetable kingdom was at a stand, and there was nothing either to engage the attention or attract the eye. Our animals had laid the ground bare for miles around the camp, and never came towards it but to drink. The axe had made a broad gap in the line of gum-trees which ornamented the creek, and had destroyed its appearance. We had to witness the gradual and fearful diminution of the water, on the possession of which our lives depended; day after day we saw it sink lower and lower, dissipated alike by the sun and the winds. From its original depth of nine feet, it now scarcely measured two, and instead of extending from bank to bank it occupied only a narrow line in the centre of the channel. Had the drought continued for a month longer than it pleased the Almighty to terminate it, that creek would have been as dry as the desert on either side. Almost heart-broken, Mr. Browne and I seldom left our tents, save to visit our sick companion. Mr. Browne had for some time been suffering great pain in his limbs, but with a generous desire to save me further anxiety carefully concealed it from me; but it was his wont to go to some acacia trees in the bed of the creek to swing on their branches, as he told me to exercise his muscles, in the hope of relaxing their rigidity.

One day, when I was sitting with Mr. Poole, he suggested the erection of two stations, one on the Red Hill and the other on the Black Hill, as points for bearings when we should leave the Depôt. The idea had suggested itself to me, but I had observed that we soon lost sight of the hills in going to the north-west; and that, therefore, for such a purpose, the works would be of little use, but to give the men occupation; and to keep them in health I employed them in erecting a pyramid of stones on the summit of the Red Hill. It is twenty-one feet at the base, and eighteen feet high, and bears 329° from the camp, or 31° to the west of north. I little thought when I was engaged in that work, that I was erecting Mr. Poole's monument, but so it was, that rude structure looks over his lonely grave, and will stand for ages as a record of all we suffered in the dreary region to which we were so long confined.

The months of May and June, and the first and second weeks of July passed over our heads, yet there was no indication of a change of weather.

Red Hill, or Mount Poole.

It had been bitterly cold during parts of this period, the thermometer having descended to 24°; thus making the difference between the extremes of summer heat and winter's cold no less than 133°.

About the middle of June I had the drays put into serviceable condition, the wheels wedged up, and every thing prepared for moving away.

Anxious to take every measure to prevent unnecessary delay, when the day of liberation should arrive, I had sent Mr. Stuart and Mr. Piesse, with a party of chainers, to measure along the line on which I intended to move when the Depôt was broken up. I had determined, as I have elsewhere informed the reader, to penetrate to the westward, in the hope of finding Lake Torrens connected with some more extensive and more central body of water; and I thought it would be satisfactory to ascertain, as nearly as possible, the distance of that basin from the Darling, and in so doing to unite the eastern and western surveys. I had assumed Sir Thomas Mitchell's position at Williorara as correct, and had taken the most careful bearings from that point to the Depot, and the position in which they fixed it differed but little from the result of the many lunars I took during my stay there. As I purpose giving the elements of all my calculations, those more qualified than myself to judge on these matters, will correct me if I have been in error; but, as the mean of my lunars was so close to the majority of the single lunars, I cannot think they are far from the truth. Be that as it may, I assumed my position at the Depôt to be in lat. 29°40'14"S. and in long. 141°29'41"E, the variation being 5°14' East. Allowing for the variation, I directed Mr. Stuart to run the chain line on a bearing of 55° to the west of north, which I intended to cut a little to the west of the park-like and grassy plain at the termination of the creek I had traced in that direction. By supplying the party with water from the camp, I enabled them to prolong the line to 30 miles.

On the 15th of June I commenced my preparations for moving; not that I had any reason so to do, but because I could not bring myself to believe that the drought would continue much longer. The felloes and spokes of the wheels of the drays had shrunk to nothing, and it was with great difficulty that we wedged them up; but the boat, which had been so long exposed to an ardent sun, had, to appearance at least, been but little injured.

As it became necessary to point out the drays that were to go with the home returning party, I was obliged to break my intentions to Mr. Poole, who I also proposed sending in charge of them. He was much affected, but, seeing the necessity of the measure, said that he was ready to obey my orders in all things. I directed Mr. Piesse to weigh out and place apart the supplies that would be required for Mr. Poole and his men, and to pack the provisions we should retain in the most compact order. On examining our bacon we found that it had lost more than half its weight, and had now completely saturated the bran in which it had been packed. Our flour had lost more than 8 per cent., and the tea in a much greater proportion.

The most valuable part of our stock were the sheep, they had kept in excellent condition, and seldom weighed less than 55lbs. or 65lbs.; but their flesh was perfectly tasteless. Still they were a most valuable stock, and we had enough remaining to give the men a full allowance; for the parties, employed on detached excursions, could only take a day or two's supply with them, and in consequence a quantity of back rations, if I may so term them, were constantly accumulating.

Mr. Poole's reduced state of health rendered it necessary that a dray should be prepared for his transport, and I requested Mr. Browne to superintend every possible arrangement for his comfort. A dray was accordingly lined with sheep-skins, and, had a flannel quilt, as the nights were exceedingly cold, and he could not be moved to a fire. I had also a swing cot made, with pullies to raise him up when he should feel disposed to change his position.

Whilst these necessary preparations were being forwarded, I was engaged writing my public despatches.

In my communication to the Governor of South Australia, I expressed a desire that a supply of provisions might be forwarded to Williorara by the end of December, about which period I hoped I should be on my return from the interior. I regretted exceedingly putting her Majesty's Government to this additional cost, but I trust a sufficient excuse will have been found for me in the foregoing pages. I would rather that my bones had been left to bleach in that desert than have yielded an inch of the ground I had gained at so much expense and trouble.

The 27th of June completed the fifth month of our detention at the Depôt, and the prospect of our removal appeared to be as distant as ever; there were, it is true, more clouds, but they passed over us without breaking. The month of July, however, opened with every indication of a change, the sky was generally overcast, and although we had been so often disappointed, I had a presentiment that the then appearances would not vanish without rain.

About this time Mr. Poole, whose health on the whole was improving, had a severe attack of inflammation, which Mr. Browne subdued with great difficulty. After this attack he became exceedingly restless, and expressed a desire to be moved from the tent in which he had so long been confined, to the underground room, but as that rude apartment was exceedingly cold at night, I thought it advisable to have a chimney built to it before he was taken there. It was not until the 12th that it was ready for him. As the men were carrying him across the camp towards the room he was destined to occupy for so short a time, I pointed out the pyramid to him, and it is somewhat singular, that the first drops of rain, on the continuance of which our deliverance depended, fell as the men were bearing him along.

Referring back to the early part of the month, I may observe that the indications of a breaking up of the drought, became every day more apparent.

It was now clear, indeed, that the sky was getting surcharged with moisture, and it is impossible for me to describe the intense anxiety that prevailed in the camp. On the morning of the 3rd the firmament was again cloudy, but the wind shifted at noon to west, and the sun set in a sky so clear that we could hardly believe it had been so lately overcast. On the following morning he rose bright and clear as he had set, and we had a day of surpassing fineness, like a spring day in England.

The night of the 6th was the coldest night we experienced at the Depôt, when the thermometer descended to 24°. On the 7th a south wind made the barometer rise to 30°180', and with it despair once more stared us in the face, for with the wind in that quarter there was no hope of rain. On the 8th it still blew heavily from the south, and the barometer rose to 30°200'; but the evening was calm and frosty, and the sky without a cloud. I may be wearying my reader, by entering thus into the particulars of every change that took place in the weather at this, to us, intensely anxious period, but he must excuse me; my narrative may appear dull, and should not have been intruded on the notice of the public, had I not been influenced by a sense of duty to all concerned.

No one but those who were with me at that trying time and in that fearful solitude, can form an idea of our feelings. To continue then, on the morning of the 9th it again blew fresh from the south, the sky was cloudless even in the direction of Mount Serle, and all appearance of rain had passed away.

On the 10th, to give a change to the current of my thoughts, and for exercise, I walked down the Depôt creek with Mr. Browne, and turning, northwards up the main branch when we reached the junction of the two creeks, we continued our ramble, for two or three miles. I know not why it was, that, on this occasion more than any other, we should have contemplated the scene around us, unless it was that the peculiar tranquillity of the moment made a greater impression on our minds. Perhaps, the death-like silence of the scene at that moment led us to reflect, whilst gazing at the ravages made by the floods, how fearfully that silence must sometimes be broken by the roar of waters and of winds. Here, as in other places, we observed the trunks of trees swept down from the hills, lodged high in the branches of the trees in the neighbourhood of the creek, and large accumulations of rubbish lying at their butts, whilst the line of inundation extended so far into the plains that the country must on such occasions have the appearance of an inland sea. The winds on the other hand had stripped the bark from the trees to windward (a little to the south of west), as if it had been shaved off with an instrument, but during our stay at the Depôt we had not experienced any unusual visitation, as a flood really would have been; for any torrent, such as that which it was evident sometimes swells the creek, would have swept us from our ground, since the marks of inundation reached more than a mile beyond our encampment, and the trunk of a large gum-tree was jambed between the branches of one overhanging the creek near us at an altitude exceeding the height of our tents.

On the 11th the wind shifted to the east, the whole sky becoming suddenly overcast, and on the morning of the 12th it was still at east, but at noon veered round to the north, when a gentle rain set in, so gentle that it more resembled a mist, but this continued all the evening and during the night. It ceased however at 10 A.M. of the 13th, when the wind shifted a little to the westward of north. At noon rain again commenced, and fell steadily throughout the night, but although the ground began to feel the effects of it, sufficient had not fallen to enable us to move. Yet, how thankful was I for this change, and how earnestly did I pray that the Almighty would still farther extend his mercy to us, when I laid my head on my pillow. All night it poured down without any intermission, and as morning dawned the ripple of waters in a little gully close to our tents, was a sweeter and more soothing sound than the softest melody I ever heard. On going down to the creek in the morning I found that it had risen five inches, and the ground was now so completely saturated that I no longer doubted the moment of our liberation had arrived.

I had made every necessary preparation for Mr. Poole's departure on the 13th, and as the rain ceased on the morning of the 14th the home returning party mustered to leave us. Mr. Poole felt much when I went to tell him that the dray in which he was to be conveyed, was ready for his reception. I did all that I could to render his mind easy on every point, and allowed him to select the most quiet and steady bullocks for the dray he was to occupy together with the most careful driver in the party. I also consented to his taking Joseph, who was the best man I had, to attend personally upon him, and Mr. Browne put up for his use all the little comforts we could spare. I cheered him with the hope of returning to meet us after we should have terminated our labours, and assured him that I considered his services on the duty I was about to send him as valuable and important as if he continued with me. He was lifted on his stretcher into the dray, and appeared gratified at the manner in which it had been arranged. I was glad to see that his feelings did not give way at this painful moment; on my ascending the dray, however, to bid him adieu, he wept bitterly, but expressed his hope that we should succeed in our enterprise.

As I knew his mind would be agitated, and that his greatest trial would be on the first day, I requested Mr. Browne to accompany him, and to return to me on the following day. On Mr. Poole's departure I prepared for our own removal, and sent Flood after the horses, but having an abundance of water everywhere, they had wandered, and he returned with them too late for me to move. He said, that in crossing the rocky range he heard a roaring noise, and that on going to the glen he saw the waters pouring down, foaming and eddying amongst the rocks, adding that he was sure the floods would be down upon us ere long. An evident proof that however light the rain appeared to be, an immense quantity must have fallen, and I could not but hope and believe that it had been general.

Before we left the Depôt Flood's prediction was confirmed, and the channel which, if the drought had continued a few days longer, would have been perfectly waterless, was thus suddenly filled up to the brim; no stronger instance of the force of waters in these regions can be adduced than this, no better illustration of the character of the creeks can be given. The head of the Depôt creek was not more than eight miles from us, its course to its junction with the main creek was not ten, yet it was a watercourse that without being aware of its commencement or termination might have been laid down by the traveller as a river. Such however is the uncertain nature of the rivers of those parts of the continent of Australia over which I have wandered. I would not trust the largest farther than the range of vision; they are deceptive all of them, the offsprings, of heavy rains, and dependent entirely on local circumstances for their appearance and existence.

Having taken all our circumstances into consideration, our heart-breaking detention, the uncertainty that involved our future proceedings, and the ceaseless anxiety of mind to which we should be subjected, recollecting also that Mr. Browne had joined me for a limited period only, and that a protracted journey might injure his future prospects, I felt that it was incumbent on me to give him the option of returning with Mr. Poole if he felt disposed to do so, but he would not desert me, and declined all my suggestion.

On the morning of the 16th I struck the tents, which had stood for six months less eleven days, and turned my back on the Depôt in grateful thankfulness for our release from a spot where my feelings and patience had been so severely tried. When we commenced our journey, we found that our progress would be slow, for the ground was dreadfully heavy, and the bullocks, so long unaccustomed to draught, shrunk from their task. One of the drays stuck in the little gully behind our camp, and we were yet endeavouring to get it out, when Mr. Browne returned from his attendance on Mr. Poole, and I was glad to find that he had left him in tolerable spirits, and with every hope of his gradual improvement.

As we crossed the creek, between the Depôt and the glen, we found that the waters, as Flood predicted, had descended so far, and waded through them to the other side. We then rode to the glen, to see how it looked under such a change, and remained some time watching the current as it swept along.

On our return to the party I found that it would be impossible to make a lengthened journey; for, having parted with two drays, we had necessarily been obliged to increase the loads on the others, so that they sank deep into the ground. I therefore halted, after having gone about four miles only.

About seven o'clock P.M. we were surprised by the sudden return of Joseph, from the home returning party; but, still more so at the melancholy nature of the information he had to communicate. Mr. Poole, he said, had breathed his last at three o'clock. This sad event necessarily put a stop to my movements, and obliged me to consider what arrangements I should now have to make.

It appeared, from Joseph's account, that Mr. Poole had not shewn any previous indications of approaching dissolution. About a quarter before three he had risen to take some medicine, but suddenly observed to Joseph that he thought he was dying, and falling on his back, expired without a struggle.

Early on the morning of this day, and before we ourselves started, I had sent Mr. Stuart and Mr. Piesse in advance with the chainers, to carry on the chaining. On the morning of the 17th, before I mounted my horse to accompany Mr. Browne to examine the remains of our unfortunate companion, which I determined to inter at the Depôt, I sent a man to recall them.

The suddenness of Mr. Poole's death surprised both Mr. Browne and myself; but the singular fairness of his countenance left no doubt on his mind but that internal haemorrhage had been the immediate cause of that event.

On the 17th the whole party, which had so lately separated, once more assembled at the Depôt. We buried Mr. Poole under a Grevillia that stood close to our underground room; his initials, and the year, are cut in it above the grave, "J.P. 1845," and he now sleeps in the desert.

The sad event I have recorded, obliged me most reluctantly to put Mr. Piesse in charge of the home returning party, for I had had every reason to be satisfied with him, and I witnessed his departure with regret. A more trustworthy, or a more anxious officer could not have been attached to such a service as that in which he was employed.

The funeral of Mr. Poole was a fitting close to our residence at the Depôt. At the conclusion of that ceremony the party again separated, and I returned to my tent, to prepare for moving on the morrow. At 9 A.M. accordingly of the 18th we pushed on to the N.W. The ground had become much harder, but the travelling was still heavy. At three miles we passed a small creek, about seven miles from the Depôt, at which I intended to have halted on leaving that place. We passed over stony plains, or low, sandy, and swampy ground, since the valleys near the hills opened out as we receded from them. On the 19th I kept the chained line, but in consequence of the heavy state of the ground we did not get on more than 8½ miles. The character of the country was that of open sandy plains, the sand being based upon a stiff, tenacious clay, impervious to water. With the exception of a few salsolæ and atriplex, the plains were exceedingly bare, and had innumerable patches of water over them, not more than two or three inches deep. At intervals pure sand hills occurred, on which there were a few stunted casuarina and mimosæ, but a good deal of grass and thousands of young plants already springing up. As the ground was still soft, I should not have moved on the 20th, but was anxious to push on. Early in the day, and at less than 18 miles from the hills, we encountered the sandy ridges, and found the pull of them much worse than over the flats. The wheels of the drays sank deep into the ground, and in straining to get them clear we broke seven yokes. Two flights of swans, and a small flight of ducks, passed over our heads at dusk, coming from the W.N.W. The brushes were full of the Calodera, but being very wild we could not procure a specimen.

The chainers had no difficulty in keeping pace with us, and on the 26th we found ourselves in lat. 29°6', having then chained 61 miles on a bearing of 55° to the west of north, as originally determined upon. Finding that I had thus passed to the south-west of the grassy plain, I halted, and rode with Flood to the eastward; when at seven miles we descended into it, and finding that there was an abundance of water in the creek (the channel we had before noticed), I returned to Mr. Browne; but as it was late in the afternoon when we regained tents, we did not move that evening, and the succeeding day being Sunday we also remained stationary. We had halted close to one of those clear patches on which the rain water lodges, but it had dried up, and there was only a little for our use in a small gutter not far distant. Whilst we were here encamped a little jerboa was chased by the dogs into a hole close to the drays; which, with four others, we succeeded in capturing, by digging for them. This beautiful little animal burrows in the ground like a mouse, but their habitations have several passages, leading straight, like the radii of a circle, to a common centre, to which a shaft is sunk from above, so that there is a complete circulation of air along the whole. We fed our little captives on oats, on which they thrived, and became exceedingly tame. They generally huddled together in a corner of their box, but, when darting from one side to the other, they hopped on their hind legs, which, like the kangaroo, were much longer than the fore, and held the tail perfectly straight and horizontal. At this date they were a novelty to us, but we subsequently saw great numbers of them, and ascertained that the natives frequented the sandy ridges in order to procure them for food. Those we succeeded in capturing were, I am sorry to say, lost from neglect.

On Monday I conducted the whole party to the new depôt, which for the present I shall call the Park, but as I was very unwilling that any more time should be lost in pushing to the west, I instructed Mr. Stuart to change the direction of the chained line to 75° to the west of south, direct upon Mount Hopeless, and to continue it until I should overtake him. In this operation Mr. Browne kindly volunteered to assist Mr. Stuart, as the loss Mr. Piesse had so reduced my strength.

By the 30th I had arranged the camp in its new position, and felt myself at liberty to follow after the chainers. Before I left, however, I directed a stock yard to be made, in which to herd the cattle at night, and instructed Davenport to prepare some ground for a garden, with a view to planting it out with vegetables—pumpkins and melons. I left the camp with Flood, at 10 A.M. on the above day, judging that Mr.

Browne was then about 42 miles a-head of me, and stopped for the night in a little sheltered valley between two sand hills, after a ride of 28 miles. The country continued unchanged. Valleys or flats, more or less covered with water, alternated with sandy ridges, on some of which there was no scarcity of grass.

We had not ridden far on the following morning when a partial change was perceptible in the aspect of the country. The flats became broader and the sand hills lower, but this change was temporary. We gradually rose somewhat from the general level, and crossed several sand hills, higher than any we had seen. These sand hills had very precipitous sides and broken summits, and being of a bright red colour, they looked in the distance like long lines of dead brick walls, being perfectly bare, or sparingly covered with spinifex at the base. They succeeded each other so rapidly, that it was crossing the tops of houses in some street; but they were much steeper to the eastward than to the westward, and successive gales appeared to have lowered them, and in some measure to have filled up the intervening flats with the sand from their summits.

The basis of the country was sandstone, on which clay rested in a thin layer, and on this clay the sandy ridges reposed.

We overtook Mr. Browne about half an hour before sunset, and all halted together, when the men had completed their tenth mile.

On the 1st of August we did not find the country so heavy or so wet as it had been. It was indeed so open and denuded of every thing like a tree or bush, that we had some difficulty in finding wood to boil our tea. In the afternoon when we halted the men had chained 46 miles on the new bearing, but as yet we could not see any range or hill to the westward.

About two hours before we halted Mr. Browne and I surprised some natives on the top of a sand hill, two of them saw us approaching and ran away, the third could not make his escape before we were upon him, but he was dreadfully alarmed. In order to allay his fears Mr. Browne dismounted and walked up to him, whilst I kept back. On this the poor fellow began to dance, and to call out most vehemently, but finding that all he could do was to no purpose he sat down and began to cry. We managed however to pacify him, so much that he mustered courage to follow us, with his two companions, to our halting place. These wanderers of the desert had their bags full of jerboas which they had captured on the hills. They could not indeed have had less than from 150 to 200 of these beautiful little animals, so numerous are they the sand hills, but it would appear that the natives can only go in pursuit of them after a fall of rain, such as that we had experienced. There being then water, the country, at other times impenetrable, is then temporarily thrown open to them, and they traverse it in quest of the jerboa and other quadrupeds. Our friends cooked all they had in hot sand, and devoured them entire, fur, skin, entrails and all, breaking away the under jaw and nipping off the tail with their teeth.

They absolutely managed before sunset to finish their whole stock, and then took their departure, having, I suppose, gratified both their appetite and their curiosity. They were all three circumcised and spoke a different language from that of the hill natives, and came, they told us, from the west.

As we advanced the country became extremely barren, and surface water was very scarce, and the open ground, entirely denuded of timber, wore the most desolate appearance. If we had hitherto been in a region destitute of inhabitants it seemed as if we were now getting into a more populous district. About noon of the 2nd, as Mr. Browne and I were riding in front of the chainers, we heard a shout to our right, and on looking in that direction saw a party of natives assembled on a sand hill, to the number of fourteen. As we advanced towards them they retreated, but at length made a stand as if to await our approach. They were armed with spears, and on Mr. Browne dismounting to walk towards them, formed themselves into a circle, in the centre of which were two old men, round whom they danced. Thinking that Mr. Browne might run some risk if he went near, I called him back, and as I really had not time for ceremonies, we rejoined the chainers, being satisfied also that if the natives felt disposed to communicate with us, they would do so of their own accord; nor was I mistaken in this, for, judging, I suppose, from our leaving them that we did not meditate any hostility, seven of their number followed us, and as Mr. Browne was at that time in advance, I gave my horse to one of the men and again went towards them, but it was with great difficulty that I got them to a parley, after which they sat down and allowed me to approach, though from the surprise they exhibited I imagine they had never seen a white man before. They spoke a language different from any I had heard, had lost two of the front teeth of the upper jaw, and had large scars on the breast. I could not gather any information from them, or satisfactorily ascertain from what quarter they came; staying with them for a short time therefore, and giving them a couple of knives I left them, and after following abreast of us, for a mile or two, they also turned to the north, and disappeared.

The night of the 2nd August was exceedingly cold, with the wind from the N.E. (an unusual quarter from which to have a low temperature) and there was a thick hoar frost on the morning of the 3rd. Why the winds should have been so cold blowing from that quarter, whence our hottest winds also came, it is difficult to say; but at this season of the year, and in this line, they were invariably so.

Near the flat on which we stopped on the evening of the 2nd there was a hill considerably elevated, above the others; which, after unsaddling and letting out the horses, Mr. Browne and I were induced to ascend. From it we saw a line of high and broken ranges to the S.S.W. but they were very distant. At three and a half miles from this point we crossed a salt water creek, having pools in it of great depth, but so clear that we could see to the bottom and wherever our feet sank in the mud, salt water immediately oozed up. There were some, box-trees growing near this creek, which came, from the north, and fell towards the ranges. At half a mile further we crossed a small fresh water creek, and intermediate between the two was a lagoon of about a mile in length, but not more than three inches in depth. This lagoon, if it might so be called, from its size only, had been filled by the recent rains; but was so thick and muddy, from being continually ruffled by the winds, that it was unfit for use. The banks of the fresh water creek were crowded with water-hens, similar to those which visited Adelaide in such countless numbers the year before I proceeded into the interior (1843). They were running about like so many fowls; but, on being alarmed, took flight and went south.

The fresh water creek (across which it was an easy jump) joined the salt water creek a little below where we struck it, and was the first creek of the kind we had seen since we left the Depôt, in a distance of more than 100 miles, and up to this point we had entirely subsisted on the surface water left by the rains. The country we now passed through was of a salsolaceous character, like a low barren sea coast. The sand hills were lower and broader than they had been, and their sides were cut by deep fissures made by heavy torrents. From a hill, about a mile from our halting place on this day, we again saw the ranges, which had been sighted the day before. South of us, and distant about a mile, there was a large dry lagoon, white with salt, and another of a similar kind to the west of it.

These changes in the character of the country convinced me that we should soon arrive at some more important one. On the 4th we advanced as usual on a bearing of 75° to the west of south, having then chained 65 miles upon it. At about three miles we observed a sand hill in front of us, beyond which no land was to be seen, as if the country dipped, and there was a great hollow. On arriving at this sand hill our further progress westward was checked by the intervention of an immense shallow and sandy basin, upon which we looked down from the place where we stood. The hills we had seen the day before were still visible through a good telescope, but we could only distinguish their outlines; in addition to them, however, there was a nearer flat-topped range, more to the northward and westward of the main range, which latter still bore S.S.W., and appeared to belong to a high and broken chain mountains. The sandy basin was from ten to twelve miles broad, but destitute of water opposite to us, although there were, both to the southward and northward, sheets of water as blue as indigo and as salt as brine. These detached sheets were fringed round with samphire bushes with which the basin was also speckled over. There was a gradual descent of about a mile and a half, to the margin of the basin, the intervening ground being covered with low scrub. My first object was, to ascertain if we could cross this feature, which extended southward, beyond the range of vision, but turned to the westward in a northerly direction, in the shape in which Mr. Eyre has laid Lake Torrens down. For this purpose Mr. Browne and I descended into it. The bed was composed of sand and clay, the latter lying in large masses, and deeply grooved by torrents of rain. There was not any great quantity of salt to be seen, but it was collected at the bottom of gutters, and, no doubt, was more or less mixed with the soil. At about four miles we were obliged to dismount; and, tying our horses so as to secure them, walked on for another mile, when we found the ground too soft for our weight and were obliged to return; and, as it was now late, we commenced a search for water, and having found a small supply in a little hollow, at a short distance from the flag, we went to it and encamped. The length of the chain line to the flag staff was 70¾ miles, which with the 61 we had measured from the Depôt, made 131¾ miles in all; the direct distance, therefore, from the Depôt to the flag staff, was about 115 miles, on a bearing of 9½° to the North of West or W.¾N.

My object in the journey I had thus undertaken, was not so much to measure the distance between the two places, as to ascertain if the country to the north-west of Lake Torrens, on the borders of which I presumed I had arrived, was practicable or not, and whether it was connected with any more central body of water. It behoved me to ascertain these two points with as little delay as possible, for the surface water was fast drying up, and we were in danger of having our retreat cut off. Whether the country was practicable or not, in the direction I was anxious to take, it was clear that I could not have penetrated as far as I then was, with the heavy drays, with any prudence.

To be more satisfied, however, as to the nature of the country to the westward, I rode towards the N.E. angle of the Sandy Basin, on the morning of the 4th, sending Mr. Stuart southwards, to examine it in that direction; but, neither of these journeys proving satisfactory, I determined on fixing the position of the hills in reference to our chained line, and then return to the Depôt, to prepare for a more extensive exploration of the N.W. interior. I found the country perfectly impracticable to the N.W., and that it was impossible to ascertain the real character of this Sandy Basin. On the other side of it the country appeared to be wooded; beyond the wood there was a sudden fall; and, as far as I could judge, this singular feature must have been connected to Spencer's Gulf, before the passage that evidently existed once between them, was filled up.

On the 5th I ran a base line from the end of the chained line to the north-west, on a bearing of 317° to the only prominent sand hill in that direction, distant from the staff 5 1/2 miles, from the extremities of which the ranges bore as follow:—

BEARINGS FROM THE FLAG STAFF AT THE TERMINATION OF THE CHAINED LINE.

To a bluff point in the main range	198·00
To the north point of the south range	188·40
To the north point	182·50
To the highest point in south range	187·00
To the flat-topped hills	231·00
To the north-west point of the lake	283·00
To the south point	158·00

BEARINGS FROM THE NORTH-WEST EXTREMITY.

To the bluff	194·30
To the north point of south range	184·00
To the south	183·00
To the flat-topped hills	176·30
To the north-west extremity of lake	275·00

The angles given by these bearings were necessarily very acute, but that could not be avoided. With the bearings, however, from a point in our chain line, 16 miles to the rear, they gave the distance of the more distant ranges as 65 miles, that of the nearer ones as 33.

Our latitude, by altitudes of Vega and Altair, on the night of the 5th of August, was 29°14'39", and 291°15'14"; by our bearings, therefore, the flat-topped hills were in lat. 29°33', and the bluff, in the centre of the distant chain, where there appeared to be a break in it, in 30°10', and in long. 139°12'.

Presuming our Depôt to have been in lat. 29°40'10" and in long. 141°30'E, and allowing 52½ miles to a degree, our long. by measurement was 139°20'E. I had ascertained the boiling point of water at our camp, about 100 feet above the level of the basin to be 212·75; which made our position there considerably below the level of the sea: but in using the instrument on the following morning in the bed of the basin itself, I unfortunately broke it. As, however, the result of the observation at our bivouac gave so usual a depression, and as, if it was correct, Lake Torrens must be very considerably below the level of the sea, I can only state that the barometer had been compared with one in Adelaide by Capt. Frome, and that, allowing for its error, its boiling point on a level with the sea had been found by him to be 212·25.

On the 6th I left the neighbourhood of this place, and stopped at 16 miles to verify our former bearings. The country appeared more desolate on our return to the camp than when we were advancing. Almost all the surface water had dried up, or consisted of stagnant mud only, so that we were obliged to push on for the Park, at which we arrived on the 8th. On the 10th we completed the year, it being the anniversary of our departure from Adelaide.

I found that every thing had gone on regularly in the camp during my absence, and that the cattle and sheep had been duly attended to. Davenport had also dug and planned out a fine garden, which he had planted with seeds, but none had as yet made their appearance above the ground.

The day after our return to the camp we were visited by two natives, who were attracted towards us by the sound of the axe. They were crossing the plain, and were still at a considerable distance when they observed Davenport pointing a telescope, on which they stopped, but on my sending a man to meet them, came readily forward. We were in hopes that we should see our old friend in the person of one of them, but were disappointed; nor would they confirm any of his intelligence, neither could they recognise any of the fish in the different plates I had shewn him. In truth, we could get nothing out of these stupid fellows; but, as we gave them plenty to eat, they proposed bringing some other natives to taste our mutton, on the following day; and, leaving us, returned, as they said, with their father and brother, the latter a fine young lad. But neither from the old man could we gather any information, as to the nature of the country before us. These people were circumcised, like many others we had seen, but were in no way disfigured by the loss of their teeth or cuts. I can say as little for their cleanliness as for their information, since they melted the fat we gave them in troughs, and drank it as if it had been so much oil, emptying what remained on their heads, rubbing the grease into their hair, and over their bodies.

I felt satisfied on mature reflection that if the country continued to any distance either to the northward or westward, such as we had found it on our recent journey, it would be highly imprudent to venture into it with the whole party. Setting aside the almost utter impossibility of pulling the drays over the heavy sand ridges by which our route would be intersected, little or no surface water now remained. The ground was becoming as dry and parched as it had been before the fall of rain. I determined therefore before I again struck the tents to examine the country to the north-west, and not incautiously to hazard the safety of the party by leading it into a region from which I might find it difficult to retreat. As soon therefore as I had run up the charts, I prepared for this journey. Our position at the new Depôt was in latitude 29°6'30", and in longitude 141°5'8", it therefore appeared to me if I ran on a bearing of 45° to the west of north, should gain the 138th meridian about the centre of the continent, and at the same time cross the Tropics at the desired point, and I felt certain that if there were any mountain chains or ranges of hills to the westward of me connected with the north-east angle of the continent I should be sure to discover them.

In preparing for this important journey, which it was evident the success of the expedition would depend, I took more than ordinary precautions. I purposed giving the charge of the camp to Mr. Stuart.—I had established it on a small sandy rise, whereon we found five or six native huts. This spot was at the northern extremity of the Park, but a little advanced into it. Immediately in front of the tents there was a broad sheet of water shaded by gum-trees, and the low land between this and the sand hills was also chequered with them. The position was in every way eligible. The open grassy field or plain stood full in view, and the men could see the cattle browsing on it, but I directed Mr. Stuart never to permit them to be without one of the men as a guard, and to have them nightly in the stockyard. In order to provide for the further security of the camp, I marked out the lines, for the erection of a stockade, wherein I directed Mr. Stuart to pitch one of the bell tents. In this tent I instructed him to deposit the arms and ammunition, and to consider it as the rallying point in the event of any attack by the natives, in which case I told him his first step would be to secure the sheep. I desired that the stockade might be commenced as soon as I left, and that it should be built of palisades 4½ feet above the ground, and arranged close together. In such a fortification I considered that the men would be perfectly safe, and as the stockyard was in a short range of the carbines I felt the cattle would be sufficiently protected.

I selected Flood, Lewis, and Joseph to accompany me, and took 15 weeks provisions. This supply required all the horses but one, for although they had so long a rest at the old Depôt they were far from being strong, since for the last three months they had lived on salsolaceous herbs, or on the shoots of shrubs, so that although apparently in good condition they had no work in them. My last instructions to Morgan were to prepare and paint the boat in the event of her being required.

CHAPTER VIII.

LEAVE THE DEPÔT FOR THE NORTH-WEST—SCARCITY WATER—FOSSIL LIMESTONE—ARRIVE AT THE FIRST CREEK—EXTENSIVE PLAINS—SUCCESSION OF CREEKS—FLOODED CHARACTER OF THE COUNTRY—POND WITH FISH—STERILE COUNTRY—GRASSY PLAINS—INTREPID NATIVE—COUNTRY APPARENTLY IMPROVES-DISAPPOINTMENTS WATER FOUND—APPEARANCE OF THE STONY DESERT NIGHT THEREON—THE EARTHY PLAIN—HILLS RAISED BY REFRACTION—RECOMMENCEMENT OF THE SAND RIDGES—THEIR UNDEVIATING REGULARITY—CONJECTURES AS TO THE DESERT—RELATIVE POSITION OF LAKE TORRENS—CONCLUDING REMARKS.

ON THE morning of the 14th Mr. Browne and I mounted our horses, and left the camp at 9 A.M., followed by the men I had selected, and crossing the grassy plain in a N.W. direction, soon found ourselves amidst sand hills and scrub.

As I have stated I had determined to preserve a course of 45° to the west of north, or in other words a north-west course, but the reader will readily believe that in such a country I had no distant object on which to rely. We were therefore obliged take fresh bearings with great precision from almost every sand-hill, for on the correctness of these bearings, together with our latitude, we had to depend for our true position. We were indeed like a ship at sea without the advantage of a steady compass.

Throughout the whole day of our departure from the camp we traversed a better country than that between it and Lake Torrens, insomuch that there was more grass. Sand ridges and flats succeeded each other, but the former were not so broken and precipitous or the latter so barren, as on our line to the westward, and about four miles from the camp we passed a pool of water to our right. At five miles we observed a new melaleuca, similar to the one I had remarked when to the north with Joseph, growing on the skirts of the flats, but the shrubs for the most part consisted of hakea and mimosæ with geum and many other minor plants. For a time the ridges were smooth on their sides, and a quantity of young green grass was springing up on them. At nine miles we crossed some stony plains, and halted after a ride of 26 miles without water.

On the 15th a strong and bitterly cold wind blew from the westward as we passed through a country differing in no material respect from that of the day before. Spinifex generally covered the sand ridges, which looked like ocean swells rising before us, and many were of considerable height. At six miles we came to a small pool of water, where we breakfasted. On leaving this we dug a hole and let the remainder of the water into it, in the hope of its longer continuance, and halted after a long journey in a valley in which there was a kind of watercourse with plenty of water, our latitude being 28°21'39". Before we left this place we cut a deep square hole, into which as before we drained the water, that by diminishing its surface we might prevent the too speedy evaporation of it, in case of our being forced back from the want of water in the interior, since that element was becoming more scarce everyday. We saw but little change in the character of the country generally as we rode through it, but observed that it was more open to the right, in which direction we passed several extensive plains. There were heaps of small pebbles also of ironstone and quartz on some of the flats we crossed. We halted at the foot of a sand hill, where there was a good deal of grass, after a vain search for water, of which we did not see a drop during the day. The night of the 17th, like the preceding one, was bitterly cold, with the wind S.W. During the early part of this day we passed over high ridges of sand, thickly covered with spinifex, and a new polygonum, but subsequently crossed some flats of much greater extent than usual, and of much better soil, but the country again fell off in quality and appearance, although on the whole the tract we had crossed on our present journey was certainly better than that we traversed in going to Lake Torrens. We halted rather earlier than usual, at a creek containing a long pond of water between two and three feet deep. The ground near it was barren, if I except the polygonum that was growing near it. The horses however found a sufficiency to eat, and we were prevented the necessity of digging at this point, in consequence of the depth of the water. We observed some fossil limestone cropping out of the ground in several places as we rode along, and the flats were on many parts covered with small rounded nodules of lime, similar to those I have noticed as being strewed over the fossil cliffs of the Murray. It appeared to me as I rode over some of the flats that the drainage was to the south, but it was exceedingly difficult in so level and monotonous a region to form a satisfactory opinion. We saw several emus in the course of the day, and a solitary crow, but scarcely any other of the feathered tribe. There was an universal sameness in the vegetation, if I except the angophora, growing on the sand hills and superseding the acacia.

On the 18th the morning was very cold, with the wind at east, and a cloudy sky. We started at eight; and after crossing three very high sandridges, descended into a plain of about three miles in breadth, extending on either hand to the north and south for many miles. At the further extremity of this plain we observed a line of box-trees, lying, or rather stretching, right across our course; but as they were thicker to the S.W. than at the point towards which we were riding, I sent Flood to examine the plain in that direction. In the mean time Mr. Browne and I rode quietly on; and on arriving at the trees, found that they were growing in the broad bed of a creek, and overhanging a beautiful sheet of water, such as we had not seen for many a day. It was altogether too important a feature to pass without further examination; I therefore crossed, halted on its west bank, and as soon as Flood turned, (who had not seen any water,) but had ascertained that just below the trees, the creek spreads over the plain, I sent him with Mr. Browne to trace it up northward, the fall of the country apparently being from that point. In the meantime we unloaded the horses, and put them out on better grass than they had had for some time. On the opposite side of the creek, and somewhat above there were two huts, and the claws of crayfish scattered about near them. There were also a wild fowl and Hæmantopus sitting on the water, either unconscious of or indifferent to our presence. This fine sheet of water was more than 60 yards broad by about 120 long, but, as far as we could judge, it was shallow.

Mr. Browne returned to me in about three hours, having traced the creek upwards until he lost its channel, as Flood had done on a large plain, that extended northwards to the horizon. He observed the country was very open in that direction and had passed another pond of water, deeper but not so large as that at which we had stopped, surprised an old native in his hut with two of his wives, from whom he learnt that there were both hills and fish to the north.

Whilst Mr. Browne was away, I debated within myself whether or not to turn from the course on which I had been running to trace this creek up. The surface water was so very scarce, that I doubted the possibility of our getting on; but was reluctant to deviate from the line on which I had determined to penetrate, and I think that, generally, one seldom gains anything in so doing. From Mr. Browne's account of the creek, its character appeared to be doubtful, so that I no longer hesitated on my onward course; but we remained stationary for the remainder of the day.

The evening of this day was beautifully fine, and during it many flights of parrots and pigeons came to the water. Of the latter we shot several, but they were very wild and wary. There was on the opposite side of the creek a long grassy flat, with box-trees growing on it, together with a new Bauhinia, which we saw here for the first time. On this grassy flat there were a number of the water-hens we had noticed on the little fresh-water creek near Lake Torrens. These birds were running about like fowls all over the grass, but although they had been so tame as to occupy the gardens and to run about the streets of Adelaide, they were now wild enough.

Mr. Browne remarked that the females he had seen were, contrary to general custom as regards that sex, deficient in the two front teeth of the upper jaw, but that the teeth of the man were entire, and that he was not otherwise disfigured, I was anxious to have seen these natives, and, as their hut was not very far from us, we walked to it in the cool of the afternoon, but they had left, and apparently gone to the N.E.; we found some mussel shells amongst the embers of some old fires near it. Our latitude at this point was 28°3'S. a distance of 86 miles from the Park.

We left on the morning of the 20th at an early hour, and after crossing that portion of the plain lying to the westward, ascended a small conifer sand hill, that rose above the otherwise level summit of the ridge. From this little sand hill we had our anticipations confirmed as to the low nature of the country to the north as a medium point, but observing another and a much higher point to the west-ward, we went to, and found that the view extended to a much greater distance from it. The country was very depressed, both to the north and north-west. The plains had almost the character of lagoons, since it was evident they were sometimes inundated, from the water mark on the sand hills by which they were partly separated from on another. Below us, on our course, there was a large plain of about eight miles in breadth; but immediately at the foot of the hill, which was very abrupt (being the terminating point of a sandy ridge of which it was the northern extremity), there was a polygonum flat. We there saw a beautiful parrot, but could not procure it. The plain we next rode across was evidently subject to floods in many parts; the soil was a mixture of sand and clay. There was a good deal of grass here and there upon it, and box-trees stunted in their growth were scattered very sparingly round about; but the country was otherwise denuded of timber. There were large bare patches on the plains, that had been full of water not long before, but too shallow to have lasted long, and were now dry. We found several small pools, however, and halted at one, after a journey of 17 miles, near some gum-trees.

The morning of the 20th was exceedingly calm, with the wind from the west, but it had been previously from the opposite point. The channel of the creek was broad, and we traced it to some distance on either hand, but it contained no water, excepting that at which we stopped; but at about two miles before we halted, Mr. Browne found a supply under some gum-trees, a little to the right of our course, where we halted on our return.

The Bauhinia, here grew to the height of 16 to 20 feet, and was a very pretty tree; the ends of its branches were covered with seed-pods, both of this and the year before: it was a flat vessel, containing four or six flat hard beans. I regretted, at this early stage of our journey, that the horses were not up to much work, although we were very considerate with them, but the truth is, that they had for about two or three months before leaving the Depôt, been living on pulpy vegetables, in which there was no strength, they nevertheless looked in good condition. They had become exceedingly tractable, and never wandered far from our fires; Flood, however, watched them so narrowly that they could not have gone far. Since the three days' rain in July, the sky was but little clouded, but we now observed that from whatever quarter the wind blew, a bank of clouds would rise in the opposite direction—if from the east, in the west, and vice versâ—but the clouds invariably came against the wind, and must consequently have been moving in an upper current.

On the 20th we commenced our journey early, that is to say, at 6 A.M.; the sky was clear, the temperature mild, and the wind in the S.E. quarter. We crossed plains of still greater extent than any we had hitherto seen; their soil was similar to that on the flats of the Darling, and vegetation seemed to suffer from their liability to inundation. The only trees now to be seen were a few box-trees along their skirts, and on the line of the creeks, which last were a perfectly new feature in the country, and surprised me greatly. The tract we passed over on this day was certainly more subject to overflow than usual. Large flats of polygonum, and plains having rents and fissures in them, succeeded those I have already described. At ten miles we intersected a creek of considerable size, but without any water, just below where we crossed its channel it spread over a large flat and is lost. Proceeding onwards, at a mile and a half, we ascended a line of sand hills, and from them descended to firmer ground than that on which we had previously travelled. At six miles we struck another creek with a broad and grassy bed, on the banks of which we halted, at a small and muddy pool of water. The trees on this creek were larger than usual and beautifully umbrageous. It appeared as if coming from the N.E., and falling to the N.W. There were many huts both above and below our bivouac, and well-trodden paths from one angle of the creek to the other. All around us, indeed, there were traces of natives, nor can there be any doubt, but that at one season of the year or other, it is frequented by them in great numbers. From a small continuous elevation our view extended over an apparently interminable plain in the line of our course. That of the creek was marked by gum-trees, and I was not without hopes that we should again have halted on it on the 21st, but we did not, for shortly after we started it turned suddenly to the west, and we were obliged to leave it, and crossed successive plains of a description similar to those we had left behind, but with little or no vegetation upon them. At about five miles we intersected a branch creek coming from the E.N.E., in which there was a large but shallow pool of water. About a mile to the westward of this channel we ascended some hills, in the composition of which there was more clay than sand, and descended from them to a firm and grassy plain of about three and a half miles in breadth. At the farther extremity we crossed a line of sand hills, and at a mile and a half again descended to lower ground, and made for some gum-trees at the western extremity of the succeeding plain, on our old bearing of 55° to the west of north. There we intersected another creek with two pools of water in it, and as there was also a sufficiency of grass we halted on its banks.

The singular and rapid succession of these water-courses exceedingly perplexed me, for we were in a country remote from any high lands, and consequently in one not likely to give birth to such features, yet their existence was a most fortunate circumstance for us. There can be no doubt but that the rain, which enabled us to break up the old Depôt and resume our operations, had extended thus far, but all the surface water had dried up, and if we had not found these creeks our progress into the interior would have been checked. In considering their probable origin, it struck me that they might have been formed by the rush of floods from the extensive plains we had lately crossed. The whole country indeed over which we had pass from the first creek, was without doubt very low, and must sometimes be almost entirely under water, but what, it may be asked, causes such inundation? Such indeed was the question I asked myself, but must say I could arrive at no satisfactory conclusion.

That these regions are subject to heavy rains I had not the slightest doubt, but could the effect of heavy rains have produced these creeks, short and uncertain in their course, rising apparently in one plain, to spread over and terminate in another, for had we gone more to the westward in our course than we did, it is probable we should never have known of the existence of any of them. I was truly thankful that we had thus fallen upon them, and considering how much our further success depended on their continuance, I began to hope that we should find them a permanent feature in the country.

About this period and two or three days previously, we observed a white bank of clouds hanging upon the northern horizon, and extending from N.E. to N.W. No wind affected it, but without in the least altering its shape, which was arched like a bow, it gradually faded away about 3 P.M. Could this bank have been over any inland waters?

At the point to which I have now brought the reader, we were in lat. 27°38'S, and in long. 140°10' by account, and here, as I have observed, as in our journey to Lake Torrens, the N.E. winds were invariably cold. On the 22nd we crossed the creek, and traversed a large plain on the opposite side that was bounded in the distance by a line of sand hills. On this plain were portions of ground perfectly flat, raised some 12 or 18 inches above its general level; on these, rhagodia bushes were growing, which in the distance looked like large trees, in consequence of the strong refraction. The lower ground of these N.W. plains had little or no vegetation upon it, but bore the appearance of land on which water has lodged and subsided; being hard and baked in some places, but cracked and blistered in others, and against the sides of the higher portions of the plain, a line of sticks and rubbish had been lodged, such as is left by a retiring tide, and from this it seemed that the floods must have been about a foot deep on the plain when it was last inundated. At 4½ miles we reached its western extremity, and ascending the line of sand hills by which it is bounded on that side, dropped down to another plain, and at six miles intersected a creek with a deep broad and grassy bed, but no water. A high row of gum-trees marked its course from a point rather from the southward of east to the north-north-west. Crossing to the opposite side we ascended another sand hill by a gradual rise, and again descended to another plain at the farther extremity of which we could indistinctly see a dark line of trees. Arriving at these after a ride of six miles, we were stopped by another creek. Its banks were too steep for the cart, and we consequently turned northward and traced it downwards for four miles before we found a convenient spot at which to halt. The ground along the creek side was of the most distressing nature; rent to pieces by solar heat, and entangled with polygonum twisted together. We passed several muddy water-holes and at length stopped at a small clear deep pond. The colour of the water, a light green, at once betrayed its quality; but fortunately for us, though brackish it was still tolerable, much better than the gritty water we had passed. There was however but little vegetation in its neighbourhood, the grass being coarse and wiry. Both on this creek and some others we had passed, we observed that the graves of the natives were made longitudinally from north to south, and not as they usually are from east to west.

The evening we stopped at this place was very fine. We had descended into the bed of the creek, Mr. Browne and I were reclining on the ground, looking at the little pond, in which the bank above was clearly reflected. On a sudden my companion asked me if I had brought a small hook with me, as he had taken it into his head that there were fish in the pond. Being unable to supply his wants, he got a pin, and soon had a rough kind of apparatus prepared, with which he went to the water; and, having cast in his bait, almost immediately pulled out a white and glittering fish, and held it up to me in triumph. I must confess that I was exceedingly astonished, for the first idea that occurred to my mind was—How could fish get into so isolated a spot? In the water-holes above us no animals of the kind could have lived. How then were we to account for their being where we found them, and for the no less singular phenomenon of brackish waters in the bed of a fresh water creek? These were exceedingly puzzling questions to me at the time,

but, as the reader will find, were afterwards explained. Mr. Browne succeeded in taking no less than thirteen fish, and seemed to think that they were identical with the silver perch of the Murray, but they appeared to me to be a deeper and a thinner fish. Although none of them exceeded six inches in length, they were very acceptable to men, who were living on five pounds of flour only a-week.

The night we stayed here was very dark, and about 11 P.M. the horses which had been turned down the creek by Flood, rushed violently past our fire, as if they had been suddenly alarmed. They were found at a distance of five miles above us the next morning, but we could never discover why they had taken fright. Their recovery detained us longer than our usual hour, but at nine we mounted and, crossing the creek at three-quarters of a mile, ascended a hill, connected with several others by sandy valleys, and saw that the creek, a little below where we crossed it, turned to the west. We could trace its course, by the trees on its bank, for several miles. From the hills we descended to a country of a very different character from that which I have been describing. As we overlooked it from the higher ground it was dark, with a snow-white patch of sand in the centre; on traversing it we found that its productions were almost entirely samphire-bushes growing on a salty soil.

The white patch we had seen from a distance was the dry bed of a shallow salt lagoon also fringed round with samphire bushes, and being in our course we crossed it. There was a fine coating of salt on its surface, together with gypsum and clay, as at Lake Torrens. The country for several miles round it was barren beyond description, and small nodules of limestone were scattered over the ground in many places. After leaving the lagoon, which though moist had been sufficiently hard to bear our weight, we passed amidst tortuous and stunted box-trees for about three miles; then crossed the small dry and bare bed of a water-course, that was shaded by trees of better appearance, and almost immediately afterwards found ourselves on the outskirts of extensive and beautifully grassed plains, similar to that on which I had fixed the Depôt, and most probably owing, like them, their formation to the overflow of the last, or some other creek we had traced. The character of the country we had previously travelled over being so very bad, the change to the park-like scene now before us was very remarkable. Like the plains at the Depôt they had gum-trees all round them, and a line of the same trees running through their centre.

Entering upon them on a north-west course, we proceeded over the open ground, and saw three figures in the distance, who proved to be gathering seeds. They did not perceive us until were so near to them that they could not escape, but stood for some time transfixed with amazement. On riding up we dismounted, and asked them by signs where there was any water, to which question they signified most energetically that there was none in the direction we were going, that it was to the west. One of these women had a jet black skin, and long curling glossy ringlets. She seemed indeed of a different race, and was, without doubt, a secondary object of consideration with her companions; who, to secure themselves I fancy, intimated to us that we might take her away; this, however, we declined doing. One of the women went on with her occupation of cleaning the grass seeds she had collected, all the time we remained, humming a melancholy dirge. On leaving them, and turning to the point where they said no water was to be found they exhibited great alarm, and followed us at a distance.

Soon after we passed close to some gum-trees and found a dry channel under a sand hill on the other side, running this down we came suddenly on two bough huts before which two or three little urchins were playing, who, the moment they saw us, popped into the huts like rabbits. Directly opposite there was a shallow puddle rather than a pool of water, and as Joseph had just met with an accident I was obliged to stop at it. I was really sorry to do so, however, for I knew our horses would exhaust it all during the night, and I was reluctant to rob these poor creatures of so valuable a store, I therefore sent Flood to try if he could find any lower down; but, as he failed, we unsaddled our horses and sat down.

The women who had kept us in sight were then at the huts, to which Mr. Browne and I walked. In addition to the women and children, there was an old man with hair as white as snow. As I have observed, there was a sand hill at the back of the huts, and as we were trying to make ourselves understood by the women a native made his appearance over it; he was painted in all the colours of the rainbow, and armed to the teeth with spear and shield. Great was the surprise and indignation of this warrior on seeing that we had taken possession of his camp and water. He came fearlessly down the hill, and by signs ordered us to depart, threatening to go for his tribe to kill us all, but seeing that his anger only made us smile, he sat down and sulked. I really respected the native's bravery, and question much if I should have shewn equal spirit in a similar situation. Mr. Browne's feelings I am sure corresponded with my own, so we got up and left him, with an intention on my part to return when I thought he had cooled down to make him some presents, but when we did so he had departed with all his family, and returned not to the neighbourhood again. We had preserved two or three of the fish, and in the hope of making the women understand us better, produced them, on which they eagerly tried to snatch them from us, but did not succeed. They were evidently anxious to get them to eat, and I mention the fact, though perhaps telling against my generosity on the occasion, to prove how rare such a feast must be to them.

As I had foreseen, our horses finished all the water in the puddle during the night, and we left at seven in the following morning, taking up our usual N.N.W. course, from which up to this point we had not deviated. We passed for about eight miles through open box-tree forest, with a large grassy flat backed by sand hills to the right. The country indeed had an appearance of improvement. There was grass under the trees, and the scenery as we rode along was really cheerful. I began to hope we were about to leave behind us the dreary region we had wandered over, and that happier and brighter prospects would soon open out, to reward us for past disappointment. Mr. Browne and I even ventured to express such anticipations to each other as we journeyed onwards. At eight miles however, all our hopes were annihilated. A wall of sand suddenly rose before us, such as we had not before seen; lying as it did directly across our course we had no choice but to ascend. For 20 miles we toiled over as distressing a country as can be imagined, each succeeding sand ridge assumed a steeper and more rugged character, and the horse with difficulty pulled the cart along. At 13 miles we crossed a salt lagoon similar to the one I have described to the S.E. of the plains on which we had last seen the natives, but larger. Near it there was a temporary cessation of the fearful country we had just passed, but it was only temporary, the sand ridges again crossed our path, and at five or seven miles from the lagoon we pulled up for the night in a small confined valley in which there was a little grass, our poor horses sadly jaded and fatigued, and our cart in a very rickety state. We could not well have been in a more trying situation, and as Mr. Browne, and Lewis (one of the men I had with me), went to examine the neighbourhood from a knoll not far off, while there was yet light, I could not but reflect on the singular fatality that had attended us. I had little hope of finding water, and doubted in the event of disappointment whether we should get any of the horses back to the Fish-pond, the nearest water in our rear. Mr. Browne was late in returning to me, but the news he had to communicate dispelled all my fears. He had, he told me, from the summit of the knoll to which he went, observed something glittering in a dark looking valley about three

miles to the N.W., and had walked down to ascertain what it was, when to his infinite delight he found that it was a pool of water, covering no small space amongst rocks and stones. It was too late to avail ourselves, however, of this providential discovery; but we were on our way to the place at an early hour. There we broke our fast, and, I should have halted for the day to repair the cart, but there was little or no grass in the valley for the horses, so that we moved on after breakfast; but coming at less than a mile to a little grassy valley in which there was likewise water, we stopped, not only to give the animals a day of rest, and to repair the cart, but to examine the country, and satisfy ourselves as to the nature of the sudden and remarkable change it had undergone. With this in view, as soon as the camp was formed, and the men set to repair the cart, Mr. Browne and I walked the extremity of a sandy ridge that bore N.N.W. from us, and was about two miles distant. On arriving at this point we saw an immense plain occupying more than one half of the horizon, that is to say, from the south round to the eastward of north. A number of sandy ridges, similar to that on which we stood, abutted upon, and terminated in this plain like so many head lands projecting into the sea. The plain itself was of a dark purple hue, from the elevated point on which we stood appeared to be perfectly level.
There was a line of low trees far away upon it to the N.E.; and to the north, at a great distance, the sun was shining on the bright point of a sand hill. The plain was otherwise without vegetation, and its horizon was like that of the ocean. In the direction I was about to proceed, nothing was to be seen but the gloomy stone-clad plain, of an extent such as I could not possibly form any just idea. Ignorant of the existence of a similar geographical feature in any other part of the world, I was at a loss to divine its nature. I could not however pause as to what was to be done, but on our return to the party prepared to cross it. I was fully aware, before leaving the old Depôt, that as soon as we got a few miles distant from the hills, I should be unable to continue my angles, and should thenceforth have to rely on bearings. So long as we were chaining there was no great fear of miscalculating position; so far then as the second Depôt, it would not be difficult for any other traveller to

follow my course. From that point, as I have already stated, I ran on a compass bearing of 25° to the west of north, or on a N.N.W. course, and adhered to it up to the point I have now led the reader, a new bearing having been taken on some object still farther in advance from every sand hill we ascended. This appeared to me to be the most satisfactory way of computing our distances and position, for the latitude necessarily correcting both, the amount of error could not be very great. I now found, on this principle, that I was in latitude 27°4'40" south, and in longitude, by account, 139°10' east.

On reaching the cart I learnt that Lewis, while wandering about, had stumbled on a fine sheet of water, in a valley about two miles to the south of us, and that Joseph and Flood had shot a couple of ducks, or I should have said widgeon of the common kind.

On the 26th I directed Flood to keep close under the sandy ridge, to the termination of which Mr. Browne and I had been, and to move into the plain on the original bearing of 25° to the west of north until I should overtake him; Mr. Browne and I then mounted and went to see the water Lewis had discovered, for which we had not had time the previous evening. It was a pretty little sequestered spot surrounded by sand hills, excepting to the N.W. forming a long serpentine canal, apparently deep, and shaded by many gum-trees; there were a numbers of ducks on the water, but too wild to allow us within shot. Both Mr. Browne and I were pleased with the spot, and could not but congratulate ourselves in having such a place to fall back upon, if we should be forced to retreat, as it had all the promise of durability for some weeks to come. We overtook the drays far upon the plains, and continued our journey for twenty miles, when I halted on a bare piece of sandy ground on which there were a few tussocks of grass, and a small puddle of water. On travelling over the plain we found it undulating, with shining hollows in which it was evident water sometimes collected. The stones, with which the ground was so thickly covered as to exclude vegetation, were of different lengths, from one inch to six, they had been rounded by attrition, were coated with oxide of iron, and evenly distributed. In going over this dreary waste the horses left no track, and that of the cart was only visible here and there. From the spot on which we stopped no object of any kind broke the line of the horizon; we were as lonely as a ship at sea, and as a navigator seeking for land, only that we had the disadvantage of an unsteady compass, without any fixed point on which to steer. The fragments covering this singular feature were all of the same kind of rock, indurated or compact quartz, and appeared to me to have had originally the form of parallelograms, resembling both in their size and shape the shivered fragments, lying at the base of the northern ranges, to which I have already had occasion to call attention.

Although the ground on which we slept was not many yards square, and there was little or nothing on it to eat, the poor animals, loose as they were, did not venture to trespass on the adamantine plain by which they were on all sides surrounded.

On the 27th we continued onwards, obliged to keep the course by taking bearings on any prominent though trifling object in front. At ten miles there was a sensible fall of some few feet from the level of the Stony Desert, as I shall henceforth call it, and we descended into a belt of polygonum of about two miles in breadth, that separated it from another feature, apparently of equal extent but of very different character. This was an earthy plain, on which likewise there was no vegetation; resembling in appearance a boundless piece of ploughed land on which floods had settled and subsided—the earth seemed to have once been mud and then dried. It had been impossible to ascertain the fall or dip of the Stony Desert, but somewhat to the west of our course on the earthy plain there were numerous channels, which as we advanced seemed to be making to a common centre towards the N.E. Here and there a polygonum bush was growing on the edge of the channels; and some of them contained the muddy dregs of what had been pools of water. Over this field of earth we continued to advance almost all day, without knowing whether we were getting still farther into it, or working our way out. About an hour before sunset, this point was settled beyond doubt, by the sudden appearance of some hills over the line of the horizon, raised above their true position by refraction. They bore somewhat to the westward of north, but were too distant for speculation upon their character. It was very clear, however, that there was a termination to the otherwise apparently boundless level on which we were in that direction, if not in any other. Our view of these hills was but transient, for they gradually faded from sight, and in less than ten minutes had entirely disappeared. Shortly afterwards some trees were seen in front, directly in the line of our course; but, as they were at a great distance, it was near sunset before we reached them; and finding they were growing close to a small channel (of which there were many traversing the plain) containing a little water, we pulled up at them for the night, more especially as just at the same moment the hills, before seen, again became visible, now bearing due north. To scramble up into the box-trees and examine them with our telescopes was but the work of a moment, still it was doubtful whether they were rock or sand. There were dark shadows on their faces, as if produced by cliffs, and anxiously did we look at them so long as they continued above the horizon, but again they disappeared and left us in perplexity. They were, however, much more distinct on the second occasion, and Mr. Browne made out a line of trees, and what he thought was grass on our side of them.

There was not a blade of anything for our horses to eat round about our solitary bivouac, so that we were obliged to fasten them to the trees, only three in number, and to the cart. There was, however, a dark kind of weed growing in the creek, and some half dozen stalks of a white mallow, the latter of which Flood pulled up and gave to the horses, but they partook sparingly of them, and kept gnawing at the bark of the trees all night long.

In reference to our movements on the morrow, it became a matter of imperative necessity to get the poor things to where they could procure some food as soon as possible; I determined, therefore, to make for the hills, whatever they might be, at early dawn. The night was exceedingly cold, the thermometer falling to freezing point. At day-break there was a heavy fog, so we did not mount until half-past six, when the atmosphere was clearer, the fog, having in some measure dispersed. We then proceeded, and for the first time since commencing the journey turned from the course 332°, or one of N.N.W. to one of due north, allowing 5° for easterly variation. My object was to gain the trees Mr. Browne had noticed, as soon as possible, but did not reach them until a quarter to ten. We then discovered that they lined a long muddy channel, in which was a good deal of water, but not a blade of vegetation anywhere to be seen. I turned back, therefore, to a small sandy rise, whereon we had observed a few tufts of grass, and allowed the animals to pick what they could. At this spot we were about a mile and a half from the hills, which stood before us, their character fully developed, and whatever hope we might have before encouraged of the probability of a change of country on this side of the desert, was at one glance dispelled. Had these hills been as barren as the wastes over which we had just passed, so as they had been of stone we should have hailed them with joy. But, no!—sandy ridges once more rose up in terrible array against us, although we had left the last full 50 miles behind, even the animals I think regarded them with dismay.

From the little rising ground on which we had stopped, we passed to the opposite side of the creek, which apparently fell to the east, and traversing a bare earthy plain, we soon afterwards found ourselves ascending one of the very hills we had been examining with so much anxiety through a glass the evening before. It was flanked on either side by other hills, that projected into and terminated on this plain, as those we had before seen terminated in the Stony Desert; and they looked, as I believe I have already remarked, like channel head-lands jutting into the sea, and gradually shutting each other out. The one we ascended was partly composed of clay and partly of sand; but the former, protruding in large masses, caused deep shadows to fall on the faces and gave the appearance of a rocky cliff to the whole formation, as viewed from a distance.

Broad and striking as were the features of the landscape over which the eye wandered from the summit of this hill, I have much difficulty in describing them.

Immediately beneath was the low region from which we had just ascended, occupying the line of the horizon from the north-east point, southwards, round to the west. Southward, and for some degrees on either side, a fine dark line met the sky; but to the north-east and south-west was a boundless extent of earthy plain. Here and there a solitary clump of trees appeared, and on the plain, at the distance of a mile to the eastward, were two moving specks, in the shape of native women gathering roots, but they saw us not, neither did we disturb them—their presence indicated that even these gloomy forbidding regions were not altogether uninhabited.

As the reader will, I have no doubt, remember, the sandy ridges on the S.E. side of the Desert were running at an angle of about 18° to the west of north, having gradually changed from the original direction of about 6° to the eastward of that point. I myself had marked this gradual change with interest, because it was strongly corroborative of my views as to the course the current I have supposed to have swept over the central parts of the continent must have taken, i.e. a course at right angles to the ridges. It is a remarkable fact that here, on the northern side of the Desert, and after an open interval of more than 50 miles, the same sand ridges should occur, running in parallel lines at the same angle as before, into the very heart of the interior, as if they absolutely were never to terminate. Here, on both sides of us, to the eastward and to the westward, they followed each other like the waves of the sea in endless succession, suddenly terminating as I have already observed on the vast plain into which they ran. What, I will ask, was I to conclude from these facts?—that the winds had formed these remarkable accumulations of sand, as straight as an arrow lying on the ground without a break in them for more than ninety miles at a stretch, and which we had already followed up for hundreds of miles, that is to say across six degrees of latitude? No! winds may indeed have assisted in shaping their outlines, but I cannot think, that these constituted the originating cause of their formation. They exhibit a regularity that water alone could have given, and to water, I believe, they plainly owe their first existence. It struck me then, and calmer reflection confirms the impression, that the whole of the low interior I had traversed was formerly a sea-bed, since raised from its sub-marine position by natural though hidden causes; that when this process of elevation so changed the state of things, as to make a continuous continent of that, which had been an archipelago of islands, a current would have passed across the central parts of it, the direction of which must have been parallel to the sandy ridges, and consequently from east to west, or nearly so—that also being the present dip of the interior, as I shall elsewhere prove. I further think, that the line of the Stony Desert being the lowest part of the interior, the current must there have swept along it with greater force, and have either made the breach in the sandy ridges now occupied by it, or have prevented their formation at the time when, under more favourable circumstances, they were thrown up on either side of it. I do not know if I am sufficiently clear in explanation, finding it difficult to lay down on paper all that crowds my own mind on this subject; neither can I, without destroying the interest my narrative may possess, now bring forward the arguments that gradually developed themselves in support of the foregoing hypothesis.

Although I had been unable to penetrate to the north-west of Lake Torrens, that basin appeared to me to have once formed part of the back waters of Spencer's Gulf; still I long kept in view the possibility of its being connected with some more central body of water. Having however gained a position so much higher to the north, and almost on the same meridian, and having crossed so remarkable a feature as the Stony Desert (which, as I suppose, was once the focus of a mighty current, to judge from its direction passing to the westward), I no longer encouraged hopes which, if realized, would have been of great advantage to me, or regretted the circumstances by which I was prevented from more fully examining the north-east and northern shores of Lake Torrens. I felt doubtful of the immediate proximity of an inland sea, although many circumstances combined to strengthen the impression on my mind that such a feature existed on the very ground over which we had made our way. I had assuredly put great credit on the statements of the solitary old man who visited the Depôt but his information as far as we could judge had turned out to be false; and I was half angry with myself for having been so credulous, well aware as I was of the exaggerations of the natives, and how little dependence can be placed on what they say.

CHAPTER IX.

FLOOD'S QUICK SIGHT—FOREST FULL OF BIRDS—NATIVE WELL—BIRDS COLLECT TO DRINK—DANGEROUS PLAIN—FLOOD'S HORSE LOST—SCARCITY OF WATER—TURN NORTHWARD—DISCOVER A LARGE CREEK—BRIGHT PROSPECTS—SUDDEN DISAPPOINTMENT—SALT LAGOON—SCARCITY OF WATER—SALT WATER CREEK—CHARACTER OF THE INTERIOR—FORCED TO TURN BACK—RISK OF ADVANCING—THE FURTHEST NORTH—RETURN TO AND EXAMINATION OF THE CREEK—PROCEED TO THE WESTWARD—DREADFUL COUNTRY—JOURNEY TO THE NORTH—AGAIN FORCED TO RETURN—NATIVES—STATION ON THE CREEK—CONCLUDING REMARKS.

REFLECTING on the singular character of the country below me, as I stood on the pointed termination of the ridge the party had just ascended, I could not but think how fortunate it was we had not found it in a wet state, for in such a case to cross it would have been impossible. I felt assured indeed, from the moment we set foot on it, that in the event of rain, while we should be in the more distant interior, return would be altogether impracticable, but we had neither time to pause on, or provide against, the consequences of any heavy fall that might have set in. I do not think that this flashed across the minds of any of the party excepting my own, who would not have been justified in leading men forward as I was doing, without weighing every probable chance of difficulty or success.

As the line of the sand ridges was nearly parallel to that of our course, we descended to a polygonum flat, and keeping the ridge upon our left, proceeded on a bearing of 342°, or on a N.N.W. course, up a kind of valley. Whilst thus riding leisurely along, Flood, whose eyes were always about him, noticed something dark moving in the bushes, to which he called our attention. It was a dark object, and was then perfectly stationary; as Flood however insisted that he saw it move, Mr.

Browne went forward to ascertain what it could be, when a native woman jumped up and ran away. She had squatted down and put a large trough before her, the more effectually to conceal her person, and must have been astonished at the quickness of our sight in discovering her. We were much amused at the figure she cut, but as she exhibited great alarm Mr. Browne refrained from following her; after getting to some distance she turned round to look at us, and then walked off at a more leisurely pace. At the distance of about four miles, the sandy ridge made a short turn, and we were obliged to cross over to the opposite side to preserve our course. On gaining the top of the ridge, we saw an open box-tree forest, and a small column of smoke rising up from amongst the trees, towards which we silently bent our steps. Our approach had however been noticed by the natives, who no doubt were at the place not a minute before, but had now fled. We then pushed on through the forest, the ground beneath our horses feet being destitute of vegetation, and the soil composed of a whitish clay, so peculiar to the flooded lands of the interior. The farther we entered the depths of the forest, the more did the notes of birds assail our ears. Cockatoos, parrots, calodera, pigeons, crows, etc., all made that solitude ring with their wild notes, and as (with the exception of the ducks on the southern side of the Stony Desert) we had not seen any of the feathered race for many days, we were now astonished at their numbers and variety. About an hour before sunset we arrived on the banks of a large creek, with a bed of couch grass, but no water. The appearance of this creek, however, was so promising that we momentarily expected to see a pond glittering before us, but rode on until sunset ere we arrived at a place, which had attracted our attention as we approached it. Somewhat to the right, but in the bed of the creek, there were two magnificent trees, the forest still extending back on either side. Beneath these trees there was a large mound of earth, that appeared to have been thrown up. On reaching the spot we discovered a well of very unusual dimensions, and as there was water in it, we halted for the night.

On a closer examination of the locality, this well appeared to be of great value to the inhabitants. It was 22 feet deep and 8 feet broad at

the top. There was a landing place, but no steps down to it, and a recess had been made to hold the water, which was slightly brackish, the rim of the basin being also incrusted with salt. Paths led from this spot to almost every point of the compass, and in walking along one to the left, I came on a village consisting of nineteen huts, but there were not any signs of recent occupation. Troughs and stones for grinding seed were lying about, with broken spears and shields, but it was evident that the inhabitants were now dispersed in other places, and only assembled here to collect the box-tree seeds, for small boughs of that tree were lying in heaps on the ground, and the trees themselves bore the marks of having been stripped. There were two or three huts in the village of large size, to each of which two smaller ones were attached, opening into its main apartment, but none of them had been left in such order as those I have already described.

It being the hour of sunset when we reached the well, the trees were crowded with birds of all kinds coming for water, and the reader may judge of the straits to which they were driven, when he learns that they dived down into so dark a chamber to procure the life-sustaining element it contained. The wildest birds of the forest were here obliged to yield to the wants of nature at any risk, but notwithstanding, they were exceedingly wary; and we shot only few cockatoos. The fact of there being so large a well at this point, (a work that must have required the united labour of a powerful tribe to complete), assured us that this distant part of the interior, however useless and forbidding to civilized man, was not without inhabitants, but at the same time it plainly indicated, that water must be scarce. Indeed, considering that the birds of the forest had powers of flight to go where they would, I could not but regard it as a most unfavourable sign, that so many had collected here. Had this well contained a sufficiency of water, it would have been of the utmost value to us, but there was not more than enough for our wants, so that, although I should gladly have halted for a day, as our horses were both ill and tired, necessity obliged me to continue my journey, and accordingly on the 29th we resumed our progress into the interior on our original course. At about a mile we broke through the forest, and entered an open earthy plain, such as I believe man never before crossed. Subject to be laid under water by the creek we had just left, and to the effects of an almost vertical sun, its surface was absolutely so rent and torn by solar heat, that there was scarcely room for the horses to tread, and they kept constantly slipping their hind feet into chasms from eight to ten feet deep, into which the earth fell with a hollow rumbling sound, as if into a grave. The poor horse in the cart had a sad task, and it surprised me, how we all at length got safely over the plain, which was between five and six miles in breadth, but we managed it, and at that distance found ourselves on the banks of another creek, in the bed of which there was plenty of grass but no water. I was however exceedingly anxious to give the horses a day's rest; for several of them were seriously griped, and had either taken something that disagreed with them, or were beginning to suffer from constant work and irregularity of food. Mr. Browne too was unwell and Lewis complaining, so that it was advisable to indulge ourselves if possible. I therefore determined to trace the creek downwards, in the hope of finding water, and at a mile came upon a shallow pond where I gladly halted, for by this time several of the horses had swollen to a great size, and were evidently in much pain.

After arranging the little bivouac our attention was turned to the horses, and Mr. Browne found it necessary to bleed Flood's horse, to allay the inflammatory symptoms that were upon him. Still however he got worse, and no remedy we had in our power to apply seemed to do him good. The poor animal threw himself down violently on the ground, and bruised himself all over, so that we were obliged to fasten him up, but as there appeared to be no fear of wandering, at sunset he was allowed to be loose. He remained near me for the greater part the night, and was last seen close to where I was lying, but in the morning was no where to be found, and although we searched for a whole day, and made extensive sweeps to get on his track we never saw him more, and concluded he had died under some bush. This was the horse we recovered on the Murray, the same that had escaped from the government paddock in Adelaide. The other animals had in some measure recovered, and the additional day of rest they got while we were searching for Flood's horse, enabled me to resume my journey on the last day of August. Our course being one of 335° to the west of north, or nearly N.N.W., and that of the sandy ridges being 340° we necessarily crossed them at a very acute angle, and the horses suffered a good deal. In the afternoon we travelled over large bare plains, of a most difficult and distressing kind, the ground absolutely yawning underneath us, perfectly destitute of vegetation, and denuded of timber, excepting here, and there, where a stunted box-tree was to be seen. While on the sand hills, the general covering which was spinifex, there were a few hakea and low shrubs. On such ground as that whereon we were travelling, it would have been hopeless to look for water, nevertheless our search was constant, but we were obliged to halt without having found any, and to make ourselves as comfortable as we could. All the surface water left by the July rain had entirely disappeared, and what now remained even in the creeks was muddy and thick. It was indeed at the best most disgusting beverage, nor would boiling cause any great sediment. Every here and there, as we travelled along, we passed some holes scooped out by the natives to catch rain, and in some of these there was still a muddy residuum; we moreover observed that the inhabitants of this desert made these holes in places the best adapted to their purpose, where if the slightest shower occurred, the water falling on hard clay would necessarily run into them.

The circumstances under which we halted in the evening of the 31st of August were very embarrassing. It was evident that the country into which we were now advancing, was drier and more difficult than the country we had left behind. It was impossible, indeed, to hope that the animals would get on, if it should continue as we had found it thus far. There were numerous high ridges of sand to the westward, in addition to those on the plains, and so full of holes and chasms were the latter, that the horses would soon have been placed *hors de combât,* if they had continued to traverse them. Moreover, I could not but foresee that unless I used great precaution our retreat would be infallibly cut off. Whatever water we had passed, since the morning we commenced our journey over the Stony Desert, was not to be depended upon for more than four or five days, and although we might reckon with some certainty on the native well in the box-tree forest, the supply it had yielded was so very small that we could not expect to obtain more from it than would suffice ourselves and one or two of the horses. Taking all these matters into consideration, I determined on once more turning to the north for a day or two, in order that by keeping along the flats, close under the ridges, I might get firmer travelling for the cart, and in the expectation, that we should be more likely to find water in thus doing, than by crossing the succession of ridges. Accordingly, on the 1st of September, we started on a course of 6° to the west of north, or a N.1/2W. course, that allowing for variation, being within 1 1/2 points of a due north course. On this we went up the flat where we had slept. By keeping close to the ridges we found, as I had anticipated, firmer ground, though the centre of the flat was still of the worst description. There were a few small box-trees to be seen as we passed along, but scarcely any minor vegetation. At about nine miles we were attracted by the green appearance of some low polygonum bushes, to which we went, and under them found two small puddles of water, that we might easily have passed. They must have been three feet deep after the rains, but were now barely five inches, and about the size of a loo table. However, we had no choice, and as the horse had suffered so much from the rickety motion of the cart, caused by the inequalities of the ground, and there was a silky kind of grass growing sparingly around, I stopped here for the rest of the day to effect necessary repairs. When, however, we came to examine the wheels, we found that so many of the spokes were shivered and had shrunk, that Lewis got on but slowly, renewing only such as were found absolutely useless; we were consequently detained at this point another day, but on the 3rd resumed our journey up the flat, and at two miles crossed a small sandy ridge into the opposite flat, and at five miles stopped at a second ridge of some height for Lewis and Joseph, who were a good way behind with the cart. On coming up, they informed us that they had fallen in with a tribe of natives, twelve in number, shortly after starting, and had remained some time with them. They were at a dirty puddle, such as we had left, and were at no great distance from our little bivouac. Joseph good-naturedly gave one of them his knife, but he could not understand a word they said.

After crossing the sand ridge, we kept on the edge of the flats, as I have said, for the sake of the horses. The ridges had now become very long, and varied in breadth from a few hundred yards to a mile. Box-trees were scattered over them, and, although generally bare, they were not altogether destitute of grass, or herbage; the ridges of sand, on the contrary, still continued unbroken, and several were covered with spinifex; but on the whole the country appeared to be improving, and the fall of waters being decidedly somewhat to the eastward of south, or towards the Stony Desert, I entertained hopes that we had crossed the lowest part of the interior, and reached the southerly drainage. We were again fortunate in coming on another pond at 20 miles, where we halted, the country round about us wearing an improved appearance. Still our situation was very precarious, and we were risking a great deal by thus pushing forward, for although I call the hollows (in which we found the water) ponds, they were strictly speaking the dregs, only of what had been such, and were thick, black, and muddy; but the present aspect of the country led us to hope for a favourable change, and on the morning of the 4th we still held our northerly course up the flat, on which we had travelled the greater part of the day before. As we advanced, it became more open and grassy, and at three miles we found a small supply of very tolerable water in the bed of a shallow watercourse. We had ridden about ten miles from the place where we had slept, and Mr. Browne and I were talking together, when Flood, who was some little distance a-head, held up his hat and called out to us. We were quite sure from this circumstance that he had seen something unusual, and on riding up were astonished at finding ourselves on the banks of a beautiful creek, the bed of which was full both of water and grass. The bank on our side was twenty feet high, and shelved too rapidly to admit of our taking the horses down, but the opposite bank was comparatively low.

Immediately within view were two large sheets of water around the margin of which reeds were growing, but nevertheless these ponds were exceedingly shallow. The direction of this fine watercourse was N. by W. and S. by E., coming from the first and falling to the last point, thus enabling us to trace it up without changing our own. A little above where we intersected its channel two small tributaries join it, (or, I am more inclined to think, two small branches go from it; for we had apparently been rising as we came up the valley, but more especially as the direction from which they appeared to come to the S.W.), was almost opposite to the course of the creek itself. On proceeding upwards we observed that there were considerable intervals, along which the channel of the creek was dry; but where such was the case, it was abundantly covered with couch grass, of which the horses were exceedingly fond. We passed several sheets of water, however, some of which had a depth of two feet, although the greater number were shallow. After following it for ten miles, we halted with brighter prospects, and under more cheering circumstances than we had any right to anticipate; but, although the creek promised so well, the valley on either side it was more than usually barren and scrubby, was bounded in, as usual, by high ridges of sand that still continued to head us in unbroken lines, and were the most prominent and prevailing feature of the interior; and although we were now within two degrees of the Tropics, our latitude at this point being 25°34'19", we had not as yet observed the slightest change in the vegetation, or anything to intimate our approach to a tropical country.

On the 5th we started on a course of 340°, the upward course of the creek. At two miles it turned to the N.E, but soon came round again to N.W., and afterwards kept a general course of 10° to the west of north. Its channel gradually contracted as we advanced, and the polygonum grew to the size of a very large bush upon its banks. At nine miles we arrived at a creek junction from the S.W. and traced it over grassy plains, on which some Bauhinia were growing, but finding that it took its rise in a kind of marsh occupying the centre of the plain into which it had led us, we turned away to the main creek. The country now became more open, and tertiary limestone shewed itself on the plains, and at a short distance from the creek a vein of milky quartz cropped out near a pretty sheet of water. As we proceeded upwards sandstone traversed its bed in several places; in some degree contracting its channel. A short time before we halted we passed a very large and long sheet of water, on which there were a good many wild fowl, so very shy, that although the brush grew close to the banks of the creek, so as to favour our creeping upon them, we could not shoot any.

Notwithstanding that the creek had thus changed its appearance from what it was where we first came upon it (its waters being muddy with less grass in its channel), we had no reason to suppose that it would disappoint our hopes; we therefore resumed our journey on the morning of the 6th, without any idea that we should meet with any check in the course of the day. As the immediate neighbourhood of this creek had become scrubby, we kept wide of it and travelled for 12 miles, on a bearing of 34°, over flats destitute of all manner of vegetation, but thinly scattered over with the box, acacia and the Bauhinia. These flats were still bounded on either side by high sandy ridges, covered with spinifex, excepting on their summits, which were perfectly bare. The view from them both to the eastward and westward was, as it were, over a sandy sea; ridge after ridge succeeding each other as far as the eye could stretch the vision. To the north the flat appeared to terminate at a low sand hill bearing 335° or N.N.W.1/2W.

When we again came on the creek, there was an abundance both of water and grass in its bed, but just above, the channel suddenly turned to the N.E. and in again keeping wide of it to avoid the inequalities of the ground, we arrived at the little sand hill that had previously bounded our view, and on ascending it, found that immediately beneath us, there was a clear small lake, covered with wild fowl. The colour of the water immediately betrayed its quality, and we found on tasting that it was too salt to drink. An extensive grassy flat extended to the westward of the lake, bounded by box-trees, and the channel of the creek still held its course to the N.E. I could not therefore but suppose, that this was a junction from that point, and therefore determined on passing to the opposite side, in anticipation that I should again come on our old friend amidst the trees. We accordingly crossed at the bottom of the little lake, and in so doing found amidst the other herbage two withered stalks of millet.

The grassy woodland continued for several miles, and as it was evidently subject to flood, we were in momentary expectation of seeing a denser mass of foliage before us, as indicating the course of the creek, but we suddenly debouched upon open plains, bounded by distant sand hills. There was not now a tree to be seen, but samphire bushes were mixed with the polygonum growing round about; as the changes however in this singular and anomalous region had been so sudden and instantaneous, I still held on my course, but the farther I advanced into the plains the more did the ground betray a salt formation.

We halted an hour after sunset, under a sand hill about 16 miles distant from the creek, without having succeeded in our search for water, for although we passed several muddy pools at which the birds still continued to drink they were too thick for our animals.

The prospect from the top of the sand hill under which we had formed our bivouac, was the most cheerless and I may add the most forbidding of any that our eyes had wandered over, during this long and anxious journey. To the west and north-west there were lines of heavy sand ridges, so steep and rugged as to deter me from any attempt to cross them with my jaded horses. To the north and north-east a dark green plain covered with samphire bushes (amidst which the dry beds of small salt lagoons, as white as snow, formed a singular and striking contrast) was to be seen extending for about eight miles. This plain was bounded by distant hills, the bright red tops of which gleamed, even in the twilight. I was here really puzzled what course to pursue, one only indeed was open to me—the north—unless I should determine to fall back on the creek; but I thought it better to advance, in the hope of being able to maintain my ground, and with the intention of halting for a few days at the first favourable point at which we should arrive, for my mind was filled with anxiety. It had pained me for some time, to see Mr. Browne daily suffering more and more, and although he continued to render me the most valuable assistance, a gloom hung over him; he seldom spoke, his hands were constantly behind him, pressing or supporting his back, and he appeared unfit to ride. My men were also beginning to feel the effects of constant exposure, of ceaseless journeying, and of poverty of food, for all we had was 5lbs. of flour and 2 oz. of tea per week; it is true we occasionally shot a pigeon or a duck, but the wildness of the birds of all kinds was perfectly unaccountable. The horses living chiefly on pulpy vegetation had little stamina, and were incapable of enduring much privation or hardship. No rain had fallen since July, nor was there any present indication of a change. Much as I desired it, I yet dreaded having to traverse such a country as that into which I was now about to plunge, in a wet state. With a soil of still tenacious clay, already soft from the moisture produced by the mixture of salt in it, I foresaw that in the event of heavy rain, I should be involved in almost inextricable difficulties, but there was no alternative.

On the morning of the 7th I sent Mr. Browne to the westward, to ascertain the nature of the country, and if by any chance he could again find the creek, and in case I had inadvertently mistaken the real creek for a tributary, I myself pushed on to the north, in the hope of intersecting it. Mr. Browne had not, however, been absent more than three-quarters of an hour, when he returned to inform me that he had been stopped by a salt creek, coming direct from the north, the bed of which was too soft for him to cross. He said that its channel was white as snow, and that every reed and blade of grass on its banks, was encrusted with salt. Under an impression that as long as I should continue in the neighbourhood of, and on a course nearly parallel to this creek, I could not hope for any favourable change, I decided on crossing it, and with that view turned to the west; but finding the bed of the creek still too soft to admit of our doing so, I traced it upwards to the north, along a sandy ridge.

As Mr. Browne had informed me, its channel was glittering white, and thickly encrusted with salt, nor was there any water visible, but on going down to examine it in several places where the salt had the appearance of broken and rotten ice, we found that there were deep pools of perfect brine underneath, on which the salt floated, to the thickness of three or four inches. The marks of flood on the side of the sand hill shewed a rise of 12 feet above its ordinary level. At about a mile and a half we descended the sand hill on which we had previously kept, and ascended another, when we saw the basin of the creek immediately below us, but quite dry, and surrounded by sand hills. Crossing just below it, we proceeded on a course of 331° over extensive plains, covered with samphire, excepting where the beds of dry salt lagoons occurred. The ground was spongy and soft, and the cart wheels consequently sank deep into it. The plain was surrounded on all sides by sand hills, and that towards which we were advancing appeared to run athwart our course instead of nearly parallel to it as heretofore. On gaining the summit, we found that other ridges extended from it in parallel lines, the ridge on which we stood forming the head of the respective valleys. A line of acacia, a species we had never found near water, was growing down the centre of each, and the fall of the country seemed again to be to the N.N.W.

Pushing down one of the valleys, the descent of which was very gradual, and keeping on such clear ground as there was, the ridges rose higher and higher on either side of us as we advanced, all grass and other vegetation disappeared, and at length both valley and sand ridge became thickly coated with spinifex.

At noon I halted, in the hope of obtaining a meridian altitude, but was disappointed, as also at night, the sky continuing obscured. At half-past two I pulled up, to consider whether or not it would be prudent to push on any farther. I calculated that we were now 34 miles from the creek, our only place of refuge. The horses had not tasted water from the early part of the day before, and we could not reasonably expect to get back to the salt lagoon under a day and a half. Our poor animals were not in a condition to endure much fatigue, although by going on steadily we had managed to get over a good deal of ground. It is, however, probable that I should not have had much consideration for them on this occasion, if other matters had not weighed on my mind and influenced my decision. My men were all three unwell, and had been so for some days prior to this, and Mr. Browne's sufferings were such that I hesitated subjecting him to exertions greater than those he was necessarily obliged to submit to, and by which I felt assured he would ultimately be overcome. The treacherous character of the disease by which he had been attacked was well understood. I had no hope of any improvement in his condition until such time as he could procure change of food. So far from this I dreaded every day that he might be laid prostrate as Mr. Poole had been, that I should have to carry him about in a state of helplessness, and that he would ultimately sink as his unfortunate companion had done. Had other considerations, therefore, not influenced me, I could not make up my mind to persevere, and see my only remaining companion perish at my side, and that, too, under the most trying, I had almost said the most appalling circumstances, for no one who has not seen the scurvy in its worst character can form an idea of it. I could not run the risk of being obliged to lay and leave one, in that gloomy desert, whose attention and kindness to me had been uniform, and whose life I knew was valuable to very many. The time has now passed, and I thank God that Mr. Browne, who embarked in this expedition in reliance on my discretion, is now restored to health and strength; but although he has regained his elasticity of spirits, and would, I have no doubt, again encounter even the same risks, he will yet remember Central Australia, and all that both of us there suffered. The question for me however was, how far I should be justified in pushing forward under the almost certainty of inextricable embarrassment. I was now within reach of water, but another fifteen miles would have put it out of my reach; and though I felt I had the power, I did not see the advantage of perseverance, with so many difficulties staring me in the face. Our distance from the creek may appear to be short; but it will be borne in mind that our horses had now been more than a year living upon dry grass and salsolaceous plants, that from the time of our leaving the Depôt, they had been ridden from sunrise to sunset; and that at night they had been tethered and confined to a certain range, within which there was not sufficient for them to eat. They had already been too long without water or food, and therefore that which would have been a trifling journey to them under ordinary circumstances, under existing ones was beyond their strength. Nevertheless, though thus convincing my understanding, I felt that it required greater moral firmness to determine me to retrace my steps than to proceed onwards.

Regarding our situation in its most favourable point of view, my advancing would have been attended with extreme risk. If I had advanced, and had found water, all would have been well for the time at least—if not, the extent of our misfortunes would only have been tested by their results. The first would have been the certain loss of all our horses, and I know not if one of us would ever have returned to the Depôt, then more than 400 miles distant, to tell the fate of his companions to those we had left there. On mature deliberation then, I resolved to fall back on the creek, and as my progress was arrested in this direction, to make that the centre of my movements, in trying, every other point where I thought there might be a chance of success.

I saw clearly indeed that there was no help for this measure. We had penetrated to a point at which water and feed had both failed. Spinifex and a new species of mesembryanthemum, with light pink flowers on a slender stalk, were the only plants growing in that wilderness, if I except a few withered acacia trees about four feet high. The spinifex was close and matted, and the horses were obliged to lift their feet straight up to avoid its sharp points. From the summit of a sandy undulation close upon our right, we saw that the ridges extended northwards in parallel lines beyond the range of vision, and appeared as if interminable. To the eastward and westward they succeeded each other like the waves of the sea. The sand was of a deep red colour, and a bright narrow line of it marked the top of each ridge, amidst the sickly pink and glaucous coloured vegetation around. I fear I have already wearied the reader by a description of such scenes, but he may form some idea of the one now placed before him, when I state, that, familiar as we had been to such, my companion involuntarily uttered an exclamation of amazement when he first glanced his eye over it. "Good Heavens," said he, "did ever man see such country!" Indeed, if it was not so gloomy, it was more difficult than the Stony Desert itself; yet I turned from it with a feeling of bitter disappointment. I was at that moment scarcely a degree from the Tropic, and within 150 miles of the centre of the continent. If I had gained that spot my task would have been performed, my most earnest wish would have been gratified, but for some wise purpose this was denied to me; yet I may truly say, that I should not thus have abandoned my position, if it had not been a measure of urgent and imperative necessity.

After what I have said, the feelings with which, on the morning of the 8th, we unloosed our horses from the bushes, to which they had all night been fastened, will easily be imagined. Just as we were about to mount, a flight of crested parroquets on rapid wing and with loud shriek flew over us, coming directly from the north, and making for the creek to which we were going—it was a singular occurrence just at that moment, and so I regarded it, for I had well nigh turned again. It proved, however, that to the very last, we had followed the line of migration with unerring precision. What would I not have given for the powers of those swift wanderers of the air? But as it was I knew not how long they had been on the wing, or how far it was to the spot where they had last rested.

We passed the salt lagoon about 10 A.M. of the 9th, and stopped at a shallow but fresh water pond, a little below it, no less thankful than our exhausted animals that we were relieved from want, and the anxiety attendant on the last few days. On passing the lagoon we saw two natives digging for roots, but did not disturb them. In the afternoon, however, Joseph and Lewis saw twenty, who exhibited some unfriendly symptoms, and would not allow them to approach. They were not armed, but carried red bags. The food of the natives here, as in other parts of the interior, appeared to be seeds of various kinds. They had even been amongst the spinifex gathering the seed of the mesembryanthemum, of which they must obtain an abundant harvest. The weather, a little before this time, had been very cold, but was now getting warmer every day. As we had been advancing northwards towards the Tropics, I was not surprised at this. The sky also was clear, generally speaking, but we had observed for the last two or three months that it was invariably more cloudy at the full of the moon than at any other period.

As our recent journey proved that in going to the westward on the 5th inst., we had wandered from the creek, and that instead of holding on in that direction, it had changed its course considerably to the eastward of north, I determined, after we should all have had a day of rest, to trace the channel upwards, in order to satisfy myself as to what became of it. On the 10th, therefore, Mr. Browne and myself with Flood, mounted our horses, with the intention of tracing it up until we should have ascertained to what point it led. We passed through some very pretty scenery in the proximity of the lagoon where it was lightly wooded, with an abundance of grass; and I could not help reflecting with how much more buoyant and pleasurable feelings we should have explored such a country, when compared with the monotonous and sterile region we had wandered over. The transition however from the rich to the barren, from the picturesque to the contrary, was instantaneous. From the grassy woodland we had been riding through, we debouched upon a barren plain without any vegetation, and after crossing a small channel, intersected a second much larger, a little beyond it. Both creeks evidently traversed different parts of a large plain to the north, to which they had no apparent inlet. There was a long tongue of sand, rather elevated, and running up into the plain, to the termination of which we rode, and then found ourselves, as it were, in the centre of an area, that was of great extent, and appeared to be bounded on all sides, excepting that by which we had entered, by sand hills. Unconnected lines of trees marked the courses of the channels traversing it in different directions, but as the evening had far advanced, and my object had been rather to look round about me than to make any lengthened excursion, we returned to our little bivouac, with the intention of devoting another day to the fuller examination of the neighbourhood.

On the following day I proceeded with the whole party to the westward, anticipating that the salt formation existing to the north-west was merely local, and that by thus turning a few degrees from the course on which we had before gone, we should altogether avoid it. I should not, however, have taken Joseph and Lewis with the cart, if I had not been somewhat apprehensive that the natives might visit the camp during my absence, and some misunderstanding be the consequence; for as we had hitherto found the country to the westward worse than at any other point, I was after all doubtful how far I should be able to push on.

We left the creek on a W. by N. course, the direction of the sandy ridges being to the N.N.W., so that we were obliged to cross them successively. I soon found that the country was infinitely worse than I expected. We had scarcely passed a kind of marsh at some little distance from the creek, when we once more crossed salty valleys, between high sandy ridges. The wind blowing fresh from the south, peppered us with showers of sand as we ascended the last, and carried the salt in the valleys like drifting snow from one end of them to the other, filling our eyes and entering the pores of the skin, so as to cause us much annoyance. Before noon we had crossed eighteen of these sandy undulations, and were on the top of another, having fairly tired the horses in the ascent, and I consequently pulled up, to wait for the cart, but the heavy nature of the country had so shaken it, that the men were obliged to stop; and on examining the spokes of the wheels, I really wondered how they could have got on so far, and expected that in another half mile every one of them would be shaken out, and the cart itself fall to the ground. The spokes had shrunk to such a degree that they did not hold in the felloes and axles by more than two or three 10ths of an inch. I felt it necessary therefore to turn back to the creek, to get new spokes of such wood as we could procure, there not being a tree of any kind visible near us; but it was late ere we got back to water, and once more took up our position on the same ground we had quitted in the morning. The country we had passed was certainly such as to deter me from making a second attempt in the same quarter, and to confirm my impression that from some cause or other the interior to the westward was worse than anywhere else. Lewis, the moment we got back to the creek, set to work in good earnest, with Joseph's assistance, to repair the cart, but it necessarily delayed us longer than prudence would have allowed; in the meantime, however, we were at least deriving benefit from rest.

On mature consideration, I thought the quarter in which we should have most chance of success would be a course a little to the east of north, for the day Mr. Browne and I rode up the creek it appeared to me that the country was more open in that direction. I thought it better, however, to make for the sandy tongue of land in the centre of the plain, in which the creek appeared to take its rise, and to be guided by circumstances both in the examination of that plain, and the course I should ultimately pursue. The cart being fit for use on the morning of the 12th we again left the creek, and at four miles on an east by north course arrived at the sand hill to which I desired to go; from that point I proceeded to the N.N.W., that appearing to be the general direction of the creek upwards; but as there were lines of box-trees on both sides of us, those to our left being denser than the right, I moved for them over a plain of about five miles in breadth, but so full of cracks and fissures that we had great difficulty in crossing it. Notwithstanding, however, that the cart fell constantly into them, we got it safely over. Not finding any water under or near the trees I turned a little to the north, keeping wide of the creek; but, coming on its channel again at five miles, I halted, because there happened to be a little grass there, and we were fortunate enough, after some perseverance, to find a muddy puddle that served the horses, however unfit for our use. From the appearance of the plain before us, I hardly anticipated success in our undertaking. We had evidently arrived near the head of the creek, and I felt assured that if the features of the country here, were similar to those of other parts of the interior, we should, between where we then were, and some distant sand hills, again find ourselves travelling over a salt formation. The evening had closed in with a cloudy sky, and the wind at W.N.W., and during the night we had two or three flying showers, but they were really in mockery of rain, nor was any vestige of it to be seen in the morning, which broke with a clear sky, and the wind from the S.E.

As soon as morning dawned we saddled our horses and made for the head of the plain, crossing bare and heavy ground until we neared the sand hills, when observing that I was leaving the creek, which I was anxious to trace up, we turned to the north-east for a line of gum-trees, but the channel was scarcely perceptible under them, and we had evidently run it out. There were only two or three solitary trees to be seen to the north, at which point the plain was bounded by sand hills. To the S.E. there was a short line of trees, from the midst of which the natives were throwing up a signal smoke, but as it would have taken me out of my way to have gone to them, I held on a N.N.W. course, and at the termination of the plain ascended a sand hill, though of no great height. From it we descended a small valley, the sides of which were covered with samphire bushes, and the bottom by the dry white and shallow bed of a salt lagoon. From this valley we passed into a plain, in which various kinds of salsolaceous productions were growing round shallow salty basins. At a little distance from these, however, we stumbled upon a channel with some tolerable water in it, hid amongst rhagodia bushes, but the horses refused to drink. This plain communicated with that we had just left, round the N.E. point of the sand hill we had crossed but there were no box-trees on it to mark the line of any creek or water; but the sand ridge forming its northern boundary was very high, and contrary to their usual lay, ran directly across our course, and as the ascent was long and gradual, so was it some time before we got to the top. The view which then presented itself was precisely similar to the one I have already described, and from which we had before been obliged to retreat. Long parallel lines of sandy ridges ran up northwards, further than we could see, and rose in the same manner on either side. Their sides were covered with spinifex, but there was a clear space at the bottom of the valleys, and as there was really no choice we proceeded down one of them, for 12 miles, and then halted.

At this point the open space at the bottom of the valleys had all closed in, and the cart, during the latter part of the journey, had gone jolting over the tufts and circles of spinifex to the great distress of the horse; grass and water had both failed, nor could I see the remotest chance of any change in the character of the country. It was clear, indeed, that until rains should fall it was perfectly impracticable; and with such a conviction on my mind, I felt that it would only be endangering the lives of those who were with me, if I persevered in advancing. I therefore once more determined to fall back upon the creek, there to hold my ground until such time as it should please God to send us rain. We re-entered the plain in which the creek rises at 3 P.M., and made for the trees, from whence the signal smoke was rising, and there came on a tolerable sized pond of water, at which we stopped for a short time, and while resting, ascertained that some natives were encamped at a little distance above us; but although we went to them, and endeavoured by signs and other means to obtain information, we could not succeed, they either did not or would not understand us; neither, although our manner must have allayed any fear of personal injury to themselves, did they evince the slightest curiosity, or move, or even look up when we left them. I cannot, however, think that such apparent indifference arises from a want of feeling, for that, on some points, they possess in a strong degree; but so it was, that the natives of the interior never approached our camps, however much we might encourage them. On leaving these people, of whom, if I recollect, there were seven, we tried to avoid the distressing plains we had crossed in the morning, and it was consequently late before we got to the creek and dismounted from our horses, after a journey of about 42 miles. The 13th thus found us beaten back by difficulties such as were not to be overcome by human perseverance. I had returned to the creek with the intention of abiding the fall of rain, and was not without hopes that it would have gladdened us, for the sky about this time was very cloudy, and anywhere else but in the low country in which we were, rain most assuredly would have fallen. As it was, the clouds passed over us without breaking.

A lunar we here obtained placed us in longitude 138°15'31"E., our latitude being 25°4'0"S. Computed from these data I deem I may fairly assume we were in 24°40'00"S., and on the 138th meridian, when we stopped on the 8th; being then 470 geographical miles to the north of Mount Arden, about 350 from Mount Hopeless, and rather more than midway between the first of those hills and the Gulf of Carpentaria. My readers will perhaps bear in mind, that the object of this expedition was limited "to ascertaining the existence and the character of a supposed chain of hills, or a succession of separate hills, trending down from N.E. to S.W. and forming a great natural division of the continent." I hope I do not take too much credit to myself, if I say that I have set that question at rest; and that, considering the nature of the country into which I penetrated, no such chain can reasonably be supposed to exist. If, indeed, any mountains had really been in the direction specified, it appears to me that I must have discovered them, but, as far as my poor opinion goes, I think the sandy ridges, both I and my readers have so much reason to hold in dread, are as extensive on one side of the Stony Desert as the other. In truth, I believe, that not only is such the case, but that the same region extends with undiminished breadth even to the great Australian Bight, which occupies a space along the south coast of the continent, as nearly as may be of equal breadth with the sea-born Desert itself; and I cannot but conclude that that remarkable wall, shewing a perpendicular front to the ocean, but sloping inwards from the coast, was thrown up simultaneously with the fossil bed of the

Murray, during the time those convulsions, by which the changes in the central parts of the continent, to which I have already called attention, were going on. But I venture to give these opinions with extreme diffidence; they may be contrary to general views on the subject. I merely record my own impressions from what I have observed, in the hope that I may assist the geologist in his inferences. The ideas I would desire to convey are clear enough in my own mind, but I must confess that I feel a great difficulty in placing them so forcibly and so clearly before my readers as I could desire.

CHAPTER X.

REFLECTIONS ON OUR DIFFICULTIES—COMMENCE THE RETREAT—EYRE'S CREEK—PASS THE NATIVE WELL—ACROSS THE STONY DESERT—FIND ANOTHER WELL WITHOUT WATER—NATIVES—SUCCESSFUL FISHING—VALUE OF SHEEP—DECIDE ON A RETREAT—PROPOSE THAT MR. BROWNE SHOULD LEAVE—HIS REFUSAL TO DESERT THE PARTY—MR. BROWNE'S DECISION—PREPARE TO LEAVE THE CAMP—REMARKS ON THE CLIMATE—AGAIN LEAVE THE DEPÔT—SINGULAR EXPLOSION—DISCOVER A LARGE CREEK—PROCEED TO THE NORTH—RECURRENCE OF SAND RIDGES—SALT WATER LAKE—AGAIN STRIKE THE STONY DESERT—ATTEMPT TO CROSS IT.

TO THAT man who is really earnest in the performance of his duty to the last, and who has set his heart on the accomplishment of a great object, the attainment of which would place his name high up in the roll of Fame; to him who had well nigh reached the topmost step of the ladder, and whose hand had all but grasped the pinnacle, the necessity must be great, and the struggle of feeling severe, that forces him to bear back, and abandon his task.

Let any man lay the map of Australia before him, and regard the blank upon its surface, and then let me ask him if it would not be an honourable achievement to be the first to place foot in its centre.
Men of undoubted perseverance and energy in vain had tried to work their way to that distant and shrouded spot. A veil hung over Central Australia that could neither be pierced or raised. Girt round about by deserts, it almost appeared as if Nature had intentionally closed it upon civilized man, that she might have one domain on the earth's wide field over which the savage might roam in freedom.

I had traced down almost every inland river of the continent, and had followed their courses for hundreds of miles, but, they had not led me to its central regions. I had run the Castlereagh, the Macquarie, the

Lachlan, the Murrumbidgee, the Hume, the Darling, and the Murray down to their respective terminations, but beyond them I had not passed—yet—I looked upon Central Australia as a legitimate field, to explore which no man had a greater claim than myself, and the first wish of my heart was to close my services in the cause of Geography by dispelling the mists that hung over it.

True it is that my friend Eyre had penetrated high up to the north of Mount Arden, and there can be no doubt but that his ardent and chivalrous spirit would have carried him far beyond the point he attained, if he had not met unconquerable difficulties. I thought that a cooler and more leisurely progress would enable me to feel my way into a country whose inhospitable character developed itself more the more it was penetrated. I had adopted certain opinions, the correctness of which I was anxious to test, and I thought the investigations I desired to make, were not only worthy the pursuit of private ambition, but deserving the attention of Her Majesty's Government. With these feelings I could not but be grateful to Lord Stanley, for having entertained my proposition, and given me an opportunity to distinguish myself. It is not because his Lordship is no longer at the head of the Colonial Office, that I should refrain from making my acknowledgments to him, and expressing the sense I entertain of the obligation under which he has laid me. It so happened that the course pointed out to me by Lord Stanley, and that in which I desired to go, were the same, and I had hoped that in following up my instructions, I should ultimately have gained the spot I so ardently desired to reach, and to have left the flag of my native country flying over it.

The feelings then with which I returned to the creek after the failure of our last attempt to penetrate to the north may well be imagined. I returned to it, as I have said, with perhaps a sullen determination to stand out the drought; but, on calm reflection, I found that I could not do so. I could not indeed hide from myself that in the course of a few days my retreat to the Depôt would unavoidably be cut off if rain should not fall. Looking to the chance of our being delayed until our provisions should be consumed, and to the fact that we could not expect to get back to the

Depôt in less than three weeks, and that I could not hope for any amendment either in Mr. Browne or my men, so long as they were confined to the scanty diet we then had. I determined on my return to the Park, thence to take out fresh hands, and to make another attempt to penetrate across the Desert in some other direction; but, as this measure, like our detention at the Depôt, would involve a great loss of time, I proposed to myself again to divide the party, and to send Mr. Browne home with all the men, except Mr. Stuart and two others. I saw no objection to such a course, and certainly did not anticipate any opposition to it on the part of my companion. I resolved then, with a due regard to his state, to retrace my steps with all possible expedition; and, accordingly, directed that everything should be prepared for our retreat on the morning of the 14th, for the sky had cleared, and all prospect of rain had again vanished. Although we were here so close to the Tropic, the climate was not oppressive. The general temperature after noon was 84°, the morning 46°. The prevailing wind was from S.S.E. to E.S.E. and it was invariably cold; at least we felt it so, and I regretted to observe, that in Mr. Browne's case it caused a renewed attack of violent pains in the muscles and joints, from which he had before been somewhat free. It is also remarkable, that up to this distant point, no material change had taken place in the character of the vegetation; with the exception of the few trees and plants I have mentioned the herbage of these sterile regions, and of the Darling were essentially the same, only with this difference, that here they were all more or less stunted, whereas, in the month of October, when we passed up the Darling, they were only just flowering, now in the month of September they had ripened their seed.

Before we commenced our journey back to the Depôt, I named this "Eyre's Creek." No doubt it is an important feature in the country where it exists. Like the other creeks, however, it rises in plains, and either terminates in such or falls into the Stony Desert. There can be no doubt, however, that to any one desiring to cross the continent to the north, Eyre's Creek would afford great facilities; and if the traveller happened fortunately to arrive on it at a favourable moment he would have every chance of success.

For twelve miles below the salt lagoon there is not a blade of grass either in the bed of the creek or on the neighbouring flats, the soil of both being a stiff cold clay. We passed this ungenial line, therefore, and encamped near a fine pool of water, where both our own wants and those of our horses, as far as feed and water went, were abundantly supplied.

In going along one of the flats, before we discovered the creek, Mr. Browne and I had chased a Dipus into a hollow log, and there secured it. This pretty animal we put into a box; but as it appeared to eat but little grass, we gave it some small birds, which it always devoured at night. Our dogs had killed one on the banks of the Darling, but had so mutilated it, that we could not preserve it. We hoped, however, to keep this animal alive, and up to the present time there was every chance of our doing so. It was an exceedingly pretty animal, of a light grey colour, having a long tail, feathered at the end, insectivorous, and not marsupial. On the 16th we turned from the creek to the south, and passed down the long flat up which we had previously come. On the following day we passed several of the hollows scraped by the natives, and in one of them found a little water, that must have accumulated in it from the drizzly showers that fell on the night of the 8th, and which might have been heavier here than with us. On the 19th we arrived at the creek where Flood's horse was lost, but could not make out any track to betray that he had been to water, and as there was not enough remaining in the pond for our use, we crossed the plain, over which we had had so much difficulty in travelling, and halted for a short time at the native well, out of which numbers of birds flew as we approached. From the Box-tree Forest we pushed on down the polygonum flat, where we had seen the native woman who had secreted herself in the bush. A whole family was now in the same place, but an old man only approached us. We were, indeed, passing, when he called to us, expressly for the purpose of telling us that the horse (Flood's) had gone away to the eastward. This native came out of his way, and evidently under considerable alarm, to tell us this, and to point out the direction in which he had gone, our stock of presents being pretty nearly exhausted, Mr. Browne, with his characteristic good nature, gave him a striped handkerchief, with which he was much pleased. As it was evident the poor horse had kept along the edge of the Desert, and as he was a wandering brute, not caring for companions, it was uncertain to what distance he had rambled, I did not, therefore, lose time by attempting to recover him. We were all of us sure that he would not face the Stony Desert, but he may still be alive, and wandering over that sterile country. We stopped for the night on the long channel near the sandy rise where we had before rested, about ten miles short of our camp, and the trees on the muddy plain; and having effected our passage across that plain and the Stony Desert, over which it was with extreme difficulty that we kept our track, found ourselves on the 22nd, in the little grassy valley, from which we had entered upon it; little water was remaining, however, at the place where we had then stopped, so that I sent over to the sequestered spot Lewis had discovered, but the water there had entirely disappeared. Flood managed to shoot a couple of ducks (Teal), of which there were four or five that flew away to the south-east. These two birds were, I may truly say, a God-send, and I beg to assure the reader they were uncommonly good.

From this valley we had to cross the heavy sand ridges which had so fatigued our horses before, and I hardly expected we should find water nearer than the Fish Pond. We therefore started early to get over the distance as soon as possible, and, as on the outward journey, had a most severe task of it. The ridges were certainly most formidable, although they were not of such size as those from which we had retreated. At six miles we crossed the salt lagoon, and late in the afternoon descended to the box-tree forest before mentioned, having the grassy plains now upon the left-hand side. The sandy ridges overlooked these plains, so that in riding along we noticed some natives, seven in number, collecting grass seeds upon them, on which alone, it appears to me, they subsist at this season of the year. However, as soon as they saw us, they all ran away in more than usual alarm, perhaps from the recollection of our misunderstanding with Mr. Popinjay. Their presence, however, assured us that there must be water somewhere about, and as on entering the plain, more to the west than before, we struck on a track. I directed Mr. Browne to run it down, who, at about half-a-mile, came to a large well similar to that in the creek on the other side of the Stony Desert, but not of the same dimensions. We had lost sight of him for some little time, when suddenly his horse made his appearance without a rider, and caused me great anxiety for the moment, for my mind immediately reverted to our sulky friend, and my fears were at once raised that my young companion had been speared; riding on, therefore, I came at length to the well, down which, to my inexpressible relief, I saw Mr. Browne, who was examining it, and who came out on my calling to him. There was not sufficient water to render it worth our while to stop; but the well being nine feet deep, shewed the succession of strata as follows: four feet of good alluvial soil; three feet of white clay; and two feet of sea sand.

I should perhaps have been more particular in the description of our interview with the old man and his family on the northern side of the earthy plain. As I have stated, he called out to us, and in order to discover what he wanted, I held Mr. Browne's horse, while he dismounted and went to him. The old native would not, however, sit down, but pointed to the S.E. as the direction in which, as far as we could understand, the horse, "cadli" (dog), as he called him, the only large four-legged brute of which he knew any thing, had gone. The poor fellow cried, and the tears rolled down his cheeks when he first met Mr. Browne, and the women chanted a most melancholy air during the time we remained, to keep the evil spirits off, I suppose; but they had nothing to fear from us, if they could only have known it. This confusion of tongues is a sad difficulty in travelling the wilds of Australia. Both the old man and the women wanted the two front teeth of the upper jaw, and as the former had worn his down almost to a level with his gums like an old horse, he looked sadly disfigured.

We halted about three miles short of the place at which we had before stopped, but as Joseph followed some pigeons to a clump of trees across the plain at about a mile distance, and there found a small pond of water, we moved over to it, and remained stationary on the following day to rest our wearied animals.

The 24th again saw us at the Fish Pond, where Mr. Browne again exhibited his skill in the gentle craft, and caught a good dish of the finny tribe. The mystery as to how these fish could have got into so isolated a spot, was not yet cleared up, and I was really puzzled on the subject.

On the 27th, as we were crossing the country between the creeks, some natives came in from the north and called out to us, in consequence of which Mr. Browne and I rode up to them. They were in a sad state of suffering from the want of water; their lips cracked, and their tongues swelled. They had evidently lingered at some place or other, until all the water, intermediate between them and the creeks had dried up. The little water we had was not sufficient to allay their thirst, so they left us, and at a sharp trot disappeared over the sand hill.

On the 29th our journey over the sandy ridges was very distressing. They appeared to me to be much more numerous, and the valleys between them much more sandy than when we first passed over them, and were thickly covered with spinifex, although grass was also tolerably abundant in the flats. At this stage of our journey, I was the only one of the party who was not ill; Mr. Browne and all the men were suffering, added to which, the men were fairly knocked up. Their labours were now, however, drawing to a close, and I was only too thankful, that I retained my strength.

We had crossed the first or Strzelecki's Creek on the 29th, and had halted that night without water. During it some of the horses broke loose and wandered back; but Flood and Joseph soon overtook and brought them back. We should have had a distance of 85 miles to travel without water, but fortunately the precaution we had taken of digging wells in going out, insured us a supply in one of them, so that our return over this last long and dry tract of country was comparatively light, and we gained the Park and joined Mr. Stuart at the stockade on the evening of the 2nd of October, after an absence of seven weeks, during which we had ridden more than 800 miles. Had it not been for the precaution of digging these wells, I do not think that two or three of the horses would have reached their journey's end. We only found water in one, it is true, but that one was of the most essential service, inasmuch as it saved several of our animals; and this is a point, I hope future travellers in such a country will bear in mind. Mr. Browne found it necessary to put all the men on the sick list, and their comrades made them as comfortable as they could, after their late fatigues.

It was a great satisfaction to me to learn that everything had gone on well at the camp during my absence; Mr. Stuart had a good report to make of all. The cattle had been duly attended to, and had become exceedingly tame and quiet. The sheep were in splendid condition, but their flesh had a peculiar flavour—and that, too, not a very agreeable one, still their value was unquestionable, for if we had been living on salt provisions, it is more than probable that half of the party would have been left in the desert. The practicability of taking a flock of sheep into the interior, had now been fully proved in our case, at all events; but I am ready to admit that they are, notwithstanding, a precarious supply, and that unless great care be taken, they may be lost. The men, however, appeared to consider them of far too great importance to be neglected, and I think that when taken, they will for that very reason be well looked after.

The stockade had been erected and really looked very well; it was built just as I had directed, with the flag flying at the entrance. I availed myself of the opportunity, therefore, to call it, "Fort Grey," after his Excellency the then Governor of South Australia.

Mr. Stuart informed me that a few natives only had visited the camp; but that on one occasion some of them appeared armed, being as they said on their way to a grand fight, four of their tribe having been killed in a recent encounter. Only the day before, however, a party had visited the camp, one of whom had stolen Davenport's blanket. He was pretty sure of the thief, however, so we did not despair of getting it back again.

I observed that when we were on Eyre's Creek, the climate and temperature were cool and agreeable. From that period the heat had considerably increased, and the thermometer now ranged from 96 to 100°. The wind having settled in its old quarter the E.S.E., in this latitude was not so cold as we had felt it in a more northerly one. Why it should have been so, it is difficult to say: we know the kind of country over which an E.S.E. wind must pass between the coast and the latitude of Fort Grey, and could not expect that it should be other than hot, but we are ignorant of the kind of country over which it may sweep higher up to the north. Can it be that there is a large body of water in that quarter? We shall soon have to record something to strengthen that supposition. About this period the sky was generally cloudy, and, as I have before remarked, in any other region it would have rained, but here only a few drops fell, no signs of which remained half an hour afterwards; the barometer, however, was very low, and it was not unreasonable to have encouraged hopes of a favourable change.

On the 3rd the natives who had visited the camp before our return, again came, together with the young boy who Davenport suspected had stolen his blanket. He charged him with the theft, therefore, and told him not to return to the tents again without it, explaining at the same time what he had said, to the other natives. The boy went away before the rest, but all of them returned the next day, and he gave up the blanket. On hearing this, I went out and praised him, and as he appeared to be sorry for his offence, I gave him a knife, in which I believe I erred, for we afterwards learnt, that the surrender of the blanket was not a voluntary act, but that he had been punished, and forced to restore it by his tribe. I cannot help thinking, however, that if the theft had not been discovered, the young rogue would have been applauded for his dexterity. I had, during my journey back to the Depôt, sat up to a late hour writing, that no delay might take place in my intended arrangements on our arrival at Fort Grey. In revolving in my own mind the state of the country, I felt satisfied that, although the water had decreased fearfully since the July rain, the road was still open for Mr. Browne to make good his retreat, but it was quite uncertain how long it might continue so. It was evident, indeed, that neither he nor myself had any time to lose, but I waited for a few days before I broke the subject to him, reluctant as I was to hasten his departure, and feeling I should often have to regret the loss of such a companion. The varied reverses and disappointments we had encountered together, and the peculiar character of the expedition, had, as far as Mr. Browne and myself were concerned, removed all restraint, and left to ourselves in that dreary wilderness, we regarded each other as friends only, who were united in a common cause, in the success of which we were almost equally interested. I knew, therefore, that the proposal I was about to make would give him pain; but I counted on his acquiescence, and as time would not admit of delay, I availed myself of an opportunity that presented itself the third day after our return, to break it to him.

As we were sitting in the tent after dinner, with our tea still before us, I said to him, "I am afraid, Browne, from what I have observed, that you have mistaken the object for which I have returned to the Depôt, and that you have been buoying yourself up with the hope that it is done preparatory to our return to Adelaide; for myself I cannot encourage any such hope for the present, at least. So far indeed from this, I have for some time been reflecting as to the most prudent course to be pursued under our present circumstances; for, I would not conceal from you the pain I have felt at the failure of our endeavours to penetrate farther than we have been able to do into the interior, neither can I conceal from myself the fact, that whatever our personal exertions, the results of our labours have not been commensurate with our expectations, and that however great our perseverance or however difficult the task we have had to perform, the world at large will alone judge of its merits by its success. In considering how we can yet retrieve our misfortunes, one only step occurs to me, and whatever pain our separation may cost us, I am sure, where the interests of the services call for it, you will readily comply with my wishes. I propose, then, your return to Adelaide, with all the party but three; that you should leave me five horses, and take with you only such provisions as you may absolutely require upon the road. By such an arrangement I might yet hold out against the drought, and ultimately succeed in doing something to make up for the past." My young friend was evidently unprepared for the proposition I had made. "You have done all you were sent out to do," he observed, "why then seek to penetrate again into that horrid desert? It is impossible that you can succeed during the continuance of the dry weather. If you now go you will never get back again; besides, have you," he asked, "made any calculations as to the means both of provisions and carriage you will require?" "That," I replied, "is for my consideration, but I have done so, and it appears to me that both are ample." "Well" said Mr. Browne, "it may be so, I do not know, but I can never consent to leave you in this dreadful desert. Ask me to do anything else, and I will do it; but I cannot and will not desert you." It was in vain that I assured him, he took a wrong view of the matter. That, as I had sent Mr. Poole home to increase my means, so I wished to send him, and that he would be rendering me as valuable, though not such agreeable service, as if he continued with me. "You know, Browne," I added, "that the eyes of the geographical world are fixed on me, and that I have a previous reputation to maintain; with you it is different." If I hoped to make any discovery I would not ask you to leave me. Believe me, I would that you shared the honour as you have shared the privations and anxieties of this desert with me; but I entertain no such hope, and would save you from further exposure. I have not seen enough of this dreary region to satisfy me as to its present condition. How then shall I satisfy others? That Stony Desert was, I believe, the bed of a former stream, but how can I speak decidedly on the little I have observed of it. No! as we have been forced back from one point, I must try another—and I hope you will not throw any impediment in the way. There is every reason why you should return to Adelaide: your health is seriously impaired, you are in constant pain, and your affairs are going to ruin; on all these considerations I would urge you to comply with my wishes." Mr. Browne admitted the truth of what I said, but felt certain that if he left, it would only be to hear of my having perished in that horrid desert—that my life was too valuable to others to be so thrown away, that he owed me too much to forsake me, and that he could not do that of which his conscience would ever after reproach him;—that his brother would attend to his interests, and that if it were otherwise, it would be no excuse for him to desert his friend—that he would acquiesce in any other arrangement, but to leave me he could not. "Well," I said, "I ask nothing unreasonable from you, nothing but what the sternness of duty calls for; and if you will not yield to friendly solicitations, I must order you home." "I cannot go," he replied; "I do not care for any pecuniary reward for my services, and will give it up: I want no pay, but desert you I will not!" The reader will better imagine than I can describe, such a scene passing in the heart of a wilderness, and under such circumstances I may not state all that passed; suffice it to say, that we at length separated, with an assurance on Mr. Browne's part, that he would consider what I had proposed, and speak to me again in the morning. The morning came, and after breakfast, he said he had endeavoured to force himself into a compliance with my wishes, but to no purpose;—that he could not leave me, and had made up his mind to take the consequences. It was in vain that I remonstrated, and I therefore ceased to importune him on a point which, however much I might regret his decision, I could not but feel that he was influenced by the most disinterested anxiety for my safety. But it became necessary to make some other arrangements; I had already been four days idle, and it was not my intention to let the week so pass over my head. Mr. Browne was too ill to accompany me again into the field. I sent, therefore, for Mr. Stuart, and told him to put up ten weeks provisions for four men—to warn Morgan and Mack that I should require them to attend me when I again left the camp—and to hold himself and them in readiness to commence the journey the day but one following; as I felt the horses required the rest I should myself otherwise have rejected.

I then sent for Mr. Browne, and told him that I proposed leaving the stockade in two days, by which time I hoped the horses would in some measure have recovered from their fatigues—that as he could not attend me, I should take Mr. Stuart with two fresh men—that in making my arrangements I found that I should be obliged to take all the horses but two, the one he rode and a weaker animal; to this, however, he would by no means consent—entreating me to take his horse also, as he felt assured I should want all the strength I could get.

No rain had as yet fallen, but every day the heat was increasing: the thermometer rising, even thus early in the season, to 98° and 100° in the shade, and the wind keeping steadily to the E.S.E. The country was so dry, and the largest pools of water had so diminished in quantity, that I doubted whether or not I should be able to get on, since as it was I should have to travel the first 86 miles without water, there being none in any other direction to the north of us. Even the large sheet in the first creek, to which I proposed going, had fearfully shrunk. But what gave me most uneasiness, was the reduced state of water on which the men and animals depended. From a fine broad sheet it was now confined within the limits of its own narrow channel, and I felt satisfied that if I should be absent many weeks, Mr. Browne would be obliged to abandon his position. Foreseeing this contingency, I arranged with him that in the event of his finding it necessary to retire, he should fall back on the little creek, near the old Depôt. That before he finally broke up the camp he should dig a hole in some favourable part of the creek into which the water he might leave would drain, so as to insure on my return as much as possible, and we marked a tree under which he was to bury a bottle, with a letter in it to inform me of his intended movements. Nothing could have been more marked or more attentive than Mr. Browne's manner to me, and I am sure he saw me mount my horse to depart with sincere regret; but the interval between the conclusion of these arrangements and the day fixed on to resume my labours soon passed over, although I deferred it to the 9th, in consequence of Flood's assuring me that the horses required the additional rest.

I had, indeed, been the more disposed to postpone the day of my departure, because I hoped, from appearances, that rain would fall, but I was disappointed. On the 6th it was very close, and heavy clouds passed over us from the N.E., our rainy quarter, towards the Mount Serle ranges, but still no rain fell on the depressed and devoted region in which we were. At eight, however, it rained slightly for about a quarter of an hour, and the horizon was black with storm clouds; all night heavy thunder rolled in the distance, both to the west and east of us; my ear caught that joyful sound as I laid on my mattress, and I fervently prayed that it might be the precursor of a fall.

I could not but hope, that, in the ordinary course of events, to revive and to support nature, the great Author of it would have blessed the land, desert as it was, with moisture at last, but I listened in vain for the pattering of rain, no drops, whether heavy or light, fell on my tent. The morning of the 7th dawned fair and clear; the sun rose in unshrouded splendour; and crossed the heavens on that day without the intervention of a cloud to obscure his disc for a moment. If then I except the rain of July, which lasted, at intervals, for three days, we had not had any for eleven months. Under the withering effects of this long continued drought, the vegetable kingdom was again at a stand; and we ourselves might be said to have been contending so long against the elements. No European in that respect had ever been more severely tried.

The day before we commenced our journey to the north it was exceedingly hot, the thermometer rose to 106° in the shade, and thus early in the season were we forewarned of what we might expect when the sun should become more vertical. In the afternoon the old man who had visited us just before we commenced our late journey, arrived in the camp with his two wives, and a nice little girl about eleven, with flowing curly hair, the cleanliness and polish of which would have done credit to the prettiest head that ever was adorned with such. They came in from the S.W., and were eagerly passing our tents without saying a word, and making for the water, when we called to them and supplied all their wants. The poor things were almost perishing from thirst, and seized the pannikins with astonishing avidity, when they saw that they contained water, and had them replenished several times. It happened also fortunately for them, that the lamb of the only ewe we had with us, and which had been dropped a few weeks before, got a *coup de soleil,* in consequence of which I ordered it to be killed, and given to the old man and his family for supper. This they all of them appeared to enjoy uncommonly, and very little of it was left after their first meal. The old man seemed to be perfectly aware that we had been out, but shook his head when I made him understand that I was going out again in the morning.

I determined, on the journey I was about to commence, to run on a due north course from the first "Strzelecki's Creek," as soon as I should reach it, and to penetrate the interior in that direction as far as circumstances might justify. As the reader will have concluded from the observations I have made, it had occurred to me that the Stony Desert had been the bed of a former stream, and I felt satisfied that if I was right in that conclusion, I should certainly strike it again. My object, therefore, was to keep at such a distance from my last course, as should leave no doubt of that fact upon my mind; it appeared to me that a due northerly course would about meet my views, and that if the Stony Desert was what I supposed it to have been, I should come upon it about two degrees to the eastward of where I had already crossed it. In pushing up to the north I also hoped that I might find a termination to the sandy ridges, although I could not expect to get into any very good country, for from what we saw to the north it was evidently much lower than that over which we had passed, and I therefore looked for a cessation of the sandy ridges we had before been so severely distressed on passing.

I shook hands with Mr. Browne about half-past eight on the morning of the 9th of October, and left the depôt camp at Fort Grey, with Mr. Stuart, Morgan and Mack, taking with me a ten weeks' supply of flour and tea. I once more struck into the track I had already twice traversed, with the intention of turning to the north as soon as I should gain Strzelecki's Creek. As we rode over the sand-hills, they appeared as nothing to me, after the immense accumulations of sand we had crossed when Mr. Browne and I were out together. We stopped short of the flat in which we had sunk the largest well on that occasion, to give the horses time to feed a little before sunset, and not to hurry them too much at starting. The day was exceedingly warm, and the wind from the N.E. A few heat-drops fell during the night, but the short thunder shower at the Depôt on the Sunday did not appear to have extended so far as where we then were. Nevertheless it would appear, that these low regions are simultaneously affected by any fall of rain; for there can be no doubt as to that of July having extended all over the desert interior, and the drizzling shower we had at the head of the northern Eyre's Creek, just as we were about to retrace our steps, having been felt the same day at the camp. I have just said that the day had been exceedingly hot, with the wind from the N.E., a quarter from whence we might naturally have expected that it would have blown warm; but I would observe, that before Mr. Browne and I passed the Stony Desert on our recent excursion, the winds from that point were unusually cold, and continued so until after we had crossed the Desert, and pushed farther up to the north, when they changed from cold to heat. I will not venture any conjecture as to the cause of this, because I can give no solution to the question, but leave it to the ingenuity of my readers, who are as well able to judge of such a fact as myself.

I would also advert to a circumstance I neglected to mention in its proper place, but which may be as forcibly done now as at the time it occurred. When Mr. Browne and I were on our recent journey to the north, after having crossed the Stony Desert, being then between it and Eyre's Creek, about nine o'clock in the morning, we distinctly heard a report as of a great gun discharged, to the westward, at the distance of half a mile. On the following morning, nearly at the same hour, we again heard the sound; but it now came from a greater distance, and consequently was not so clear. When I was on the Darling, in lat. 30°, in 1828, I was roused from my work by a similar report; but neither on that occasion, or on this, could I solve the mystery in which it was involved. It might, indeed, have been some gaseous explosion, but I never, in the interior, saw any indication of such phenomena.

We were obliged to fasten up our horses to prevent them from straying for water, and had, therefore, nothing to do but to saddle them on the morning of the 10th, and started at six. Our journey the day before had been 33 miles: this day we rode about 36, to the little muddy creek the reader will, I have no doubt, call to mind. In it, contrary to my expectation, we found a small supply of water, though difficult to get; and I halted at it, therefore, for the night, and reached the Strzelecki Creek about half-past ten on the morning of the 11th, in which I was rejoiced to find that the water was far from being exhausted. Turning northwards up the creek, I halted about half-past one at the upper pool, about seven miles from the first. As far as this point the lay of the sand ridges was N.N.E. and S.S.W.

As Mr. Browne had stated to me, the country to the north was much more open from the point at which we now were than to the west. A vast plain, indeed, met the horizon in the first direction, and as we rode up it on the 12th, we observed that it was bounded at irregular distances, varying from three to six miles, on either side of us, by low sand hills. The whole plain was evidently subject to flood, and the travelling in some places was exceedingly heavy. We had ridden from early dawn until the sun had sunk below the horizon, without seeing any apparent termination to this plain, or the slightest; indication of water. Just as it was twilight we got on a polygonum flat; there being a little sand hill on one side of it, under which I determined to stop for the night.

While the men were tethering the horses on the best part of the flat, where there happened to be a little green grass, Mr. Stuart and I walked up the sand hill; but in the obscure light then prevailing, we could not see any thing distinctly. It appeared, however, that the country before us was traversed by a belt either of forest or of scrub; there was a long dark line running across the country but we could not make out what it was, so that we descended to our little bivouac full of hope, and anxious for the morning dawn to satisfy ourselves as to what we had been looking at. Day had scarcely broke when we were again on the hill; and as objects became clearer, saw a broad belt of gum-trees extending from the southward of east to the north-west. It was bounded on either side by immense plains, on which were here and there ridges of sand, but at a great distance from each other. There was another small sand hill distant four miles, and an apparently high and broken chain of mountains was visible to the N.E., distant more than 50 miles. The trees were not more than three miles from us, and were denser and seemingly larger than any we had seen; and although we could not see any water glittering amidst the foliage, yet I could not but hope that we were on the eve of some important discovery. There were likewise mountains in the distance, with broken lofty peaks, exactly resembling the Mount Serle chain, and I ventured to hope that I had at length found a way to escape from the gloomy region to which we had been so long confined. Descending from our position we pushed for a dark mass of foliage to the N.E., and shortly after crossing the dry bed of a lagoon, found ourselves riding through an open box-tree forest, amidst an abundance of grass. At half a mile further we were brought up by our arrival on the banks of a magnificent channel. There was a large sheet of water to our left, covered with wild fowl. Flooded gum-trees of large size grew on its banks, and its appearance was altogether imposing. I stood looking in admiration on the broad mirror so close to me, and upon a sight so unusual; and I deeply regretted at that moment that Mr. Browne was not with me to enjoy the gratification of such a scene.

We dismounted and turned our horses out to feed on the long grass in the bed of this beautiful creek, and whilst Morgan prepared breakfast, Mr. Stuart and Mack took their guns and knocked over the ducks, that were, I suppose, never used to be so taken in; but the remainder would not stand fire long, and flew off to the eastward. As they passed, however, I snatched up a carbine, and, without taking any aim, discharged it into the midst of them, and brought one of their number down—the only bird I had shot for many years.

After giving the horses a good feed and a good rest, I crossed the channel of the creek to ascend the little hill I had seen from our morning position, that by taking bearings of the distant ranges from both, I might arrive at their approximate distance from me. From this little hill the prospect was much the same as from the first, only that the distant ranges seemed to be still higher, and there was a long line either of water or mirage at their base, and we now appeared to be in a belt of wood, for the hill on which we stood, rose in the midst of the trees, and our eyes wandered over the tops of them to the distant plains. We descended from it northwards, but had not gone half a mile, when we were again stopped by another creek, still broader and finer than the first. The breadth of its channel was more than 200 yards, its banks were from fifteen to eighteen feet high, and it had splendid sheets of water: both above and below us. The natives, whose broad and well beaten paths leading from angle to angle of the creek we had crossed on our approach to it, had fired the grass, and it was now springing up in the bed of the most beautiful green. I determined, therefore, to stay where I was until the following day, to give my animals the food and rest they so much required, and myself time for reflection. We accordingly dismounted, and turned the horses out, and it was really a pleasure to see them in clover.

The whole bed of the creek was of a vivid green, excepting where gravel had been deposited in it, but the animals kept on the grass, close to the water's edge. As we had approached the creek through a belt of wood, so it extended on the other side for a considerable distance into the plains, but the soil was not so good as in the neighbourhood of the first channel we had crossed, since bushes of rhagodia were growing underneath the trees, as indicative of a slight mixture of salt in the earth. The appearance of the creek, however, embosomed as it was in wood, was very fine, more especially the upward view of it, where there was a splendid sheet of water, in the centre of which the branches of a huge tree appeared reflected, the trunk being completely hid. About a quarter of a mile above us a tributary joins the main branch from the eastward, that when flooded must have a fall of three or four feet, and something of the character of a Canadian rapid.

When I sat down beside the waters of the beautiful channel to which Providence in its goodness had been pleased to direct my steps, I felt more than I had ever done in my life, the responsibility of the task I had undertaken. When I left the Depôt I had determined on keeping a northerly course into the interior, for the reasons I have already assigned; but knowing the state of the country as I did, and the little chance there was of finding water on its parched and yawning surface, I now hesitated whether I should persevere in my first determination, or proceed in the examination of this new feature, and of the mountain ranges to the N.E. both of which I had every reason to hope would lead me out of the present fearful desert into better country. Any one perhaps less experienced than myself in the treacherous character of the most promising river of the Australian Continent would have acted differently. It would in all probability have occurred to them to trace the creek, either upwards or downwards, in the hope of its leading to something better. It was clear, however, that the first channel I had crossed, was a branch only of that upon which I was resting, and by which the plains I had traversed on approaching it were laid under water, and I felt assured that if my conclusion as to the Stony Desert was correct, I should derive no advantage in tracing the creek downwards, since I knew it would either terminate in extensive grassy plains as I had found other creeks to do, or be lost on the broad surface of the Stony Desert. Taking every thing into consideration, I had resolved on turning to the eastward, to examine the upward course of the creek, believing it more than probable that it would lead me into the hills, but, as I was weighing these things in my mind, the sky became suddenly overcast and a thunder-storm passed over us, which for the short half hour it continued was of unusual violence, filling all the little hollows on the plains, and chequering them over with sheets of water. The road northwards being thus thrown open to me, I returned to my original purpose, and determined on the morrow to pursue a northerly course directly into the interior, in the hope that ere the surface water left by the thunder-storm should be dried up, I might reach such another creek as the one I was about to quit, or find some other such permanent place of safety; leaving the examination of the upper branches of the creek, and of the mountain ranges to the period of my return. Accordingly on the morning of the 13th, we left our position, crossing to the proper right bank of the creek, and breaking through the nearer box tree forest, traversed open plains, the soil of which was principally sand, but there was an abundance of grass upon them, and they were somewhat elevated above the more alluvial flats near the creek. At 2½ miles we crossed a large tributary from the N.E., the main branch trended to the N.W., and we kept the belt of trees in view as we rode along, during the greater part of the day. At seven miles we descended a little from the grassy plains to a flooded plain of considerable extent, but again rose from it to the sandy level, and finding a small puddle of rain water at 36 miles I halted.

As I was about to trust entirely to the supply of water left by the recent storm, and knew not to what distance it had extended, I felt it necessary to take every precaution to insure our retreat. We worked, therefore, by the light of the moon, and dug a square into which we drained all the water that remained after the horses had satisfied themselves in the morning, but the quantity was so small that I scarcely hoped to derive any advantage from it on our return; and it was really the zeal of Morgan and Mack that induced me to allow them to finish it. Warm as the weather had been at Fort Grey, the night was bitterly cold, with the wind from the S.S.E. We left this, our first well, at early dawn, riding across a continuation of the same grassy and sandy land as that we had journeyed over the day before, only that it had many bare patches upon it full of water, the undersoil being a red clay. The same kind of tree we had seen to the eastward, between the old Depôt and the Darling, and which I had there taken to be a species of Juglans, prevailed hereabouts in sheltered places.

The creek line of trees was still visible to our left, so that it must have come up a little more to the north. We crossed several native paths leading to it: the impression of an enormous foot was on one of them. At eight miles we descended to a flooded plain, scattered over with stunted box-trees, the greater number being dead, and I may remark that we generally found such to be the case on lands of a similar description; a fact, it appears to me, that can only be accounted for from the long- continued drought to which these unhappy regions are subject. These flooded plains are generally torn to pieces by cracks of four, six, and eight feet deep, of a depth, indeed, far below that at which I should imagine trees draw their support, but the box-tree spreads its roots very near the surface of the ground, having, I suppose, no prominent tap root, and can therefore receive no moisture from such a soil as that in which we so often found it in premature decay; the excess of moisture at one time, and the want of it at another, must be injurious to trees and plants of all kinds, and this circumstance may be a principal cause of the deficiency of timber in the interior of Australia. From the level, we ascended to sandy and grassy plains as before, but they were now bounded by sandy ridges of a red colour, and partly covered with spinifex. I really shuddered at the re- appearance of those solid waves which I had hoped we had left behind, but such was not the case. At six miles we arrived at the base, and ascending one of them, found that it was flanked on both sides by others; the space between the ridges being occupied by the white and dry beds of salt lagoons. The reader will, I am sure, sympathise with me in these repeated disappointments, for the very aspect of these dreaded deposits, if I may so call them, withered hope. To whatever point of the compass I turned, whether to the west, to the north, or to the east, these heart-depressing features existed to damp the spirits of my men, and irresistibly depress my own; but it was not for me to repine under such circumstances, I had undertaken a task, and in the performance of it had to take the country as it laid before me, whether a Desert or an Eden. Still whatever moral convictions we may have, we cannot always control our feelings. The direction of the ridges was nearly north and south, somewhat to the westward of the first point, so that at a distance of more than two degrees to the eastward they almost preserved their parallelism. We rode along the base of a ridge for about three miles, but as on ascending it to take a survey, I observed that at about a mile beyond, it terminated, and that the dry bed of the lagoon to our right passed into a plain of great breadth immediately in front, the character and appearance of which was very doubtful, and as it was now sunset, and we had journeyed upwards of 34 miles, I halted for the night at another puddle, rather larger than the last, but with sorry feed for the horses. At this place we dug our second well, by moonlight, as we had dug the first, and laid down on the ground to rest, fatigued, I candidly admit, both in mind and body.

The day had been exceedingly cold, as was the night, and on the following morning with the wind at S.S.E., and a clear and cloudless sky the temperature still continued low. At about a mile from where we had bivouacked, we arrived at the termination of the sandy ridge, and descended into the plain I had been reluctant to traverse in the uncertain light of evening. It proved firm, however, though it was evidently subject to floods. Samphire, salsolæ, and mesembryanthemum, were growing on it, and one would have supposed from its appearance that it was a sea marsh. Mr. Stuart shot a beautiful ground parrot as we were crossing it, on a bearing of 345°, or little more than a N. and by W. course. At 6½ miles we ascended some heavy sandy ridges, without any regularity in their disposition, but lying in great confusion. Toiling over these, at seven or eight miles farther we sighted a fine sheet of water, bearing N. and distant about two miles. At another mile I altered my course to 325°, to pass to the westward of this new feature, which then proved to be a lake about the size of Lake Bonney, that is to say from 10 to 12 miles in circumference. The ridge by which we had approached it terminated suddenly and directly over it; to our right there were other ridges terminating in a similar manner, with rushy flats between them; eastward the country was dark and very low; to the north there was a desert of glittering white sand in low hillocks, scattered over with dwarf brush, and on it the heat was playing as over a furnace. Immediately beneath me to the west there was a flat leading to the shore of the lake, and on the western side a bright red sand hill, full eighty feet high, shut out the view in that quarter. This ridge was not altogether a mile and a half in length, and behind it there were other ridges of the same colour bounding the horizon with edges as sharp as icebergs.

I did not yet know whether the waters of the lake were salt or fresh, although I feared they were salt. Looking on it, however, I saw clearly that it was very shallow; a line of poles ran across it, such as are used by the natives for catching wild fowl, of which there were an abundance, as well as of hematops on the water. As soon as we descended from the sand ridge we got on a narrow native path, that led us down to a hut, about 100 yards from the shore of the lake.

As we approached the water, the effluvia from it was exceedingly offensive, and the ground became a soft, black muddy sand. On tasting it we found that the water was neither one thing or the other, neither salt nor fresh, but wholly unfit for use. Close to its margin there was a broad path leading to the eastward, or rather round the lake; and under the sand ridge to the west, were twenty-seven huts, but they had long been deserted, and were falling to decay. Nevertheless they proved that the waters of the lake were sometimes drinkable, or that the natives had some other supply of fresh water at no great distance, from whence they could easily come to take wild fowl, nor could I doubt such place would be the creek.

Notwithstanding that the water was so bad, I tried several places by digging, but invariably came to salt water, oozing through black mud, and I therefore presumed that a good deal of rain must have fallen hereabouts, to have tempered the water of the lake so much; which it struck me would otherwise have been quite saline. From the point where we first came down upon it, we traversed a flat beach covered with a short coarse rush, having the high red sand hill, of which I have spoken, to our left; before us a vast extent of low white sand, and to the eastward an extremely dark and depressed country. I was really afraid of entering on the scorching sands in our front, for we were now full 90 miles from the creek, and it was absolutely necessary, before I should exceed that distance, to find a more permanent supply of water than the wells we had dug on our way out. In order to ascertain the nature of the country more satisfactorily, however, I ascended the rugged termination of the sandy ridge, close to which we had been riding, and was induced, from what I then saw, to determine on a course somewhat to the west of north, since a due north course was evidently closed upon me; for I now saw that the country in that direction was hopeless, as well as in an easterly direction; but although I stood full 80 feet above the lake, I could not distinguish any thing like a hill on the distant horizon. To the westward, as a medium point, there were a succession of sandy ridges, similar to that on which I stood; but to the S.W. there seemed to be an interval of plain. As the thunder storm had reached as far as the place where we last slept, I did not doubt but that it had also reached the lake, and on consideration determined to keep as northerly a course as circumstances would permit, in pushing into a country in which I was meeting new difficulties every hour. Descending, therefore, on a bearing of 340°, I went to a distance of six miles before coming to a small puddle at which I was glad to halt, it being the only drinkable water we had seen. Here we dug a third well, although, like the first, there was but little chance of benefiting by it. It behoved me therefore to be still more careful in increasing my distance from the creek, so that on the morning of the 17th I thought it prudent to search for some, and as the country appeared open to the south, I turned to that point in the hope of success.

We crossed some low sand hills to a swamp in which there was a good deal of surface water, but none of a permanent kind. We then crossed the N.W. extremity of an extensive grassy plain, similar to those I have already described, but infinitely larger. It continued, indeed, for many miles to the south, passing between all the sandy points jutting into it; and so closely was the Desert allied to fertility at this point, and I may say in these regions, that I stood more than once with one foot on salsolaceous plants growing in pure sand, with the other on luxuriant grass, springing up from rich alluvial soil. At two miles and a quarter from the swamp, striking a native path we followed it up to the S.W., and, at three-quarters of a mile, we reached two huts that had been built on a small rise of ground, with a few low trees near them. Our situation was too precarious to allow of my passing these huts without a strict search round about, for I was sure that water was not far off; and at length we found a small, narrow, and deep channel of but a few yards in length, hid in long grass, at a short distance from them. The water was about three feet deep, and was so sheltered that I made no doubt it would last for ten days or a fortnight. Grateful for the success that had attended our search, I allowed the horses to rest and feed on the grass for a time; but it was of the kind from which the natives collect so much seed, and though beautiful to the eye, was not relished by our animals. The plains extended for miles to the south and south-east, with an aspect of great luxuriance and beauty; nor could I doubt they owed their existence to the final overflow of the large creek we had all along marked trending down to this point. Such, indeed, I felt from the first, even when I looked on its broad and glittering waters, would sooner or later be its termination, or that it would expend itself, less usefully, on the Stony Desert. As yet, however, there was no indication of our approach to that iron region. The plains were surrounded on all sides by lofty ridges of sand, and the whole scene bore ample testimony to the comparative infancy, if I may so express myself, of the interior. We next pursued a N.N.W. course into the interior, and soon left the grassy plains, crossing alternate sand ridges and flats on a bearing of 346°, the whole country having a strong resemblance to that between Sydney and Botany Bay in New South Wales. On one of the ridges we surprised a native, who ran from us in great terror, and with incredible speed. About noon we crossed a plain, partly covered with stones and partly bare, and at the further extremity of it passed through a gorge between two sand hills into another plain that was barren beyond description, with only salsolaceous herbs. It had large white patches of clay on it, the shallow receptacles of rain water, but they were all dry. The plain was otherwise covered with low salsolæ, excepting on the higher ground, on which samphire alone was growing. It was surrounded on all sides by sand hills of a fiery red, and not even a stunted hakea was to be seen. From this plain we again crossed alternate sand hills and flats, the former covered with spinifex, the latter being quite denuded of all vegetation; but one of the horses at last knocking up, I was obliged to halt in this gloomy region, at the only puddle of rain water we had seen since leaving the grassy plain. I was sure, however, from the change that had taken place, and the character of the country around us, that we were approaching that feature, the continuance of which, in order to elucidate its probable origin, it had been a principal object in my present journey to ascertain. I felt so convinced on this point, that I could not have returned to Adelaide without having satisfied my mind on the subject. I might, indeed, have had general ideas as to the past state of the depressed interior, from what I had already seen of it; but the Stony Desert was the key to disclose the whole—and although I feared again to tread its surface, its existence so far away to the eastward of where I had first been on it, would at least tend to confirm my impressions as to what it had been.

It was clear, indeed, from the character of the country through which we had just passed, that we were again approaching the salt formation; more especially when, from the highest ground near us, I observed its generally dark aspect, and that there was the dry bed of a large salt lagoon directly in our course. We here dug a fourth well: the water was extremely muddy and thick, for the basin in which it was contained was very shallow, and the wind constantly playing on its surface raised waves that had stirred up the mud; but as there was more water than usual, I hoped that by deepening, it might settle. This was nothing new to us, for not only on our journey to Lake Torrens and to the N.W., had we subsisted on similar beverage, but the water at the Depôt at Fort Grey was half mud, and perfectly opaque. However, it was a matter of necessity to retain it here if possible, and we therefore took the best measures in our power to do so.

On the 19th we resumed our journey on the former bearing, the wind blowing keen from the south. At about a mile and a half we reached the salt lagoon, as it appeared to be in the distance, but which proved to be rather a flooded plain. It was about two miles broad, and three and three-quarters long, and was speckled over rather than covered with salt herbs. At this time, also, we had an immense barren plain to our left, bounded around, but more particularly to the north, by sand hills; over these we toiled for nine miles, when at their termination the centre of the plain bore 176° the east of north, or nearly south. At five miles and a half further, having previously crossed a small stony plain, succeeded by sand ridges and valleys, both covered with spinifex, we ascended a pointed hill that lay directly in our course, and from it beheld the Stony Desert almost immediately below our feet. I must acknowledge, that coming so suddenly on it, I almost lost my breath. It was apparently unaltered in a single feature: herbless and treeless, it occupied more than one half of the visible horizon, that is to say, from 10° east of north, westward round to south. As to the eastward, so here the ridges we had just crossed abutted upon it, and as many of them were lower than the line of the horizon, they looked like sea dunes, backed by storm clouds, from the dusky colour of the plain.

After surveying this gloomy expanse of stone-clad desert we looked for some object on the N.W. horizon upon which to move across it, but none presented itself, excepting a very distant sand hill bearing 308°, towards which I determined to proceed. We accordingly descended to the plain, and soon found ourselves on its uneven surface. There was a narrow space destitute of stones at the base of the sand hill, stamped all over with the impressions of natives' feet. From eighty to one hundred men, women and children must have passed along there; and it appeared to me that this had been a migration of some tribe or other during the wet weather, but it was very clear those poor people never ventured on the plain itself.

Descended from our high position, we could no longer see the sand hill just noticed, but held on our course by compass like a ship at sea, being two hours and forty minutes in again sighting it; and reaching it in somewhat less than an hour afterwards, calculated the distance at thirteen miles. As we approached, it looked like an island in the midst of the ocean; but we found a large though shallow sheet of water amongst the stones under it, for which we were exceedingly thankful. From this point we crossed to another sand hill that continued northerly further than we could see, having the Desert on either hand. Our horses beginning to flag, I halted at five on the side of the ridge, near a small puddle that had only water enough for them to drink off at once.

The morning of the 20th was bitterly cold, with the wind at S.S.E., and I cannot help thinking that there are extensive waters in some parts of the interior, over which it came: the thermometer at 42°. We started on a course of 335° for a distant sandy peak rising above the general line of the horizon. At a mile, one of the horses fortunately got bogged in a little narrow channel just like that in the grassy plain; I say fortunately, for we might otherwise have passed the water it contained without knowing it, so completely was it shaded. In looking along the channel more closely, we discovered a little pool about three yards long and one broad, but deep. At this we breakfasted and watered the horses, and then pushed on. The lodgement of this water had been caused by local drainage, and was evident from the green feed round about. Here again it appeared we had occasion to be thankful, for on this supply I hoped we might safely calculate for a week at least, so that we still held on our course with more confidence, keeping at the base of the ridge, and passing an extent of five miles through an open box-tree forest, every tree of which was dead. The whole scene being one of the most profound silence and marked desolation, for here no living thing was to be seen.

At nine miles we ascended the ridge, and from it the Desert appeared to be interminable from N. to N.E., but a few distant sand hills now shewed themselves to the eastward of the last mentioned point. We then descended into a valley of sand and spinifex, and at four miles and a half ascended an elevated peak in a sandy ridge lying in our way. From this, the view to the north-west was over a succession of sand hills. The point we stood upon, as well as the ridge, was flanked southwards by an immense plain of red sand and clay, and to the N.E. by a similar but smaller plain. Crossing a portion of the great plain, at four miles and a half we ascended another peak, and then traversed a narrow valley crossing from it into a second valley, down which we travelled for six miles.

At that distance it was half a mile in breadth, and there was a little verdure near some gum-trees, but no water. As we were searching about, a cockatoo, (Cacatua Leadbeateri) flew over the sand hill to our right, and pitched in the trees; we consequently crossed to the opposite side and halted for the night, where there was a good deal of green grass for the horses, but no water in the contiguous valley.

CHAPTER XI.

THE HORSES—ASCEND THE HILLS—IRRESOLUTION AND RETREAT—HORSES REDUCED TO GREAT WANT—UNEXPECTED RELIEF—TRY THE DESERT TO THE N.E.—FIND WATER IN OUR LAST WELL—REACH THE CREEK—PROCEED TO THE EASTWARD—PLAGUE OF FLIES AND ANTS—SURPRISE AN OLD MAN—SEA-GULLS AND PELICANS—FISH—POOL OF BRINE—MEET NATIVES—TURN TO THE N.E.—COOPER'S CREEK TRIBE, THEIR KINDNESS AND APPEARANCE—ATTEMPT TO CROSS THE PLAINS—TURN BACK—PROCEED TO THE NORTHWARD—EFFECTS OF REFRACTION—FIND NATIVES AT OUR OLD CAMP AND THE STORES UNTOUCHED—COOPER'S CREEK, ITS GEOGRAPHICAL POSITION.

I HAD taken all the horses, with the exception of one, out with me on this journey, and as they will shortly bear a prominent part in this narrative, I will make some mention of them. My own horse was a grey—for which reason I called him Duncan. I had ridden him during the whole period of my wanderings, and think I never saw an animal that could endure more, or suffered less from the want of water; he was aged, and a proof, that in the brute creation as well as with mankind, years give a certain stamina that youth does not possess. This animal, as the reader will believe, knew me well, as indeed did all the horses, for I had stood by to see them watered many a time. Mr. Stuart rode Mr. Browne's horse, a little animal, but one of great endurance also; Mack used a horse we called the Roan, a hunter that had been Mr. Poole's. Morgan rode poor Punch, whose name I have before had occasion to mention, and who, notwithstanding subsequent rest, had not recovered from the fatigues of his northern excursion. Besides these we had four pack horses:—Bawley, a strong and compact little animal, with a blaze on the forehead, high spirited, with a shining coat, and having been a pet, was up to all kind of tricks, but was a general favourite, and a nice horse—the other was Traveller, a light chesnut, what the hunter would call a washy brute, always eating and never fat;—the Colt, so called from his being young, certainly unequal to such a journey as that on which he was taken; and Slommy, another aged horse. During the summer, Traveller had had a great discharge from the nose, and I was several times on the point of ordering him to be shot, under an apprehension that his disease was the glanders; but, although the colt and my own horse contracted it, I postponed my final mandate, and all recovered; however, he continued weak. At this time they were unshod, and had pretty well worn their hoofs down to the quick, insomuch that any inequality in the ground made them limp, and it was distressing to ride them; but, notwithstanding, they bore up singularly against the changes and fatigues they had to go through.

From a small rising ground near where we stopped in the valley, on the occasion of which I am speaking, and in the obscure light of departing day we saw to the N.N.W. a line of dark looking hills, at the distance of about ten or twelve miles, but we could not discover tree or bush upon them, all we could make out was that they were dark objects above the line of horizon, and that the intervening country seemed to be as dark as they were. The weather had changed from cold to hot, the wind having flown from S. to the N.E., and the day and night were exceedingly warm. I was sorry to observe, too, that the horses had scarcely touched the grass on which, for their sakes, I had been tempted to stop, and that they were evidently suffering from the previous day's journey of from 34 to 36 miles, that being about the distance we had left the water in the grassy valley. Before mounting, on the morning of the 21st, Mr. Stuart and I went to see if we could make out more than we had been able to do the night before, what kind of country was in front of us, but we were disappointed, and found that we should have to wait patiently until we got nearer the hills to judge of their formation. About half a mile below where we had slept, the valley led to the N.N.E., and on turning, we found it there opened at once upon the Stony Desert; but the hills were now hid from us by sandy undulations to our left, and even when we got well into the plain we could hardly make out what the hills were. As we neared them, however, we observed that they were nothing more than high sand hills, covered with stones even as the desert itself, to their tops. That part of it over which we were riding also differed from any other portion, in having large sharp-pointed water-worn rocks embedded in the ground amongst the stones, as if they had been so whilst the ground was soft. There was a line of small box-trees marking the course of a creek between us and the hills, and a hope that we should find water cheered us for a moment, but that ray soon vanished when we saw the nature of its bed. We searched along it for about half an hour in vain, and then turned to the hills and ascended to the top of one of the highest, about 150 feet above the level of the plain. From it the eye wandered hopelessly for some bright object on which to rest. Behind us to the south-east lay the sand hills we had crossed, with the stony plain sweeping right round them, but in every other direction the dark brown desert extended. The line of the horizon was broken to the north-west and north by hills similar to the one we had ascended; but in those directions not a blade of grass, not a glittering spot was to be seen.

At this point, which I have placed in lat. 25°54' and in long. 139°25', I had again to choose between the chance of success or disaster, as on the first occasion; if I went on and should happen to find water, all for the time would be well, if not, destruction would have been inevitable. I was now nearly 50 miles from water, and feared that, as it was, some of my horses would fall before I could get back to it, yet I lingered undecided on the hill, reluctant to make up my mind, for I felt that if I thus again retired, it would be a virtual abandonment of the task undertaken. I should be doing an injustice to Mr. Stuart and to my men if I did not here mention that I told them the position we were placed in, and the chance on which our safety would depend if we went on. They might well have been excused if they had expressed an opinion contrary to such a course, but the only reply they made was to assure me that they were ready and willing to follow me to the last. After this, I believe I sat on the hill for more than half an hour with the telescope in my hand, but there was nothing to encourage me onwards; our situation, however, admitted not of delay. I might, it is true, have gone on and perished with all my men; but I saw neither the credit nor the utility of such a measure. I trust the reader will believe that I would not have shrunk from any danger that perseverance or physical strength could have overcome; that indeed I did not shrink from the slow fate, which, as far as I could judge, would inevitably have awaited me if I had gone on; but that in the exercise of sound discretion I decided on falling back. The feeling which would have led me onwards was similar to that of a man who is sensible of having committed an error, yet is ashamed to make an apology, and who would rather run the risk of being shot, than of having the charge of pusillanimity fixed upon him; but I have never regretted the step I took, and it has been no small gratification to me to find that the Noble President of the Royal Geographical Society, Lord Colchester, when addressing the members of that enlightened body, in its name presenting medals to Dr. Leichhardt and myself, for our labours in the cause of Geography, alluded to and approved "the prudence with which further advance was abandoned, when it could only have risked the loss of those entrusted to my charge."

We slowly retraced our steps to the valley in which we had slept, and I stopped there for half an hour, but none of the horses would eat, with the exception of Traveller, and he certainly made good use of his time. The others collected round me as I sat under a tree, with their heads over mine, and my own horse pulled my hat off my head to engage my attention. Poor brute! I would have given much at that moment to have relieved him, but I could not. We were all of us in the same distress, and if we had not ultimately found water must all have perished together. Finding that they would not eat, we saddled and proceeded onwards, I should say backwards—and at 10 P.M. we were on the sand ridges. At the head of the valley Traveller fell dead, and I feared every moment that we should lose the colt. At one I stopped to rest the horses till dawn, and then remounted, but Morgan and Mack got slowly on, so that I thought it better to precede them, and if possible to take some water back to moisten the mouth of their horses, and I accordingly went in advance with Mr. Stuart. I thought we should never have got through the dead box-tree forest I have mentioned, however we did so about 11 A.M., and made straight for the spot where we expected to relieve both ourselves and our horses, but the water was gone. Mr. Stuart poked his fingers into the mud and moistened his lips with the water that filled the holes he had made, but that was all. We were yet searching for water when Morgan and Mack appeared, but without the colt; fortunately they had descended into the valley higher up, and had found a little pool, which they had emptied, under an impression that we had found plenty; and were astonished at hearing that none any longer remained. In this situation, and with the apparent certain prospect of losing my own and Mr. Browne's horse, and the colt which was still alive when the men left him, not more than a mile in the rear, we continued our search for water, but it would have been to no purpose. Suddenly a pigeon topped the sand hill—it being the first bird we had seen—a solitary bird—passing us like lightning, it pitched for a moment, and for a moment only, on the plain, about a quarter of a mile from us, and then flew away. It could only have wetted its bill, but Mr. Stuart had marked the spot, and there was water. Perhaps I ought to dwell for a moment on this singular occurrence, but I leave it to make its own impression on the reader's feelings. I was enabled to send back to the colt, and we managed to save him, and as there was a sufficiency of water for our consumption, I determined to give the men a day of rest, and to try if I could find a passage across the Desert a little to the eastward of north, and with Mr. Stuart proceeded in that direction on the morning of the 24th; but at 3 P.M. we were out of sight of all high land. The appearance of the Desert was like that of an immense sea beach, and large fragments of rock were imbedded in the ground, as if by the force of waters, and the stones were more scattered, thus shewing the sandy bed beneath and betwixt them. The day was exceedingly hot, and our horses' hoofs were so brittle that pieces flew off them like splinters when they struck them against the stones. We were at this time about sixteen or seventeen miles from the sand hill where we had left the men. The Desert appeared to be taking a northerly direction, and certainly was much broader than further to the westward, making apparently for the Gulf of Carpentaria; nor could I

Sunset on the Murray.

Ana-Branch of the Darling

The Depot Glen.

Lake Torrens.

View from Stanley Range.

Strzelecki's Creek.

Port Adelaide.

King William Street.

The Murray River.

Mount Bryan.

doubt but that there had once been an open sea between us and it. We reached our little bivouac at 9 P.M. both ourselves and our horses thoroughly wearied, and disappointed as we had been, I regretted that I had put the poor things to unnecessary hardships. Perhaps I was wrong in having done so, but I could not rest. Our latitude here was 26°26' and our long. by account 139°21'. In the morning we crossed the remaining portion of the Desert, as I had determined on making the best of my way to the creek, and passing the sandy ridges reached our first water (the 4th going out), about sunset or a little before. Water still remained, but it was horridly thick, and in the morning smelt so offensive that it was loathsome to ourselves and the animals. Our great, indeed our only, dependence then was on the water in the little channel on the grassy plain; at this we arrived late on the afternoon of the 25th. Another day and we should again have been disappointed: the water on which I had calculated for a fortnight was all but gone. In the morning we drained almost the last drop out of the channel. We were now about 92 miles from the creek, without the apparent probability of relief till we should get to it, for it seemed hopeless to expect that we should find any water in the wells we had dug. Crossing the grassy plains on an east-north-east course, we passed the salt lake about 10 A.M. to our left, and ran along the sandy ridges between it and our encampment of the 15th, where we had made our second well, at 6 P.M., but it was dry and the bottom cracked and baked.

I would gladly have given my poor horses a longer rest than prudence would have justified, but we had not time for rest. At 8 we again mounted, and went slowly on; and when darkness closed around us lit a small lamp, and one of us walking in front led the way for the others to follow; thus tracking our way over those dreary regions all night long, we neared our last remaining well, 36 miles distant from the creek, just as morning dawned. Objects were still obscure as we approached the spot where our hopes rested, for our horses could hardly drag one foot after the other. Mr. Stuart was in front, and called to me that he saw the little trees under whose shade we had slept; soon after he said he saw something glittering where the well was, and immediately after shouted out, "Water, water." It is impossible for me to record all this without a feeling of more than thankfulness to the Almighty Power that guided us. At this place we were still 180 miles from Fort Grey; and if we had not found this supply, it is more than probable the fate of our horses would have sealed our own. As it was we joyfully unsaddled, and, after watering, turned them out to feed. Singular it was that the well on which we had least dependence, and from which we had been longest absent, should thus have held out—but so it was. At 9 we resumed our journey, there being about half a gallon a-piece for the horses just before we started; but although this, and the short rest they had, had relieved them, they got on slowly; and it was not until after midnight of the 27th, A.M. indeed of the 28th, that we reached the creek, with two short of our complement of horses, the Roan and the Colt both having dropped on the plains, but fortunately at no great distance, so that we recovered them in the course of the day. It will naturally be supposed that, arrived at a place of safety, we here rested for a while; but my mind was no sooner relieved from one cause for anxiety, than it was filled with another. If I except the thunder-storm which had enabled me to undertake my late journey from the creek, no rain had fallen, the weather had suddenly become oppressively hot, with a sky as clear as ether. I had still the mountain range to the N.E. to examine, and the upper branches of the creek, and in this necessary survey I knew no time was to be lost. Indeed I doubted if my return to the Depôt was not already shut out, by the drying up of the water in Strzelecki's Creek, although I hoped Mr. Browne still held his ground; but not only was I anxious on these heads, but as to our eventual retreat from these heartless regions. I would gladly have rested for a few days, for I was beginning to feel weak. From the 20th of July, and it was now the last day but two of October, I had been in constant exercise from sunrise to sunset; and if I except the few days I had rested at the Depôt, had slept under the canopy of heaven. My food had been insufficient to support me, and I had a malady hanging upon me that was slowly doing its work; but I felt that I had no time to spare, and, as I could not justify indulgence to myself, so on the 29th we commenced our progress up the creek, but halted at six miles on a beautiful sheet of water, and with every promise of success. In the course of the day we passed a singularly large grave. It was twenty-three feet long, and fourteen broad. The boughs on the top of it were laid so as to meet the oval shape of the mound itself, but the trees were not carved, nor were there any walks about it, as I had seen in other parts of the continent.

Native Grave.

Before we commenced our journey up the creek, I determined to secrete all the stores I could, in order to lighten the loads of the horses as much as possible, for they were now almost worn out; but it was difficult to say where we should conceal them, so as to be secure from the quick eyes of the natives. At first I thought my best plan would be to dig a hole and bury them, and then to light a fire, so as to obliterate the marks; but I changed my purpose, and placed them under a rhagodia bush, a short distance from the creek, and arranged some boughs all round it. In this place I hoped they would escape observation, for there were one or two things I should have exceedingly regretted to lose.

The weather had been getting warmer and warmer, and it had at this time become so hot that it was almost intolerable, worse indeed than at this season the previous year. The 30th was a day of oppressive heat, and the flies and mosquitoes were more than usually troublesome. I have not said much of these insects in the course of this narrative, for after all they are secondary objects only; but it is impossible to describe the ceaseless annoyance of these and a small ant. The latter swarmed in myriads in the creek and on the plains, and what with these little creatures at night, and the flies by day, we really had no rest. I continually wore a veil, or I could not have attended to our movements, or performed my duties. It is probable that being in the neighbourhood of water they were more numerous, but here they were a perfect plague, and in our depressed and wearied condition we, perhaps, felt their attacks more than we should otherwise have done. We commenced our journey at seven, and crossing the creek at three-quarters of a mile, ascended a small sand hill upon its proper left bank. Where we had crossed the channel was perfectly dry, but from the sand hill another magnificent sheet of water stretched away to the south-east as far as we could see.

From this point the creek appeared to be bounded by forest land, partly scrubby and partly grassed. To the south there were flats seemingly subject to floods, and lightly timbered, and beyond these were low sand hills. To the S.W. a high line of trees marked the course of a tributary from that quarter. To the north the country was exceedingly sandy and low, as well as to the east; and the direction of the sand ridges was only 5° to the west of north, so that from this point to our extreme west they gradually alter their line 17°, as in 138° of longitude they ran 22° to the west of north. I was not able to take more than one bearing from the hill I had ascended, to a remarkable flat-topped hill nearly N.E. I now crossed the creek on an east course, and traversed sandy plains, and low undulations, there being a tolerable quantity of grass on both and at four miles changed the route a little to

Cooper's Creek.

the northward for a small conical sand hill, from which the flat-topped hill bore 41°, and from it some darker hills were visible, somewhat more to the eastward, and as they appeared to be different from the sand ridges, I again changed my course for them, and crossing the bed of the creek at four miles, ascended a small stony range trending to the eastward, the creek being directly at their base. Following up its proper left bank I ascended another part of the range at three miles and a half, from which the flat-topped hill bore 24°, and the last hill I had ascended 239°. The channel of the creek had been dry for several miles, but we now saw a large sheet of water bearing due east, distant two miles, to which we made our way, and then stopped. From the top of this range the creek seemed to pass over extensive and bare plains in many branches, southward there were some stony hills, treeless and herbless, like those nearer to us. I was fairly driven down to the valley by the flies, as numerous on the burning stones on the top of the hill as anywhere else, and I left a knife and a pocket handkerchief behind me.

Notwithstanding the magnificent sheet of water we were now resting near, I began thus early to doubt the character of this creek. It had changed so often during the day, at one place having a broad channel, at another splitting into numerous small ones, having a great portion of its bed dry, and then presenting large and beautiful reaches to view, that I hardly knew what opinion to form of it; I also observed that it was leading away from the hills and taking us into a low and desolate region, almost as bad as that to the westward; however, time alone was to prove whether I was right in my surmises.

In the afternoon two natives made their appearance on the opposite side of the water, and I walked over to them, as I could not by any signs induce them to come to us. They were not bad looking men, and had lost their two front teeth of the upper jaw. To one I gave a tomahawk, and a hook to the other, but when I rose to depart, they gave them both back to me, and were astonished to find that I had intended them as presents. Seeing, I suppose, that we intended them no injury, these men in the morning went on with their ordinary occupations, and swimming into the middle of the water began to dive for mussels. They looked like two seals in the water with their black heads, and seemed to be very expert: at all events they were not long in procuring a breakfast.

Notwithstanding the misgivings I had as to the creek, the paths of the natives became wider and wider as we advanced. They were now as broad as a footpath in England, by a road side, and were well trodden; numerous huts of boughs also lined the creek, so that it was evident we were advancing into a well peopled country, and this circumstance raised my hopes that it would improve. As, however, our horses had no longer a gallop in them, we found it necessary to keep a sharp look out; although the natives with whom we had communicated, did not appear anxious to leave the place as they generally are to tell the news of our being on the creek to others above us.

On the 31st we started at 7 A.M., and at a mile and a half found ourselves at the termination of the stony ranges to our left. They fell back to the north, and a larger plain succeeded them. At two miles we crossed a small tributary, and passed over a stony plain, from which we entered an open box-tree forest extending far away to our left. At five miles and a half we found ourselves again on the banks of the creek, where it had an upper and a lower channel, that is to say, it had a lower channel for the stream, and an upper one independently of it. In the lower bed there was a little water, and we therefore stopped for a short time, the day being exceedingly hot. While here we saw a native at some water a little lower down, mending a net, but did not call to him. On resuming our journey we kept in the upper channel, and had not ridden very far when we saw a native about 150 yards ahead of us, pulling boughs. On getting nearer we called out to him, but to no purpose. At the distance of about 70 yards, we called out again, but still he did not hear, perhaps because of the rustling of the boughs he was breaking down. At length he bundled them up, and throwing them over his shoulder, turned from us to cross to the lower part of the creek, when suddenly he came bolt up against us. I cannot describe his horror and amazement, down went his branches, out went his hands, and trembling from head to foot, he began to shout as loud as he could bawl. On this we pulled up, and I desired Mr. Stuart to dismount and sit down. This for a time increased the poor fellow's alarm, for he doubtless mistook man and horse for one animal, and he stretched himself out in absolute astonishment when he saw them separate. When Mr. Stuart sat down, however, he stood more erect, and he gradually got somewhat composed. His shouting had brought another black, who had stood afar off, watching the state of affairs, but who now approached. From these men I tried to gather some information, and my hopes were greatly raised from what passed between us, insomuch that one of the men could not help expressing his hope that we were now near the long sought for inland sea.

On my seeking to know, by signs, to what point the creek would lead us, the old man stretched out his hand considerably to the southward of east, and spreading out his fingers, suddenly dropped his hand, as if he desired us to understand that it commenced, as he shewed, by numerous little channels uniting into one not very far off. On asking if the natives used canoes, he threw himself into the attitude of a native propelling one, which is a peculiar stoop, in which he must have been practised. After going through the motions, he pointed due north, and turning the palm of his hand forward, made it sweep the horizon round to east, and then again put himself into the attitude of a native propelling a canoe. There certainly was no mistaking these motions. On my asking if the creek went into a large water, he intimated not by again spreading out his hand as before and dropping it, neither did he seem to know anything of any hills. The direction he pointed to us, where there were large waters, was that over which the cold E.S.E. wind I have noticed, must have passed. This poor fellow was exceedingly communicative, but he did not cease to tremble all the while we were with him. After leaving him, the creek led us up to the northward of east, and we cut off every angle by following the broad and well beaten paths crossing from one to the other. At three miles I turned to ascend a conical sand hill, from whence the country appeared as follows: to the north were immense plains, with here and there a gum-tree on them; they were bounded in the distance by hills that I took to be the outer line of the range we purposed visiting; to the eastward the ground was undulating and woody; and southward, the prospect was bounded by low stony elevations, or a low range. The course of the creek was now north-east, in the direction of two distant sand hills. We now ran along it for seven miles, under an open box-tree forest, varying in breadth from a quarter of a mile to two miles; the creek frequently changed from a broad channel to a smaller one, but still having splendid sheets of water in it. At length, as we pushed up, it became sandy, and the lofty gum-trees that had ornamented it, gradually disappeared. Nevertheless we encamped on a beautiful spot.

The 1st of November broke bright and clear over us. Started at seven, the poor horses scarcely able to draw one leg after the other, the Roan having worn his hoof down the quick was exposed and raw, and he walked with difficulty. At a mile and a half we ascended an eminence, and to the eastward, saw a magnificent sheet of water to which we moved, and at five miles reached a low stony range, bounding the creek to the north; having ridden along a broad native path the whole of that distance, close to the edge of the above mentioned water. There were large rocks in the middle of it, and pelicans, one swan, several sea-gulls, and a number of cormorants on its bosom, together with many ducks, but none would let us within reach. We next ran on a bearing of 75°, or nearly east, along a large path, crossing numerous small branches of the creek, with deep and sandy beds, and occasionally over small stony plains. At noon we were at some distance from the creek, but then went towards it. The gum-trees were no longer visible, but melaleucas, from fifteen to twenty feet high, lined its banks like a copse of young birch. We now observed a long but somewhat narrow sheet of water, to which we rode; our suspicions as to its quality being roused by its colour, and the appearance of the melaleuca. It proved, as we feared, to be slightly brackish, but not undrinkable. Near the edge of the water, or rather about four or five feet from it, there was a belt of fine weeds, between which and the shore there were myriads of small fish of all sizes swimming, similar to those we had captured to the westward, in the fourth or O'Halloran's Creek. Here then was not only the clue as to how fish got into that isolated pond, but a proof of the westerly fall of the interior, since there was now no doubt whatever, but that the whole of the country Mr. Browne and I had traversed, even to the great sand hills on this side the Stony Desert, was laid under water, and by the overflow of this great creek filled the several creeks, and inundated the several plains that we had crossed. By so unexpected a fact, was this material point discovered. The Roan, at this time, could hardly walk, and not knowing when or at what distance we might again find water, or what kind of water it would be, I stopped on reaching the upper end of this pool, but even there it had a nasty taste, nor were any fish to be seen; a kind of weed covered the bed of the creek, and it looked like an inlet of the sea.

I was exceedingly surprised that we had not seen more natives, and momentarily expected to come on some large tribe, but did not, and what was very singular, all the paths were to the right, and none on the southern bank of the creek.

The weather continued intensely hot, and the flies swarmed in hundreds of thousands. The sky was without a cloud, either by day or night, and I could not but be apprehensive as to the consequences if rain should not fall; it was impossible that the largest pools could stand the rapid evaporation that was going on, but I did not deem it right to unburden my mind, even to Mr. Stuart, at this particular juncture.

On the morning of the 2nd of November the horses strayed for the first time, and delayed us for more than two hours, and we were after all indebted to three natives for their recovery, who had seen them and pointed out the direction in which they were. It really was a distressing spectacle to see them brought up, but their troubles and sufferings were not yet over. The Roan was hardly able to move along, and in pity I left him behind to wander at large along the sunny banks of the finest watercourse we had discovered.

Starting at 10 A.M. we crossed the creek, and traversed a large sandy plain, intersected by numerous native paths, that had now become as wide as an ordinary gravel walk. From this plain we observed a thin white line along the eastern horizon. The plain itself was also of white sand, and had many stones upon it, similar in substance and shape to those on the Stony Desert, but there was, notwithstanding, some grass upon it. A little above where we had slept, we struck a turn or angle of the creek where there was a beautiful sheet of water, but of a deep indigo blue colour. This was as salt as brine, insomuch that no animal could possibly have lived in it, and we observed water trickling into it from many springs on both sides. At four miles when we again struck the creek, after having crossed the plain, the water was perfectly fresh and sweet in a large pool close to which we passed. Here again there were several sea-gulls sitting on the rocks in the water, and a good many cormorants in the trees, yet I do not think there were any fish in this basin; I have no other reason for so thinking, however, than that we never saw any, either swimming in the water or rising to its surface in the coolness of evening on the sheets of fresh water. There might, however, have been fish of large size in the deep pools of this creek, although I would observe that I had two reasons for believing otherwise. The first was, that, the meshes of the nets used by the natives, of which we examined several hanging in the trees, were very small, and that among the fish bones at the natives' fires, we never saw any of a larger size than those we had ourselves captured, and it was evident that at this particular time, it was not the fishing season. I was led to think, that the water in which we noticed so many swimming about, was sacred, and that it is only when the creek overflows, that the fish are generally distributed along its whole line, that the natives take them. Certainly, to judge from the smooth and delicate appearance of the weeds round that sheet of water the fish were not disturbed.

We had been riding for some time on the proper left bank of the creek, but I at length crossed to the right and altered my course to E.S.E., but shortly afterwards ran due east across earthy plains covered with grass in tufts and very soft, but observing that I had got outside of the native tracks, and that there was no indication of the creek in front, I turned to the S.E. and at five miles struck a small sandy channel which I searched in vain for water; I therefore left it, crossing many similar channels still on a S.E. course; but observing that they all had level sandy beds, I gave up the hope of finding water in them and turned to the south, as the horses were not in a condition to suffer from want. At about two miles I ascended a sand hill, but could not see any thing of the creek; it was now getting late and two of the horses were hardly able to get along. Had we halted then, there was not a tree or a bush to which we could have tethered our animals, anxious too to get them to water I turned to the west, and at a mile got on a native path, that ultimately led me to the creek, and we pulled up at a small pond, where there was better feed than we had any right to expect.

We had hardly arranged our bivouac, when we heard a most melancholy howling over an earthen bank directly opposite to us, and saw seven black heads slowly advancing towards us. I therefore sent Mr. Stuart to meet the party and bring them up. The group consisted of a very old blind man, led by a younger one, and five women. They all wept most bitterly, and the women uttered low melancholy sounds, but we made them sit down and managed to allay their fears. It is impossible to say how old the man was, but his hair was white as snow, and he had one foot in the grave.

These poor creatures must have observed us coming, and being helpless, had I suppose thought it better to come forward, for they had their huts immediately on the other side of the bank over which they ventured. We gave the old man, a great coat, as the most useful present, and he seemed delighted with it. I saw that it was hopeless to expect any information from this timid party, so I made no objection to their leaving us after staying for about half an hour. Our latitude here, by an altitude of Jupiter, was 27°47'S.; our longitude by account 141°51'E.

The plains we had crossed during the day were very extensive, stretching from the north-west, to the south-east, like an open sea. They were thinly scattered over with box-trees, and comprised hundreds of thousands of acres of flooded grassy land. It is worthy of remark that none of these plains existed to the south of the creek, in which quarter the country was very barren, neither were there any native paths. We were at this time in too low a position to see any of the mountain ranges of which I have spoken. As the old native with the boughs had told us, the creek led us to the southward of east, and consequently away from them, and I feared that his further information would prove correct, and that we should soon arrive at its commencement.

The morning of the 3rd of November was as cloudy as the night of the 2nd had been, during which it blew violently from the N.W., and a few heat-drops fell, but without effect on the temperature. One of the horses got bogged in attempting to drink, and Mack's illness made it nine before we mounted and resumed our journey up the creek, on a N.N.E. course, but it gradually came round to north. At six miles we crossed the small and sandy bed of a creek coming from the stony plains to the south, and beneath a tree, near two huts, observed a large oval stone. It was embedded in the ground, and was evidently used by the natives for pounding seeds. We now proceeded along a broad native path towards some gum-trees, having stony undulating hills upon our right. Underneath the trees there was a fine deep pool in the channel of the creek, which had again assumed something of its original shape; but as we were in an immense hollow or bowl, and the view was very limited, I branched off to the hills, then not more than half a mile distant. From their summit the country to the south and south-west appeared darkly covered with brush; to the west, there were numerous stony undulations; northward and to the east were immense grassy plains, with many creeks, all making for a common centre upon them. In the near ground to the south-east the surface of the country was of fine white sand, partly covered with salsolaceous plants, with small fragments of stone, and patches of more grassy land. There was no fixed point on which to take a bearing, nor could we see anything of the higher ranges, now to the north-west of us.

In returning to the creek, we observed a body of natives to our left. They were walking in double file, and approaching us slowly. I therefore pulled up, and sent Mr. Stuart forward on foot, following myself with his horse. As he neared them the natives sat down, and he walked up and sat down in front of them. The party consisted of two chiefs and fourteen young men and boys. The former sat in front and the latter were ranged in two rows behind. The two chiefs wept as usual, and in truth shed tears, keeping their eyes on the ground; but Mr. Stuart, after the interview, informed me that the party behind were laughing at them and sticking their tongues in their cheeks. One of the chiefs was an exceedingly tall man, since he could not have measured less than six feet three inches, and was about 24 years of age. He was painted with red ochre, and his body shone as if he had been polished with Warren's best blacking. His companion was older and of shorter stature. We soon got on good terms with them, and I made a present of a knife to each. They told us, as intelligibly as it was possible for them to do, that we were going away from water; that there was no more water to the eastward, and, excepting in the creek, none anywhere but to the N.E. I had observed, indeed, that the native paths had altogether ceased on the side of the creek on which we then were (the south or left bank), and the chief pointed that fact out to me, explaining that we should have to cross the creek at the head of the water, under the trees, and get on a path that would lead us to the N.E. On this I rose up and mounting my horse, riding quietly towards it, descended into the bed of the creek, in which the natives had their huts, but their women and children were not there. The two chiefs and the other natives had followed, but, the former only crossed the creek and accompanied us. We almost immediately struck on the native path which, as my tall friend had informed me, led direct to the N.E.

I was not at first aware, what object our new friends had in following or rather accompanying us; but, at about a mile and a half, we came to a native hut at which there was an old man and his two lubras. The tall young man introduced him to us as his father, in consequence of which I dismounted, and shook hands with the old gentleman, and, as I had no hatchet or knife to give him, I parted my blanket and gave him half of it. We then proceeded on our journey, attended as before, and at a mile, came on two huts, at which there were from twelve to fifteen natives. Here again we were introduced by our long-legged friend, who kept pace with our animals with ease, and after a short parley once more moved on, but were again obliged to stop with another tribe, rather more numerous than the last, who were encamped on a dirty little puddle of water that was hardly drinkable; however, they very kindly asked us to stay and sleep, an honour I begged to decline. Thus, in the space of less than five miles, we were introduced to four different tribes, whose collective numbers amounted to seventy-one. The huts of these natives were constructed of boughs, and were of the usual form, excepting those of the last tribe, which were open behind, forming elliptic arches of boughs, and the effect was very pretty.

These good folks also asked us to stop, and I thought I saw an expression of impatience on the countenance of my guide when I declined, and turned my horse to move on. We had been riding on a sandy kind of bank, higher than the flooded ground around us. The plains extended on either side to the north and east, nor could we distinctly trace the creek beyond the trees at the point we had crossed it, but there were a few gum-trees separated by long intervals, that still slightly marked its course. When we left the last tribe, we rode towards a sand hill about half a mile in front, and had scarcely gone from the huts when our ambassadors, for in such a light I suppose I must consider them, set off at a trot and getting ahead of us disappeared over the sand hill. I was too well aware of the customs of these people, not to anticipate that there was something behind the scene, and I told Mr. Stuart that I felt satisfied we had not yet seen the whole of the population of this creek; but I was at a loss to conjecture why they should have squatted down at such muddy puddles, when there were such magnificent sheets of water for them to encamp upon, at no great distance; however, we reached the hill soon after the natives had gone over it, and on gaining the summit were hailed with a deafening shout by 3 or 400 natives, who were assembled in the flat below. I do not know, that my desire to see the savage in his wild state, was ever more gratified than on this occasion, for I had never before come so suddenly upon so large a party. The scene was one of the most animated description, and was rendered still more striking from the circumstance of the native huts, at which there were a number of women and children, occupying the whole crest of a long piece of rising ground at the opposite side of the flat.

I checked my horse for a short time on the top of the sand hill, and gazed on the assemblage of agitated figures below me, covering so small a space that I could have enclosed the whole under a casting net, and then quietly rode down into the flat, followed by Mr. Stuart and my men, to one of whom I gave my horse when I dismounted, and then walked to the natives, by whom Mr. Stuart and myself were immediately surrounded.

Had these people been of an unfriendly temper, we could not by any possibility have escaped them, for our horses could not have broken into a canter to save our lives or their own. We were therefore wholly in their power, although happily for us perhaps, they were not aware of it; but, so far from exhibiting any unkind feeling, they treated us with genuine hospitality, and we might certainly have commanded whatever they had. Several of them brought us large troughs of water, and when we had taken a little, held them up for our horses to drink; an instance of nerve that is very remarkable, for I am quite sure that no white man, (having never seen or heard of a horse before, and with the natural apprehension the first sight of such an animal would create,) would deliberately have walked up to what must have appeared to them most formidable brutes, and placing the troughs they carried against their breast, have allowed the horses to drink, with their noses almost touching them. They likewise offered us some roasted ducks, and some cake. When we walked over to their camp, they pointed to a large new hut, and told us we could sleep there, but I had noticed a little hillock on which there were four box-trees, about fifty yards from the native encampment, on which, foreseeing that we could go no farther, I had already determined to remain, and on my intimating this to the natives they appeared highly delighted; we accordingly went to the trees, and unsaddling our animals turned them out to feed. When the natives saw us quietly seated they came over, and brought a quantity of sticks for us to make a fire, wood being extremely scarce.

The men of this tribe were, without exception, the finest of any I had seen on the Australian Continent. Their bodies were not disfigured by any scars, neither were their countenances by the loss of any teeth, nor were they circumcised. They were a well-made race, with a sufficiency of muscular development, and stood as erect as it was possible to do, without the unseemly protrusion of stomach so common among the generality of natives. Of sixty-nine who I counted round me at one time, I do not think there was one under my own height, 5 feet 10¾ inches, but there were several upwards of 6 feet. The children were also very fine, and I thought healthier and better grown than most I had seen, but I observed here, as elsewhere amongst smaller tribes, that the female children were more numerous than the males, why such should be the case, it is difficult to say. Whilst, however, I am thus praising the personal appearance of the men, I am sorry to say I observed but little improvement in the fairer sex. They were the same half-starved unhappy looking creatures whose condition I have so often pitied elsewhere.

These were a merry people and seemed highly delighted at our visit, and if one or two of them were a little forward, I laid it to the account of curiosity and a feeling of confidence in their own numbers. But a little thing checked them, nor did they venture to touch our persons, much less to put their hands into our pockets, as the natives appear to have done, in the case of another explorer. It is a liberty I never allowed any native to take, not only because I did not like it, but because I am sure it must have the effect of lowering the white man in the estimation of the savage, and diminishing those feelings of awe and inferiority, which are the European's best security against ill treatment. The natives told us, that there was no water to the eastward, and that if we went there we should all die. They explained that the creek commenced on the plains, by spreading out their fingers as the old man had done, to shew that many small channels made a large one, pointing to the creek, and they said the water was all gone to the place we had come from; meaning, to the lower part of it. On asking them by signs, if the creek continued beyond the plains, they shook their heads, and again put their extended hand on the ground, pointing to the plain. They could give us no account of the ranges to which I proposed going, any more than others we had asked. On inquiring, if there was any water to the north-west a long discussion took place, and it was ultimately decided that there was not. I could understand, that several of them mentioned the names of places where they supposed there might be water, but it was evidently the general opinion that there was none. Neither did they appear to know of any large waters, on which the natives had canoes, in confirmation of the old man's actions. On this interesting and important point they were wholly ignorant.

The smallness of the water-hole, on which these people depended, was quite a matter of surprise to me, and I hardly liked to let the horses drink at it, in consequence. At sunset all the natives left us (as is their wont at that hour), and went to their own encampment; nor did one approach us afterwards, but they sat up to a late hour at their own camp, the women being employed beating the seed for cakes, between two stones, and the noise they made was exactly like the working of a loom factory. The whole encampment, with the long line of fires, looked exceedingly pretty, and the dusky figures of the natives standing by them, or moving from one hut to the other, had the effect of a fine scene in a play. At 11 all was still, and you would not have known that you were in such close contiguity to so large an assemblage of people. When I laid down, I revolved in my own mind what course I should pursue in the morning. If the account of the natives was correct, it was clear that my further progress eastward, was at an end. My horses, indeed, were now reduced to such a state, that I foresaw my labours were drawing to a close. Mack, too, was so ill, that he could hardly sit his animal, and although I did not anticipate anything serious in his case, anything tending to embarrass was now felt by us. Mr. Stuart and Morgan held up well, but I felt myself getting daily weaker and weaker. I found that I could not rise into my saddle with the same facility, and that I lost wind in going up a bank of only a few feet in height. I determined, however, on mature consideration, to examine the plain, and to satisfy myself before I should turn back, as to the fact of the creek commencing upon it. Accordingly, in the morning, we saddled and loaded our horses, but none of the natives came to us until we had mounted; when they approached to take leave, and to persuade us not to go in the direction we proposed, but to no purpose. The pool from which they drew their supply of water, was in the centre of a broad shallow grassy channel, that passed the point of the sand hill we had ascended, and ran up to the northward and westward; we were, therefore, obliged to cross this channel, and soon afterwards got on the plains. They were evidently subject to flood, and were exceedingly soft and blistered; the grass upon them grew in tufts, not close, so that in the distance, the plains appeared better grassed than they really were. At length, we got on a polygonum flat of great size, in the soil of which our horses absolutely sunk up to the shoulder at every step. I never rode over such a piece of ground in my life, but we managed to flounder through it, until at length we got on the somewhat firmer but still heavy plain. It was very clear, however, that our horses would not go a day's journey over such ground. It looked exactly as I have described it—an immense concavity, with numerous small channels running down from every part, and making for the creek as a centre of union; nor, could we anywhere see a termination to it. Had the plain been of less extent, I might have doubted the information of the natives; but, looking at the boundless hollow around me, I did not feel any surprise that such a creek even as the one up which we had journeyed, should rise in it, and could easily picture to myself the rush of water there must be to the centre of the plain, when the ground has been saturated with moisture.

The day being far advanced, whilst we were yet pushing on, without any apparent termination to the heavy ground over which we were riding, I turned westward at 2 P.M., finding that the attainment of the object I had in view, in attempting to cross the plain, was a physical impossibility. We reached the water, at which the blind native visited us, a little after sunset, and were as glad as our poor animals could have been, when night closed in upon us, and our labours.

On the 5th, we passed the old man's camp, in going down the creek, instead of crossing the plains as before, and halted at the junction of a creek we had passed, that came from the north, and along the banks of which I proposed turning towards the ranges. On the morning of the 6th we kept the general course of this tributary, which ran through an undulating country of rocks and sand. Its channel was exceedingly capacious, and its banks were high and perpendicular, but everything about it, was sand or gravel. Its bed was perfectly level, and its appearance at once destroyed the hope of finding water in it.

The ground over which we rode, was, as I have stated, a mixture of gravel and rocks, and our horses yielded under us at almost every step as they trod on the sharp pointed fragments. At eight miles we reached the outer line of hills, as they had appeared to us in the distance, and entered a pass between two of them, of about a quarter of a mile in width. At this confined point there were the remains and ravages of terrific floods. The waters had reached from one side of the pass to the other, and the dead trunks of trees and heaps of rubbish, were piled up against every bush.

There was not a blade of vegetation to be seen either on the low ground or on the ranges, which were from 300 to 400 feet in height, and were nothing more than vast accumulations of sand and rocks. At a mile, we arrived at the termination of the pass, and found ourselves at the entrance of a barren, sandy valley, with ranges in front of us, similar to those we had already passed. I thought it advisable, therefore, to ascend a hill to my left, somewhat higher than any near it, to ascertain, if possible, the character of the northern interior. The task of clambering to the top of it however, was, in my then reduced state, greater than I expected, and I had to wait a few minutes before I could look about me after gaining the summit. I could see nothing, after all, to cheer me in the view that presented itself. To the northward was the valley in which the creek rises, bounded all round by barren, stony hills, like that on which I stood; and the summits of other similar hills shewed themselves above the nearer line. To the east the apparently interminable plains on which we had been, still met the horizon, nor was anything to be seen beyond them. Westward the outer line of hills continued backed by others, in the outlines of which we recognised the peaks and forms of the apparently lofty chain we first saw when we discovered the creek. Thus, then, it appeared, that I had been entirely deceived in the character of these hills, and that it had been the effect of refraction in those burning regions, which had given to these moderate hills their mountain-like appearance.

Satisfied that my horses had not the strength to cross such a country, and that in it I had not the slightest chance of procuring the necessary sustenance for them, I turned back to Cooper's Creek, and then deemed it prudent to travel quietly on towards the place at which we first struck it, and had subsequently left our surplus stores.

In riding amongst some rocky ground, we shot a new and beautiful little pigeon, with a long crest. The habits of this bird were very singular, for it never perched on the trees, but on the highest and most exposed rocks, in what must have been an intense heat; its flight was short like that of a quail, and it ran in the same manner through the grass when feeding in the evening. We reached our destination on the evening of the 8th, and were astonished to see how much the waters had shrunk from their previous level. Such an instance of the rapid diminution of so large a pool, made me doubt whether I should find any water in Strzelecki's Creek to enable me to regain the Depôt.

As we descended from the flats to cross over to our old berth, we found it occupied by a party of natives, who were disposed to be rather troublesome, especially one old fellow, whose conduct annoyed me exceedingly. However, I very soon got rid of them; and after strolling for a short time within sight of us, they all went up the creek; but I could not help thinking, from the impertinent pertinacity of these fellows, that they had discovered my magazine, and taken all the things, more especially as they had been digging where our fire had been, so that, if I had buried the stores there as intended, they would have been taken.

As soon as the natives were out of sight, Mr. Stuart and I went to the rhagodia bush for our things. As we approached, the branches appeared just as we had left them; but on getting near, we saw a bag lying outside, and I therefore concluded that the natives had carried off everything. Still, when we came up to the bush, nothing but the bag appeared to have been touched, all the other things were just as we left them, and, on examining the bag, nothing was missing. Concluding, therefore, that the natives had really discovered my store, but had been too honest to rob us, I returned to the creek in better humour with them; but, a sudden thought occurring to Mr. Stuart, that as there was an oil lamp in the bag, a native dog might have smelt and dragged it out of its place, we returned to the bush, to see if there were any impressions of naked feet round about it, but with the exception of our own, there were no tracks save those of a native dog. I was consequently obliged to give Mr. Stuart credit for his surmise, and felt somewhat mortified that the favourable impression I had received as to the honesty of the natives had thus been destroyed. They had gone up the creek on seeing that I was displeased, and we saw nothing more of them during the afternoon; but on the following morning they came to see us, and as they behaved well, I gave them a powder canister, a little box, and some other trifles; for after all there was only one old fellow who had been unruly, and he now shewed as much impatience with his companions as he had done with us, and I therefore set his manner down to the score of petulance.

At 10 A.M. on the 9th we prepared to move over to the branch creek, as I really required rest and quiet, and knew very well that as long as I remained where I was, we should be troubled by our sable friends, who, being sixteen in number, would require being well looked after. Before we finally left the neighbourhood, however, where our hopes had so often been raised and depressed, I gave the name of Cooper's Creek to the fine watercourse we had so anxiously traced, as a proof of my great respect for Mr. Cooper, the Judge of South Australia. I am not conversant in the language of praise, but thus much will I venture to say, that whether in his public or private capacity, Mr. Cooper was equally entitled to this record of my feelings towards him. I would gladly have laid this creek down as a river, but as it had no current I did not feel myself justified in so doing. Had it been nearer the located districts of South Australia, its discovery would have been a matter of some importance. As it is we know not what changes or speculations may lead the white man to its banks. Purposes of utility were amongst the first objects I had in view in my pursuit of geographical discovery; nor do I think that any country, however barren, can be explored without the attainment of some good end. Circumstances may yet arise to give a value to my recent labours, and my name may be remembered by after generations in Australia, as the first who tried to penetrate to its centre. If I failed in that great object, I have one consolation in the retrospect of my past services. My path amongst savage tribes has been a bloodless one, not but that I have often been placed in situations of risk and danger, when I might have been justified in shedding blood, but I trust I have ever made allowances for human timidity, and respected the customs and prejudices of the rudest people. I hope, indeed, that in this my last expedition, I have not done discredit to the good opinion Sir C. Napier, an officer I knew not, was pleased to entertain of me. Most assuredly in my intercourse with the savage, I have endeavoured to elevate the character of the white man. Justice and humanity have been my guides, but while I have the consolation to know that no European will follow my track into the Desert without experiencing kindness from its tenants, I have to regret that the progress of civilized man into an uncivilized region, is almost invariably attended with misfortune to its original inhabitants.

I struck Cooper's Creek in lat. 27°44', and in long. 140°22', and traced it upwards to lat. 27°56', and long. 142°00'. There can be no doubt but that it would support a number of cattle upon its banks, but its agricultural capabilities appear to me doubtful, for the region in which it lies is subject evidently to variations of temperature and seasons that must, I should say, be inimical to cereal productions; nevertheless I should suppose its soil would yield sufficient to support any population that might settle on it.

CHAPTER XII.

CONTINUED DROUGHT—TERRIFIC EFFECT OF HOT WIND—THERMOMETER BURSTS—DEATH OF POOR BAWLEY—FIND THE STOCKADE DESERTED—LEAVE FORT GREY FOR THE DEPÔT—DIFFERENCE OF SEASONS—MIGRATION OF BIRDS—HOT WINDS—EMBARRASSING POSITION—MR. BROWNE STARTS FOR FLOOD'S CREEK—THREE BULLOCKS SHOT—COMMENCEMENT OF THE RETREAT—ARRIVAL AT FLOOD'S CREEK—STATE OF VEGETATION—EFFECTS OF SCURVY—ARRIVE AT ROCKY GLEN—COMPARISON OF NATIVE TRIBES—HALT AT CARNAPAGA—ARRIVAL AT CAWNDILLA—REMOVAL TO THE DARLING—LEAVE THE DARLING—STATE OF THE RIVER—OPPRESSIVE HEAT—VISITED BY NADBUCK—ARRIVAL AT MOORUNDI.

BY HALF past eleven of the 9th November we had again got quietly settled, and I then found leisure to make such arrangements as might suggest themselves for our further retreat. To insure the safety of the animals as much as possible, I determined to leave all my spare provisions and weightier stores behind, and during the afternoon we were engaged making the loads as compact and as light as we could.

It was not, however, the fear of the water in Strzelecki's Creek having dried up, that was at this moment the only cause of anxiety to me, for I thought it more than probable that Mr. Browne had been obliged to retreat from Fort Grey, in which case I should still have a journey before me to the old Depôt of 170 miles or more, under privations, to the horses at least, of no ordinary character; and I had great doubts as to the practicability of our final retreat upon the Darling. The drought had now continued so long, and the heat been so severe, that I apprehended we might be obliged to remain another summer in these fearful solitudes. The weather was terrifically hot, and appeared to have set in unusually early.

Under such circumstances, and with so many causes to render my mind anxious, the reader will believe I did not sleep much. The men were as restless as myself, so that we commenced our journey before the sun had risen on the morning of the 10th of November, to give the horses time to take their journey leisurely. Slowly we retraced our steps, nor did I stop for a moment until we had got to within five miles of our destination, at which distance we saw a single native running after us, and taking it into my head that he might be a messenger from Mr. Browne, I pulled up to wait for him, but curiosity alone had induced him to come forward. When he got to within a hundred yards, he stopped and approached no nearer. This little delay made it after sunset before we reached the upper pool (not the one Mr. Browne and I had discovered), and were relieved from present anxiety by finding a thick puddle still remaining in it, so that I halted for the night, Slommy, Bawley, and the colt had hard work to keep up with the other horses, and it really grieved me to see them so reduced. My own horse was even now beginning to give way, but I had carried a great load upon him.

As we approached the water, three ducks flew up and went off down the creek southwards, so I was cheered all night by the hope that water still remained at the lower pool, and that we should be in time to benefit by it. On the 11th, therefore, early we pushed on, as I intended to stop and breakfast at that place before I started for the Depôt. We had scarcely got there, however, when the wind, which had been blowing all the morning hot from the N.E., increased to a heavy gale, and I shall never forget its withering effect. I sought shelter behind a large gum-tree, but the blasts of heat were so terrific, that I wondered the very grass did not take fire. This really was nothing ideal: everything, both animate and inanimate, gave way before it; the horses stood with their backs to the wind, and their noses to the ground, without the muscular strength to raise their heads; the birds were mute, and the leaves of the trees, under which we were sitting, fell like a snow shower around us. At noon I took a thermometer, graduated to 127°, out of my box, and observed that the mercury was up to 125°. Thinking that it had been unduly influenced, I put it in the fork of a tree close to me, sheltered alike from the wind and the sun. In this position I went to examine it about an hour afterwards, when I found that the mercury had risen to the top of the instrument, and that its further expansion had burst the bulb, a circumstance that I believe no traveller has ever before had to record. I cannot find language to convey to the reader's mind an idea of the intense and oppressive nature of the heat that prevailed. We had reached our destination however before the worst of the hot wind set in; but all the water that now remained in the once broad and capacious pool to which I have had such frequent occasion to call the attention of the reader, was a shining patch of mud nearly in the centre. We were obliged to dig a trench for the water to filter into during the night, and by this means obtained a scanty supply for our horses and ourselves.

About sunset the wind shifted to the west, a cloud passed over us, and we had heavy thunder; but a few drops of rain only fell. They partially cooled the temperature, and the night was less oppressive than the day had been. We had now a journey of 86 miles before us: to its results I looked with great anxiety and doubt. I took every precaution to fortify the horses, and again reduced the loads, keeping barely a supply of flour for a day or two. Before dawn we were up, and drained the last drop of water, if so it could be called, out of the little trench we had made, and reserving a gallon for the first horse that should fall, divided the residue among them. Just as the morning was breaking, we left the creek, and travelled for 36 miles. I then halted until the moon should rise, and was glad to see that the horses stood it well. At seven we resumed the journey, and got on tolerably well until midnight, when poor Bawley, my favourite horse, fell; but we got him up again, and abandoning his saddle, proceeded onwards. At a mile, however, he again fell, when I stopped, and the water revived him. I now hoped he would struggle on, but in about an hour he again fell. I was exceedingly fond of this poor animal, and intended to have purchased him at the sale of the remnants of the expedition, as a present to my wife. We sat down and lit a fire by him, but he seemed fairly worn out. I then determined to ride on to the Depôt, and if Mr. Browne should still be there, to send a dray with water to the relief of the men. I told them, therefore, to come slowly on, and with Mr. Stuart pushed for the camp. We reached the plain just as the sun was descending, without having dismounted from our horses for more than fifteen hours, and as we rode down the embankment into it, looked around for the cattle, but none were to be seen. We looked towards the little sandy mound on which the tents had stood, but no white object there met our eye; we rode slowly up to the stockade, and found it silent and deserted. I was quite sure that Mr. Browne had had urgent reasons for retiring. I had indeed anticipated the measure: I hardly hoped to find him at the Fort, and had given him instructions on the subject of his removal, yet a sickening feeling came over me when I saw that he was really gone; not on my own account, for, with the bitter feelings of disappointment with which I was returning home, I could calmly have laid my head on that desert, never to raise it again. The feeling was natural, and had no mixture whatever of reproach towards my excellent companion.

We dismounted and led our horses down to water before I went to the tree under which I had directed Mr. Browne to deposit a letter for me. A good deal of water still remained in the channel, but nevertheless a large pit had been dug in it as I had desired. I did not drink, nor did Mr. Stuart, the surface of the water was quite green, and the water itself was of a red colour, but I believe we were both thinking of any thing but ourselves at that moment. As soon as we had unsaddled the horses, we went to the tree and dug up the bottle into which, as agreed upon, Mr. Browne had put a letter; informing me that he had been most reluctantly obliged to retreat; the water at the Depôt having turned putrid, and seriously disagreed with the men; he said that he should fall back on the old Depôt along the same line on which we had advanced, and expressed his fears that the water in Strzelecki's Creek would have dried, on the permanence of which he knew our safety depended. Under present circumstances the fate of poor Bawley, if not of more of our horses, was sealed. Mr. Stuart and I sat down by the stockade, and as night closed in lit a fire to guide Morgan and Mack on their approach to the plain. They came up about 2 P.M. having left Bawley on a little stony plain, and the colt on the sand ridges nearer to us, and in the confusion and darkness had left all the provisions behind; it therefore became necessary to send for some, as we had not had anything for many hours. The horses Morgan and Mack had ridden were too knocked up for further work, but I sent the latter on my own horse with a leather bottle that had been left behind by the party, full of water for poor Bawley, if he should still find him alive. Mack returned late in the afternoon, having passed the Colt on his way to the Depôt, towards which he dragged himself with difficulty, but Bawley was beyond recovery; he gave the poor animal the water, however, for he was a humane man, and then left him to die.

We had remained during the day under a scorching heat, but could hardly venture to drink the water of the creek without first purifying it by boiling, and as we had no vessel until Mack should come up we had to wait patiently for his arrival at 7 P.M. About 9 we had a damper baked, and broke our fast for the first time for more than two days. While sitting under a tree in the forenoon Mr. Stuart had observed a crow pitch in the little garden we had made, but which never benefited us, since the sun burnt up every plant the moment it appeared above the ground. This bird scratched for a short time in one of the soft beds, and then flew away with something in his bill. On going to the spot Mr. Stuart scraped up a piece of bacon and some suet, which the dogs of course had buried. These choice morsels were washed and cooked, and Mr. Stuart brought me a small piece of bacon, certainly not larger than a dollar, which he assured me had been cut out of the centre and was perfectly clean. I had not tasted the bacon since February, nor did I now feel any desire to do so, but I ate it because I thought I really wanted it in the weak state in which I was.

Perhaps a physician would laugh at me for ascribing the pains I felt the next morning to so trifling a cause, but I was attacked with pains at the bottom of my heels and in my back. Although lying down I felt as if I was standing balanced on stones; these pains increased during the day, insomuch that I anticipated some more violent attack, and determined on getting to the old Depôt as soon as possible; but as the horses had not had sufficient rest, I put off my journey to 5 P.M. on the following day, when I left Fort Grey with Mr. Stuart, directing Mack and Morgan to follow at the same hour on the following day, and promising that I would send a dray with water to meet them. I rode all that night until 3 P.M. of the 17th, when we reached the tents, which Mr. Browne had pitched about two miles below the spot we had formerly occupied. If I except two or three occasions on which I was obliged to dismount to rest my back for a few minutes we rode without stopping, and might truly be said to have been twenty hours on horseback.

Sincere I believe was the joy of Mr. Browne, and indeed of all hands, at seeing us return, for they had taken it for granted that our retreat would have been cut off. I too was gratified to find that Mr. Brown was better, and to learn that everything had gone on well. Davenport had recently been taken ill, but the other men had recovered on their removal from the cause of their malady.

When I dismounted I had nearly fallen forward. Thinking that one of the kangaroo dogs in his greeting had pushed me between the legs, I turned round to give him a slap, but no dog was there, and I soon found out that what I had felt was nothing more than strong muscular action brought on by hard riding.

As I had promised I sent Jones with a dray load of water to meet Morgan and Mack, who came up on the 19th with the rest of the horses. Mr. Browne informed me that the natives had frequently visited the camp during my absence. He had given them to understand that we were going over the hills again, on which they told him that if he did not make haste all the water would be gone. It now behoved us therefore to effect our retreat upon the Darling with all expedition. Our situation was very critical, for the effects of the drought were more visible now than before the July rain, no more indeed had since fallen, and the water in the Depôt creek was so much reduced that we had good reason to fear that none remained anywhere else. On the 18th I sent Flood to a small creek, between us and the pine forest, but he returned on the following day with information that it had long been dry. Thus then were my fears verified, and our retreat to the Darling apparently cut off. About this time too the very elements, against which we had so long been contending, seemed to unite their energies to render our stay in that dreadful region still more intolerable. The heat was greater than that of the previous summer; the thermometer ranging between 110° and 123° every day; the wind blowing heavily from N.E. to E.S.E. filled the air with impalpable red dust, giving the sun the most foreboding and lurid appearance as we looked upon him. The ground was so heated that our matches falling on it, ignited; and, having occasion to make a night signal, I found the whole of our rockets had been rendered useless, as on being lit they exploded at once without rising from the ground.

I had occasion—in the first volume of this work—to remark that I should at a future period have to make some observations on the state of the vegetation at this particular place; there being about a month or six weeks difference between the periods of the year when we first arrived at, and subsequently returned to it. When we first arrived on the 27th of January, 1845, the cereal grasses had ripened their seed, and the larger shrubs were fast maturing their fruit; the trees were full of birds, and the plains were covered with pigeons—having nests under every bush. At the close of November of the same year—that is to say six weeks earlier—not an herb had sprung from the ground, not a bud had swelled, and, where the season before the feathered tribes had swarmed in hundreds on the creek, scarcely a bird was as now to be seen. Our cattle wandered about in search for food, and the silence of the grave reigned around us day and night.

Was it instinct that warned the feathered races to shun a region in which the ordinary course of nature had been arrested, and over which the wrath of the Omnipotent appeared to hang? Or was it that a more genial season in the country to which they migrate, rendered their desertion of it at the usual period unnecessary? Most sincerely do I hope that the latter was the case, and that a successful destiny will await the bold and ardent traveller* who is now crossing those regions.

On the 20th I sent Flood down the creek to ascertain if water remained in it or the farther holes mentioned by the natives, thinking that in such a case we might work our way to the eastward; but on the

* Dr. Leichhardt had started to cross the Continent some time before.

23rd he returned without having seen a drop of water from the moment he left us. The deep and narrow channel I had so frequently visited, and which I had hoped might still contain water, had long been dry, and thus was our retreat cut off in that quarter also. There was apparently no hope for us—its last spark had been extinguished by this last disappointment; but the idea of a detention in that horrid desert was worse than death itself.

On the morning of the 22nd the sky was cloudy and the sun obscure, and there was every appearance of rain. The wind was somewhat to the south of west, the clouds came up from the north, and at ten a few drops fell; but before noon the sky was clear, and a strong and hot wind was blowing from the west: the dust was flying in clouds around us, and the flies were insupportable.

At this time Mr. Stuart was taken ill with pains similar to my own, and Davenport had an attack of dysentery.

On the 23rd it blew a fierce gale and a hot wind from west by north, which rendered us still more uncomfortable: nothing indeed could be done without risk in such a temperature, and such a climate. The fearful position in which we were placed, caused me great uneasiness; the men began to sicken, and I felt assured that if we remained much longer, the most serious consequences might be apprehended.

On the 24th, Mr. Browne went with Flood to examine a stony creek about 16 miles to the south, and on our way homewards. We had little hope that he would find any water in it, but if he did, a plan had suggested itself, by which we trusted to effect our escape. It being impossible to stand the outer heat, the men were obliged to take whatever things wanted repair, to our underground room, and I was happy to learn from Mr. Stuart, who I sent up to superintend them, that the natives had not in the least disturbed Mr. Poole's grave.

On the 25th Mr. Browne returned, and returned unsuccessful: he could find no water any where, and told me it was fearful to ride down the creeks and to witness their present state.

We were now aware that there could be no water nearer to us than 118 miles, i.e. at Flood's Creek, and even there it was doubtful if water any longer remained. To have moved the party on the chance of finding it would have been madness: the weather was so foreboding, the heat so excessive, and the horses so weak, that I did not dare to trust them on such a journey, or to risk the life of any man in such an undertaking. I was myself laid up, a helpless being, for I had gradually sunk under the attack of scurvy which had so long hung upon me. The day after I arrived in camp I was unable to walk in a day or two more, my muscles became rigid, my limbs contracted, and I was unable to stir; gradually also my skin blackened, the least movement put me to torture, and I was reduced to a state of perfect prostration. Thus stricken down, when my example and energies were so much required for the welfare and safety of others, I found the value of Mr. Browne's services and counsel. He had already volunteered to go to Flood's Creek to ascertain if water was still to be procured in it, but I had not felt justified in availing myself of his offer. My mind, however, dwelling on the critical posture of our affairs, and knowing and feeling as I did the value of time, and that the burning sun would lick up any shallow pool that might be left exposed, and that three or four days might determine our captivity or our release, I sent for Mr. Browne, to consult with him as to the best course to be adopted in the trying situation in which we were placed, and a plan at length occurred by which I hoped he might venture on the journey to Flood's Creek without risk. This plan was to shoot one of the bullocks, and to fill his hide with water. We determined on sending this in a dray, a day in advance, to enable the bullock driver to get as far as possible on the road, we then arranged that Mr. Browne should take the light cart, with 36 gallons of water, and one horse only; that on reaching the dray, he should give his horse as much water as he would drink from the skin, leaving that in the cart untouched until he should arrive at the termination of his second day's journey, when I proposed he should give his horse half the water, and leaving the rest until the period of his return, ride the remainder of the distance he had to go. I saw little risk in this plan, and we accordingly acted upon it immediately: the hide was prepared, and answered well, since it easily contained 150 gallons of water. Jones proceeded on the morning of the

27th, and on the 28th Mr. Browne left me on this anxious and to us important journey, accompanied by Flood. We calculated on his return on the eighth day, and the reader will judge how anxiously those days passed. On the day Mr. Browne left me, Jones returned, after having deposited the skin at the distance of 32 miles.

On the eighth day from his departure, every eye but my own was turned to the point at which they had seen him disappear. About 3 P.M., one of the men came to inform me that Mr. Browne was crossing the creek, the camp being on its left bank, and in a few minutes afterwards he entered my tent. "Well, Browne," said I, "what news? Is it to be good or bad?" "There is still water in the creek," said he, "but that is all I can say. What there is is as black as ink, and we must make haste, for in a week it will be gone." Here then the door was still open, a way to escape still practicable, and thankful we both felt to that Power which had directed our steps back again ere it was finally closed upon us; but even now we had no time to lose: to have taken the cattle without any prospect of relief until they should arrive at Flood's Creek, would have been to sacrifice almost the whole of them, and to reduce the expedition to a condition such as I did not desire. The necessary steps to be taken, in the event of Mr. Browne's bringing back good tidings, had engaged my attention during his absence, and with his assistance, that on which I had determined was immediately put into execution. I directed three more bullocks to be shot, and their skins prepared; and calculated that by abandoning the boat and our heavier stores, we might carry a supply of water on the drays, sufficient for the use of the remaining animals on the way. Three bullocks were accordingly killed, and the skins stripped over them from the neck downwards, so that the opening might be as small as possible.

The boat was launched upon the creek, which I had vainly hoped would have ploughed the waters of a central sea. We abandoned our bacon and heavier stores, the drays were put into order, their wheels wedged up, their axles greased, and on the 6th of December, at 5 P.M., we commenced our retreat, having a distance of 270 miles to travel to the Darling, and under circumstances which made it extremely uncertain how we should terminate the journey, since we did not expect to find any water between Flood's Creek and the Rocky Glen, or between the Rocky Glen and the Darling itself. The three or four days preceding our departure had been quite overpowering, neither did there seem to be a likelihood of any abatement of the heat when we left the Depôt. At 5 A.M. of the morning of the 7th, having travelled all night, I halted to rest the men and animals. We had then the mortification to find one of the skins was defective, and let out the water at an hundred different pores. I directed the water that remained in the skin to be given to the stock rather than that it should be lost; but both horses and bullocks refused it. During the first part of the night it was very oppressive; but about an hour after midnight the wind shifted to the south, and it became cooler. We resumed our journey at 7, and did not again halt until half past 12 P.M. of the 8th, having then gained the Muddy Lagoon, at which the reader will recollect we stopped for a short time after breaking through the Pine forest about the same period the year before; but as there was nothing for the animals to eat, I took them across the creek and put them upon an acre or two of green feed along its banks. I observed that the further we advanced southwards, the more forward did vegetation appear; Mr. Browne made the same remark to me on his return from Flood's Creek, where he found the grasses ripe, whereas at the Depôt Creek the ground was still perfectly bare.

About 3 A.M. we had a good deal of thunder and lightning, and at 7 the wind shifted a point or two to the eastward of south. Notwithstanding the quarter from which the wind blew, heavy clouds came up from the west, and about 11 we had a misty rain with heavy thunder and lightning. The rain was too slight to leave any puddles, but it moistened the dry grass, which the animals greedily devoured.

On leaving the creek we kept for about eight miles on our old track, but at that distance turned due south for two hills, the position of which Mr. Browne had ascertained on his recent journey, and by taking this judicious course avoided the Pine ridges altogether. We were, however, obliged to halt, as the moon set, in the midst of an open brush, but started again at day-break on the morning of the 9th.

Before we left the creek, near the Muddy Lagoon, all the horses and more than one half of the bullocks had drank plentifully of the water in the hides, in consequence of which they got on tolerably well. On resuming our journey we soon cleared the remainder of the scrub, and got into a more open sandy country, but the travelling on it was good; and at 20 minutes to two we halted within a mile of the hills towards which we had been moving, then about 26 miles from Flood's Creek. Being in great pain I left Mr. Browne at half-past three P.M., and reached our destination at midnight. Two hours afterwards Mr. Browne came up with the rest of the party. So we completed our first stage without the loss of a single animal, but had it not been for the slight rain that fell on the morning of the 8th, and the subsequent change of temperature, none of our bullocks could have survived the journey thus far.

As it had occupied three nights and two days, it became necessary to give both men and animals a day of rest. I could not however be so indulgent to Mr. Browne or to Flood. The next place at which we hoped to find water, was at the Rocky Gully at the foot of the ranges, distant 49 miles, if water failed us there, neither had Mr. Browne or Flood any reasonable expectation that we should procure any until we gained the Darling itself, then distant 150 miles. Mr. Browne was himself suffering severely from attacks of scurvy, but he continued with unwearied zeal to supply my place. On the 11th, at one P.M, he left me for the hills, but before he started we arranged that he should return and meet me half way whether he succeeded in finding water or not, and in order to ensure this I proposed leaving the Creek on the 13th.

As Mr. Browne had informed me, we found the vegetation much more forward at this place than we had hitherto seen it, still many of the grasses were invisible, not having yet sprung up, but there was a solitary stool of wheat that had been accidentally dropped by us and had taken root, which had 13 fine heads upon it quite ripe. These Mr. Browne gathered, and, agreeably to my wishes, scattered the seed about in places where he thought it would be most likely to grow. There was also a single stool of oats but it was not so fine as the wheat. On the 12th, at 2 P.M., Flood suddenly returned, bringing information that Mr. Browne had unexpectedly found water in the lower part of a little rocky creek in our way, distant 18 miles, and that he was gone on to the Rocky Gully. On receiving this intelligence I ordered the bullocks to be yoked up, and we started for the creek at which we had left the cart on our outward journey, at 7 P.M. It was blowing heavily at the time from the S.W. and large clouds passed over us, but the sky cleared as the wind fell at midnight. We reached our destination at 3 A.M. of the 13th. Here I remained until half-past six when we again started and gained the Horse-cart Creek at half-past twelve. Here, as at Flood's Creek, we found a large plant of mustard and some barley in ear and ripe, where few of the native grasses had more than made their appearance out of the ground.

Stopping to rest the animals for half an hour, I went myself to the little branch creek, on which the reader will recollect our cattle depended when we were last in this neighbourhood, and where I had arranged to meet Mr. Browne, who arrived there about half an hour before me. He had again been successful in finding a large supply of water in the Rocky Gully, and thought that rain must have fallen on the hills.

At 4 the teams again started, but I was too unwell to accompany them immediately. I had in truth lost the use of my limbs, and from the time of our leaving the Depôt had been lifted in and out of the cart; constant jolting therefore had greatly fatigued me, and I found it necessary to stop here for a short time after the departure of the drays. At half-past six however, we followed and overtook the party about five miles from the gully, where we halted at 3 A M. of the 14th.

Mr. Browne had found a large party of natives at the water, who had been very kind to him, and many of them still remained when we came up. He had observed some of them eating a small acid berry, and had procured a quantity for me in the hope that they would do me good, and while we remained at this place he good-naturedly went into the hills and gathered me a large tureen full, and to the benefit I derived from these berries I attribute my more speedy recovery from the malady under which I was suffering. We were now 116 miles from the Darling, and although there was no longer any doubt of our eventually reaching it, the condition in which we should do so, depended on our finding water in the Coonbaralba pass, from which we were distant 49 miles. In the evening I sent Flood on ahead to look for water, with orders to return if he succeeded in his search. In consequence of the kindness of the natives to Mr. Browne I made them some presents and gave them a sheep, which they appeared to relish greatly. They were good-looking blacks and in good condition, speaking the language of the Darling natives.

It was late on the 15th before we ascended the ranges; but, as I had only a limited distance to go it was not of much consequence, more especially as I purposed halting at the little spring, in the upper part of the Rocky Gully, at which Morgan and I stopped on a former occasion, when Mr. Browne and Flood were looking for a place by which we could descend from the hills to the plains of the desert interior. Mr. Browne took the short cut up the gully with the sheep; but when I reached the glen he had not arrived, and as he did not make his appearance for some time I became anxious, and sent after him, but he had only been delayed by the difficulty of the road, along which he described the scenery as very bold and picturesque.

We had not up to this time experienced the same degree of heat that prevailed at the Depôt. The temperature since the thunder on the 8th had been comparatively mild, and on ascending the hills we felt a sensible difference. I attributed it, however, to our elevated position, for we had on our way up the country experienced the nature of the climate of the Darling. We could not decidedly ascertain the fact from the natives, but as they were at this place in considerable numbers, both Mr. Browne and myself concluded that the river had not been flooded this year; neither had the season been the same as that of the former year, for it will be remembered that at the period the party crossed the ranges, a great deal of rain had fallen, in so much that the wheels of the drays sunk deep into the ground; but now they hardly left an impression, as they moved over it; and although more rain might have fallen on the hills than in the depressed region beyond them, it was clear that none had fallen for a considerable length of time in this neighbourhood.

Mr. Browne saw five or six rock Wallabies as he was coming up the glen, and said they were beautiful little animals. He remarked that they bounded up the bold cliffs near him with astonishing strength and activity; in some places there were basaltic columns, resting on granite, 200 and 300 feet high.

Flood returned at 4 A.M. having found water, though not of the best description, in the pass. His horse had, however, drank plentifully of it, so that I determined on pushing from that point to Cawndilla, hoping by good management to secure the cattle reaching it in safety.

Considering the distance we had to go we started late, but the bullocks had strayed down the creek, and it took some time to drive them over such rugged ground.

I preceded the party in the cart, leaving Mr. Browne in charge of the drays, and crossing the ranges descended into the pass two hours after sunset. We passed a brackish pool of water, and stopped at a small well, at which there were two native women. The party came up about two hours after midnight, the men and animals being greatly fatigued, so that it was absolutely necessary to remain stationary for a day. Our retreat had been a most harassing one, but it admitted of no hesitation. Though we had thus far, under the blessing of Providence, brought every thing in safety, and had now only one more effort to make, Cawndilla was still distant 69 miles, between which and our position there was not a drop of water.

One of the women we found here, came and slept at our fire, and managed to roll herself up in Mr. Browne's blanket, who, waking from cold, found that his *fair* companion had uncovered him, and appropriated the blanket to her own use. The natives suffer exceedingly from cold, and are perfectly paralysed by it, for they are not provided with any covering, neither are their huts of a solidity or construction such as to protect them from its effects. About noon a large tribe joined us from the S.W. and we had a fine opportunity to form a judgment of them, when contrasted with the natives of the Desert from which we had come. Robust, active, and full of life, these hill natives were every way superior to the miserable half-starved beings we had left behind, if

I except the natives of Cooper's Creek. During the day they kept falling in upon us and in the afternoon mustered more than one hundred strong, in men, women, and children. As they were very quiet and unobtrusive I gave them a couple of sheep, with which they were highly delighted, and in return, they overwhelmed our camp at night with their women.

I mentioned in a former part of this work, that Mr. Browne and I had succeeded in capturing a Dipus, when journeying to the N.W. We had subsequently taken another, and had kept them both for some time, but one died, and the other springing out of its box was killed by the dogs. From the habits of this animal I did not expect to succeed in taking it home, but I had every hope that some Jerboas, of which we had five, would outlive the journey, for they thrived well on the food we gave them. I was, however, quite provoked at this place to find that two of them had died from the carelessness of the men throwing the tarpauline over the box, and so smothering them. The survivors were all but dead when looked at, and I feared we should lose them also.

As the morning of the 19th dawned, and distant objects became visible, the plains of the Darling gradually spread out before us. We commenced our journey to Cawndilla at half-past 7, and travelled down the creek until 2 P.M., when we halted for two hours during the heat of the day at Carnapaga. At 4 we resumed our journey, and again stopped for an hour on the little sand hill at the lower part of the creek, to enable the men to take some refreshment. At quarter-past 8 we turned from the creek and travelled all night by the light of a lamp, and at daylight were 18 miles from Cawndilla. We had kept upon our former tracks, on which the cattle had moved rapidly along, but they now began to flag. Mr. Browne was in front of the party with Mr. Stuart, but he suddenly returned, and coming up to my cart gave me a letter he had found nailed up to a tree by Mr. Piesse. This letter was to inform me of his arrival on the banks of the Williorara on the 6th of the month, of his having been twice on the road in the hope of seeing us, and sent natives to procure intelligence of us, who returned in so exhausted a state, that he had given up all expectation of our being able to cross the hills. He stated that we should find a barrel of water a little further on, together with a letter from head quarters, but had retained all other letters until he should see me; nevertheless, he had the gratification to tell me that he had seen Mrs. Sturt the day before he left Adelaide, and that she was well. About a mile further on, we found the barrel of water, and relieved our suffering horses, and thus benefited by the prudent exertions of Mr. Piesse. Nothing, indeed, appeared to have escaped the anxious solicitude of that zealous officer to relieve our wants.

I reached Cawndilla at 9 A.M. and stopped on the banks of the Williorara at the dregs of a water-hole, about six inches deep, it being all that remained in the creek, but I was too much fatigued to push on to the Darling, a further distance of seven miles, where Mr. Piesse then was. The drays came up a little after noon; the cattle almost frantic from the want of water. It was with difficulty the men unyoked them, and the moment they were loose they plunged headlong into the creek and drank greedily of the putrid water that remained.

Amongst the letters I now received was one from the Colonial Secretary, informing me, that supplies had been forwarded to the point I had specified, according to the request contained in my letter of July; that my further suggestions had been acted upon, and that the Governor had availed himself of Mr. Piesse's services again, to send him in charge of the party: thus satisfied that he was on the Darling, I sent Mr. Browne and Mr. Stuart in advance, to apprise him of our approach.

On their arrival at his camp Mr. Piesse lost no time in repairing to me, and I shall not readily forget the unaffected joy he evinced at seeing me again. He had maintained a friendly intercourse with the natives, and had acquitted himself in a manner, as creditable to himself, as it had been beneficial to me.

Mr. Piesse was the bearer of numerous letters from my family and friends, and I was in some measure repaid for the past, by the good intelligence they conveyed: that my wife and children were well, and the colony was in the most flourishing condition—since, during my absence, that stupendous mine had been discovered, which has yielded such profit to the owners—and the pastoral pursuits of the colonists were in an equally flourishing condition. Mr. Browne, too, received equally glad tidings from his brother, who informed him of his intention to meet the party on its way homeward.

On the 21st I moved over to the Darling; and found a number of natives at the camp, and amongst them the old Boocolo of Williorara, who was highly delighted at our return.

Mr. Piesse had constructed a large and comfortable hut of boughs—which was much cooler than canvass. In this we made ourselves comfortable, and I hoped that the numerous and more generous supplies of eatables and drinkables than those to which we had been accustomed would conduce to our early restoration to health. I could not but fancy that the berries Mr. Browne had procured for me, and of which I had taken many, were beginning to work beneficially although I was still unable to move. As I proposed remaining stationary until after Christmas Day, I deemed it advisable to despatch messengers with letters for the Governor, advising him of my safety, and to relieve the anxiety of my family and friends. Mr. Browne accordingly made an agreement with two natives, to take the letter-bag to the ana-branch of the Darling, and send it on to Lake Victoria by other natives, who were to be rewarded for their trouble. For this service our messengers were to receive two blankets and two tomahawks, and the bag being closed they started off with it. I had proposed to Mr. Browne to be himself the bearer of it, but he would not leave me, even now. In order, therefore, to encourage the messengers, I gave them in advance the tomahawks they were to have received on their return. Our tent was generally full of natives; some of them very fine young men, especially the two sons of the Boocolo. Topar made his appearance two or three days after our arrival, but Toonda was absent on the Murray: the former, however, having been detected in attempting a theft, I had him turned out of the tent and banished the camp. The old Boocolo came daily to see us, and as invariably laid down on the lower part of my mattrass.

On the 23rd I sent Mr. Stuart to verify his former bearings on Scrope's Range, and Mr. Browne kindly superintended the chaining of the distance between a tree I had marked on the banks of the Darling and

Sir Thomas Mitchell's last camp. This tree was about a quarter of a mile below the junction of the Williorara, and had cut on it, [G.A.E., Dec. 24, 1843] the distance between the two points was three miles and 20 chains.

The 25th being Christmas Day, I issued a double allowance to the men, and ordered that preparations should be made for pushing down the river on the following morning. About 2 P.M. we were surprised at the return of our two messengers, who insisted that they had taken the letter-bag to the point agreed upon, although it was an evident impossibility that they could have done so. I therefore evinced my displeasure and refused to give them the blankets—for which, nevertheless, they greatly importuned me. Mr. Browne, however, explained to the Boocolo why I refused, and charged the natives with having secreted it somewhere or other. On this there was a long consultation with the natives, which terminated in the Boocolo's two sons separating from the others, and talking together for a long time in a corner of my hut; they then came forward and said, that my decision was perfectly just, for that the men had not been to the place agreed upon, but had left the bag of letters with a tribe on the Darling, and therefore, that they had been fully rewarded by the present of the tomahawks. This decided opinion settled the dispute at once, and the parties quietly acquiesced.

I had, as stated, been obliged to turn Topar out of my tent, and expel him the camp for theft, but at the same time Mr. Browne explained to the natives why I did so, and told them that I should in like manner expel any other who so transgressed, and they appeared fully to concur in the justice of my conduct. There is no doubt indeed but that they punish each other for similar offences, although perhaps the moral turpitude of the action is not understood by them.

The Darling at this time had ceased to flow, and formed a chain of ponds. The Williorara was quite dry from one end to the other, as were the lagoons and creeks in the neighbourhood. The natives having cleared the river of the fish that had been brought down by the floods, now subsisted for the most part on herbs and roots of various kinds, and on the caterpillar of the gum-tree moth, which they procured out of the ground with their switches, having a hook at the end. I do not think they could procure animal food in the then state of the country, there being no ducks or kangaroos in the neighbourhood, in any great quantity at all events.

I thus early began to feel the benefit of a change of diet in the diminished rigidity of my limbs, and therefore entertained great hopes that I should yet be able to ride into Adelaide. The men too generally began to recover from their fatigues, but both Mr. Browne and Mr. Stuart continued to complain of shooting pains in their limbs. The party and the animals however being sufficiently recruited to enable us to resume our progress homewards, we broke up our camp at the junction of the Williorara on the 26th of the month as I had proposed, under more favourable circumstances than we could have expected, the weather being beautifully fine and the temperature pleasant. When I was carried out of my tent to the cart, I was surprised to see the verdure of that very ground against the barrenness of which I had had to declaim the preceding year; I mean the flats of the Williorara, now covered with grass, and looking the very reverse of what they had done before; so hazardous is it to give an opinion of such a country from a partial glimpse of it. The incipient vegetation must have been brought forth by flood or heavy rains.

We passed two tribes of natives, with whom we staid for a short time as the old Boocolo was with us. Amongst these, natives we did not notice the same disproportion in the sexes as in the interior, but not only amongst these tribes but with those of Williorara and Cawndilla, we observed that many had lost an eye by inflammation from the attacks of flies. I was really surprised that any of them could see, for most assuredly it is impossible to conceive anything more tormenting than those brutes are in every part of the interior.

On the 27th we passed two of our old encampments, and halted after a journey of 16 miles in the close vicinity of a tribe of natives, about fifty in number, the majority of whom were boys as mischievous as monkeys, and as great thieves too, but we reduced them to some kind of order by a little patience. The Darling had less water than in the previous year before the flood, but its flats were covered with grass, of which hundreds of tons might have been cut, so that our cattle speedily began to improve in condition.

About this time the weather was exceedingly oppressive, and heavy thunder-clouds hung about, but no rain fell.

Our journey on the 28th was comparatively short. We passed the location of another tribe during the day, and recovered our letter-bag, which had been left by our messengers with a native belonging to it. Here the old Boocolo left us and returned to Williorara.

The last days of 1845 and the few first of 1846 were exceedingly oppressive, and the heat was almost as great as in the interior itself. On the 5th of January we crossed over from the Darling to its ancient channel, and on the 6th Mr. Browne left for Adelaide. On the 8th I reached Lake Victoria, where I learnt that our old friend Nadbuck had been speared by a native, whose jealousy he had excited, but that his wound was not mortal. He was somewhere on the Rufus, which I did not approach, but made a signal fire in the hope that he would have seen it, and, had they not been spoiled, I should have thrown up a rocket at night. However Nadbuck heard of our return, and made a successful effort to get to us, and tears chased each other down the old man's cheeks when he saw us again. Assuredly these poor people of the desert have the most kindly feelings—for not only was his reception of us such as I have described, but the natives one and all exhibited the utmost joy at our safety, and cheered us on every part of the river.

It blew very heavily on the night of the 10th, but moderated towards the morning, and the day turned out cooler than usual. The lagoons of the Murray were full of fish and wild fowl, and my distribution of all the hooks and lines I had brought back enabled my sable friends to capture an abundance of the former without going into the water, and they very soon appreciated the value of such instruments.

On the 13th I left Mr. Piesse in charge of the party, and pushed on to Moorundi, and arrived at the settlement, into which I was escorted by the natives raising loud shouts, on the 15th. Here my kind friends made me as comfortable as they could. Mr. Eyre had gone to England on leave of absence, and Mr. Nation was filling his appointment as Resident.

On the 17th I mounted my horse for the first time since I had been taken ill in November, and had scarcely left Moorundi when I met my good friends Mr. Charles Campbell and Mr. A. Hardy in a carriage to convey me to Adelaide. I reached my home at midnight on the 19th of January, and, on crossing its threshold, raised my wife from the floor on which she had fallen, and heard the carriage of my considerate friends roll rapidly away.

Mr. Eyre's house at Moorundi.

CHAPTER XIII.

REMARKS ON THE SEASON—DRY STATE OF THE ATMOSPHERE—THERMOMETRICAL OBSERVATIONS—WINDS IN THE INTERIOR—DIRECTION OF THE RANGES—GEOLOGICAL OBSERVATIONS—NON-EXISTENCE OF ANY CENTRAL CHAIN—PROBABLE COURSE OF THE STONY DESERT—WHETHER CONNECTED WITH LAKE TORRENS—OPINIONS OF CAPTAIN FLINDERS—NO INFORMATION DERIVED FROM THE NATIVES—THE NATIVES—THEIR PERSONAL APPEARANCE—DISPROPORTION BETWEEN THE SEXES—THE WOMEN—CUSTOMS OF THE NATIVES—THEIR HABITATIONS—FOOD—LANGUAGE—CONCLUSION.

HAVING thus brought my narrative to a conclusion I shall trespass but little more on the patience of the reader. It appears to me that a few observations are necessary to clear some parts, and to make up for omissions in the body of my work. I have written it indeed under considerable disadvantage; for although I have in a great measure recovered from the loss of sight consequent on my former services, I cannot glance my eye so rapidly as I once did over such a voluminous document as this journal; and I feel that I owe it to the public, as well as to myself, to make this apology for its imperfections.

There were two great difficulties against which, during the progress of the expedition, I had to contend. The one was, the want of water; the other, the nature of the country. That it was altogether impracticable for wheeled carriages of any kind, may readily be conceived from my description; and in the state in which I found it, horses were evidently unequal to the task. I cannot help thinking that camels might have done better; not only for their endurance, but because they carry more than a horse. I should, undoubtedly, have been led to try those animals if I could have procured them; but that was impossible. Certain however it is, that I went into the interior to meet with trials that scarcely camels could have borne up against; for I think there can be no doubt, from the facts I have detailed, that the season, during which this expedition was undertaken, was one of unusual dryness; but although the arid state of the country contributed so much to prevent its movements, I question whether, under opposite circumstances, it would have been possible to have pushed so far as the party succeeded in doing. Certainly, if the ground had been kept in a state of constant saturation, travelling would have been out of the question; for the rain of July abundantly proved how impracticable any attempt to penetrate it under such circumstances would have been.

It is difficult to say what kind of seasons prevail in Central Australia. That low region does not, as far as I can judge, appear to be influenced by tropical rains, but rather to be subject to sudden falls. That the continent of Australia was at one time more humid than it now is, appears to be an admitted fact; the marks of floods, and the violence of torrents (none of which have been witnessed), are mentioned by every explorer as traceable over every part of the continent; but no instance of any general inundation is on record: on the contrary the seasons appear to be getting drier and drier every year, and the slowness with which any body exposed to the air decomposes, would argue the extreme absence of moisture in the atmosphere. It will be remembered that one of my bullocks died in the Pine Forest when I was passing through it in December, 1844. In July, 1845, when Mr. Piesse was on his route home from the Depôt in charge of the home returning party, he passed by the spot where this animal had fallen; and, in elucidation of what I have stated, I will here give the extract of a letter I subsequently received from him from India. Speaking of the humidity of the climate of Bengal, he says: "It appears to me that heat alone is rather a preservative from decomposition; of which I recollect an instance, in the bullock that died in the march through the Pine scrub on the 1st of January, 1845. When I passed by the spot in the following July, the carcase was dried up like a mummy, and was in such a perfect state of preservation as to be easily recognised."

No stronger proof, I apprehend, could have been adduced of the dryness of the atmosphere in that part of the interior, or more corroborative of the intensity of heat there during the interval referred to; but the singular and unusual effects it had on ourselves, and on every thing around was equally corroborative of the fact. The atmosphere on some occasions was so rarified, that we felt a difficulty in breathing, and a buzzing sensation on the crown of the head, as if a hot iron had been there.

There were only two occasions on which the thermometer was noticed to exceed the range of 130° in the shade, the solar intensity at the same time being nearly 160°. The extremes between this last and our winter's cold, when the thermometer descended to 24° was 133°. I observe that Sir Thomas Mitchell gives the temperature at the Bogan, in his tent at 117° and when exposed to the wind at 129°; but I presume that local causes, such as radiation from stones and sand, operated more powerfully with us than in his case. Whilst we were at the Depôt about May, the water of the creek became slightly putrid, and cleared itself like Thames water; and during the hotter months of our stay there, it evaporated at the rate of nearly an inch a day, as shewn by a rod Mr. Browne placed in it to note the changes, but the amount varied according to the quiescent or boisterous state of the atmosphere. It will readily be believed that in so heated a region the air was seldom still; to the currents sweeping over it we had to attribute the loathsome and muddy state of the water on which we generally subsisted after we left that place, for the pools from which we took it were so shallow as to be stirred up to the consistency of white-wash by the play and action of the wind on their surfaces. During our stay at the Depôt the barometer never rose above 30·260, or fell below 29·540.

From December, 1844, to the end of April of the following year, the prevailing winds were from E.N.E. to E.S.E., after that month they were variable, but westerly winds predominated. The south wind was always cold, and its approach was invariably indicated by the rise of the barometer.

The rain of July commenced in the north-east quarter and gradually went round to the north-west; but more clouds rose from the former point than from any other. The sky generally speaking was without a speck, and the dazzling brightness of the moon was one of the most distressing things we had to endure when out in the bush. It was impossible indeed to shut out its light which ever way one turned, and its irritating effects were remarkable.

It will be observable to those who cast their eyes over the chart of South Australia that the range of mountains between St. Vincent's Gulf and the Murray river runs up northwards into the interior. In like manner the ranges crossed by the Expedition also ran in the same direction. The Black Rock Hill, so named by Captain Frome, is in lat. 32°45' and in the 139th meridian, and is the easternmost of the chain to which it belongs. Mount Gipps on the Coonbaralba range is in lat. 31°52' and in long. 141°41', but from that point the ranges trend somewhat to the westward of south, and consequently, may run nearer to that (of which the Black Rock Hill forms so prominent a feature) than we may suppose, but there is a distance of nearly 150 miles of country still remaining to be explored, before this point can be decided. Nevertheless, it is more than probable the two chains are in some measure connected, especially as they greatly resemble each other in

their classification. They are for the most part composed of primary igneous rocks, amongst which there is a general distribution of iron, and perhaps of other metals. The iron ore, however, that was discovered during the progress of the Expedition, of which Piesse's Knob is a remarkable specimen, was of the purest kind.

It was, as has been found in South Australia, a surface deposit, protruding or cropping out of the ground in immense clean blocks. This ore was highly magnetic; the veins of the metal run north and south, the direction of the ranges, as did a similar crop on the plains at the S.E. base of the ranges. Generally speaking there was nothing bold or picturesque in the scenery of the Barrier Range, but the Rocky Glen and some few others of a similar description were exceptions. As the Barrier Range ran parallel to the coast ranges, so there were other ranges to the eastward of the Barrier Range, running parallel to it, and they were separated by broad plains, partly open and partly covered with brush. The general elevation of the ranges was about 1,200 feet above the level of the sea, but some of the hills exceeded 1,600. Mount Lyell was 2,000; Mount Gipps 1,500; Lewis's Hill 1,000: but the general elevation of the range might be rather under than over what I have stated. It appears to me that the whole of the geological formation of this portion of the continent is the same, and that all the lines of ranges terminate in the same kind of way to the north, that is to say, in detached flat-topped hills of compact or indurated quartz shewing white and abrupt faces. So terminated the Coonbaralba Range, and so Mr. Eyre tells us did the Mount Serle Range, and so terminated the range we saw to the westward of Lake Torrens.

That they exhibit evidences of a past violent commotion of waters, I think any one who will follow my steps and view them, will be ready to admit.

That the range of hills I have called "Stanley's Barrier Range," and that all the mountain chains to the eastward and westward of it, were once so many islands I have not the slightest doubt, and that during the primeval period, a sea covered the deserts over which I wandered; but it is impossible for a writer, whatever powers of description he may have, to transfer to the minds of his readers the same vivid impressions his own may have received, on a view of any external object.

From the remarks into which I have thus been led, as well as those which have escaped me in the course of this narrative, it will be seen that the impressions I had received as to the past and present state of the continent were rather strengthened than diminished, on my further knowledge of its internal structure.

It is true, that I did not find an inland sea as I certainly expected to have done, but the country as a desert was what I had anticipated, although I could not have supposed it would have proved of such boundless extent.

Viewing the objects for which the Expedition was equipped, and its results, there can, I think, be no doubt, as to the non-existence of any mountain ranges in the interior of Australia, but, on the contrary, that its central regions are nearly if not quite on a sea level, and that the north coast is separated from the south as effectually as if seas rolled between them. I have stated my opinion that that portion of the desert which I tried to cross continues with undiminished breadth to the Great Australian Bight, and I agree with Captain Flinders, in supposing that if an inland sea exists any where, it exists underneath and behind that bank, (speaking from seaward). It would, I think, be unreasonable to suppose that such an immense tract of sandy desert, once undoubtedly a sea-bed, should immediately contract; considering, indeed, the sterile character of the country to the north of Gawler's Range, to the westward of Port Lincoln, and along the whole of the south coast of Australia, nearly to King George's Sound, I must confess I have no hope of any inland fertile country. I am aware it is the opinion of some of my friends that the Stony Desert may communicate with Lake Torrens. Such may have been and still may be the case—I will not argue the contrary, or answer for the changes in so extraordinary a region. I only state my own ideas from what I observed, strengthened by my view of the position I occupied, when at my farthest north; we will therefore refer to that position, and to the position of Lake Torrens, and see how far it is probable, that a large channel, such as I have described the Stony

Desert to be, should turn so abruptly, as it must do to connect itself with that basin; the evident fall of the interior, as far as that fact could be ascertained, being plainly from east to west.

The western shore of Lake Torrens, as laid down by Mr. Eyre, is in 137°40' or thereabouts. Its eastern shore in 141° of longitude. Its southern extremity being in lat. 28½°. My position was in 138° of long. and 24°40' of latitude. I was therefore within 20 miles as far to the westward of the western-most part of Lake Torrens, and was also 250 geographical miles due north of it. To gain Lake Torrens, the Stony Desert must turn at a right angle from its known course, and in such case hills must exist to the westward of where I was, for hills alone could so change the direction of a current, but the whole aspect of the interior would argue against such a conclusion. I never lost sight of the probability of Lake Torrens being connected with some central feature, until my hopes were destroyed by the nature of the country I traversed, nor do I think it probable that in so level a region as that in which I left it, there is any likelihood of the Stony Desert changing its direction so much as to form any connection with the sandy basin to which I have alluded. Nevertheless it may do so. We naturally cling to the ideas we ourselves have adopted, and it is difficult to transfer them to the mind of another. In reference however to what I had previously stated, I would give the following quotation from Flinders. His impressions from what he observed while sailing along the coast, in a great measure correspond with mine when travelling inland, the only point we differ upon is as to the probable origin of the great sea-wall, which appeared to him to be of calcareous formation, and he therefore concluded that it had been a coral reef raised by some convulsion of nature. Had Capt. Flinders been able to examine the rock formation of the Great Australian Bight, he would have found that it was for the most part an oolitic limestone, with many shells imbedded in it, similar in substance and in formation to the fossil bed of the Murray, but differing from it in colour.

"The length of these cliffs from their second commencement is 33 leagues, and that of the level bank from New Cape Paisley, where it was first seen from the sea, no less than 145 leagues. The height of this extraordinary bank is nearly the same throughout, being nowhere less by estimation than 400 feet, not anywhere more than 600. In the first 20 leagues the rugged tops of some inland mountains were visible over it, but during the remainder of its long course, the bank was the limit of our view.

"This equality of elevation for so great an extent, and the evidently calcareous nature of the bank, at least in the upper 200 feet, would bespeak it to have been the exterior line of some vast coral reef, which is always more elevated than the interior parts, and commonly level with high water mark. From the gradual subsiding of the sea, or perhaps from some convulsion of nature, this bank may have attained its present height above the surface, and however extraordinary such a change may appear, yet when it is recollected that branches of coral still exist, upon Bald Head, at the elevation of 400 feet or more, this supposition assumes a degree of probability, and it would farther seem that the subsiding of the waters has not been at a period very remote, since these frail branches have yet neither been all beaten down nor mouldered away by the wind and weather.

If this supposition be well founded, it may with the fact of no other hill or object having been perceived above the bank in the greater part of its course, assist in forming some conjecture as to what may be within it, which cannot as I judge in such case, be other than flat sandy plains or water. The bank may even be a narrow barrier between an interior and the exterior sea, and much do I regret the not having formed an idea of this probability at the time, for notwithstanding the great difficulty and risk, I should certainly have attempted a landing upon some part of the coast, to ascertain a fact of so much importance."

Had there been any inland ranges they would have been seen by that searching officer from the ocean, but it is clear that none exists; for Mr. Eyre in his intercourse with the natives, during his journey from South Australia to King George's Sound, elicited nothing from them that led him to suppose that there were any hills in the interior, or indeed that an inland sea was to be found there; even the existence of one may reasonably be doubted, and it may be that the country behind the Great Australian Bight is, as Captain Flinders has conjectured, a low sandy country, formed by a channel of 400 or 500 miles in breadth, separating the south coast of the continent from the west and north ones. Although I did not gain the direct centre of the continent there can be very little doubt as to the character of the country round it. The spirit of enterprise alone will now ever lead any man to gain it, but the gradual development of the character of the yet unexplored interior will alone put an end to doubts and theories on the subject. The desert of Australia is not more extensive than the deserts in other parts of the world. Its character constitutes its peculiarity, and that may lead to some satisfactory conclusion as to how it was formed, and by what agent the sandy ridges which traverse it were thrown up. I would repeat that I am diffident of my own judgment, and that I should be indebted to any one better acquainted with the nature of these things than I to point out wherein I am in error.

It remains for me, before I close this part of my work, to make a few observations on the natives with whom we communicated beyond the river tribes. Mr. Eyre has given so full and so accurate an account of the natives of the Murray and Darling that it is needless for me to repeat his observations. I would only remark that I attribute our friendly intercourse with them to the great influence he had framed over them by his judicious conduct as Resident Protector at the Murray. I fully concur with him in the good that resulted from the establishment of a post on that river, for the express purpose of putting a stop to the mutual aggression of the overlanders and natives upon each other. I have received too many kindnesses at the hands of the natives not to be interested in their social welfare, and most fully approved the wise policy of Captain Grey, in sending Mr. Eyre to a place where his exertions were so eminently successful.

In another place I may be led to make some remarks on the condition of the natives of South Australia, but at present I have only to observe upon that of the natives of the distant interior with whom no white man had ever before come in contact.

If I except the tribe upon Cooper's Creek, on which they are numerous, the natives are but thinly scattered over the interior, as far as our range extended. The few families wandering over those gloomy regions may scarcely exceed one hundred souls. They are a feeble and diminutive race when compared to the river tribes, but they have evidently sprung from the same parent stock, and local circumstances may satisfactorily and clearly account for physical differences of appearance. Like the tribes of the Darling and the Murray, and indeed like the aborigines of the whole continent, they have the quick and deep set eye, the rapidly retiring forehead, and the great enlargement of the frontal sinus, the flat nose and the thick lip. It is quite true that many have not the depression of the head so great, but in such cases I think an unusual proportion of the brain lies behind the ear. In addition, however, to the above physiognomical resemblances, they have the same disproportion between the upper region of the body and the lower extremities, the same prominent chest, and the same want of muscular development, and in common with all the natives I have seen, their beards are strong and stand out from the chin, and their hair the finest ornament they possess, only that they destroy its natural beauty by filth and neglect, is both straight and curly. Their skins are nearly of the same hue; nor did we see any great difference, excepting in one woman, whose skin was of a jet black. Two young women, however, were noticed who had beautiful glossy ringlets, of which they appeared to be exceedingly proud, and kept clean, as if they knew their value. Both Mr. Browne and myself observed a great disparity of numbers in the male and female children, there being an excess of the latter of nearly two to one, and in some instances of a still greater disproportion. This fact was also obvious both to Mr. Stuart and myself in the tribe on Cooper's Creek, in which the number of female children greatly exceeded that of the male, though there were more adult men than women. The personal appearance of the men of this tribe, as I have already stated, was exceedingly prepossessing—they were well made and tall, and notwithstanding that my long-legged friend was an ugly fellow, were generally good looking. Their children in like manner were in good condition and appeared to be larger than I had remarked elsewhere, but with the women no improvement was to be seen. Thin, half-starved and emaciated they were still made to bear the burden of the work, and while the men were lounging about their fires, and were laughing and talking, the women were ceaselessly hammering and pounding to prepare that meat, of which, from their appearance, so small a proportion fell to their share. As regards the treatment of their women, however, I think I have observed that they are subjected to harsher treatment when they are members of a large tribe than when fewer are congregated together. Both parents are very fond of and indulgent to their children, and there is no surer way of gaining the assistance of the father, or of making a favourable impression on a tribe than by noticing the children.

I think that generally speaking the native women seldom have more than four children, or if they have, few above that number arrive at the age of puberty. There are, however, several reasons why the women are not more prolific; the principal of which is that they suckle their young for such a length of time, and so severe a task is it with them to rear their offspring that the child is frequently destroyed at its birth; and however revolting to us such a custom may be, it is now too notorious a fact to be disputed.

The voices of the natives, generally speaking, are soft, especially those of the women. They are also a merry people and sit up laughing and talking all night long. It is this habit, and the stars so constantly passing before their eyes, which enables them to know when they are likely to have rain or cold weather, as they will point to any star and tell you that when it shall get up higher then the weather will be cold or hot.

These primitive people have peculiar customs and ceremonies in their intercourse with strangers, and on first meeting preserve a most painful silence; whether this arises from diffidence or some other feeling it is difficult to say, but it is exceedingly awkward; but, however awkward or embarrassing it may be, there can be no doubt as to the policy and necessity of respecting it. The natives certainly do not allow strangers to pass through their territory without permission first obtained, and their passions and fears are both excited when suddenly intruded upon. To my early observation of this fact, and to my forbearing any forced interview, but giving them time to recover from the surprise into which my presence had thrown them, I attribute my success in avoiding any hostile collision. I am sure, indeed, whatever instances of violence and murder may be recorded of them, they are naturally a mild and inoffensive people.

It is a remarkable fact that we seldom or ever saw weapons in the hands of any of the natives of the interior, such as we did see were similar to those ordinarily used by natives of other parts of the continent. Their implements were simple and rude, and consisted chiefly of troughs for holding water or seeds, rush bags, skins, stones, etc. The native habitations, at all events those of the natives of the interior, with the exception of the Cooper's Creek tribe, had huts of a much more solid construction than those of the natives of the Murray or the Darling, although some of their huts were substantially built also. Those of the interior natives however were made of strong boughs with a thick coating of clay over leaves and grass. They were entirely impervious to wind and rain, and were really comfortable, being evidently erections of a permanent kind to which the inhabitants frequently returned. Where there were villages these huts were built in rows, the front of one hut being at the back of the other, and it appeared to be a singular but universal custom to erect a smaller hut at no great distance from the large ones, but we were unable to detect for what purpose they were made, unless it was to deposit their seeds; as they were too small even for children to inhabit. At the little hut to the north of the ranges, from which the reader will recollect we twice frightened away a poor native, we found a very large spear, apparently for a canoe, which I brought to the camp. This spear could not possibly have been used as a weapon, for it was too heavy, but on shewing it subsequently to some natives, they did not intimate that it was a canoe spear. It may be thought that having been in the interior for so many months I ought to have become acquainted with many of the customs and habits of the people inhabiting it, but it will have been seen that they seldom came near us.

The custom of circumcision generally prevailed, excepting with the Cooper's Creek tribe, but you would meet with a tribe with which that custom did not prevail, between two with which it did.

As regards their food, it varies with the season. That which they appeared to me to use in the greatest abundance were seeds of various kinds, as of grasses of several sorts, of the mesembryanthemum, of the acacia and of the box-tree; of roots and herbs, of caterpillars and moths, of lizards and snakes, but of these there are very few. Besides these they sometimes take the emu and kangaroo, but they are never so plentiful as to constitute a principal article of food. They take ducks when the rains favour their frequenting the creeks and lagoons, exactly as the natives of other parts of Australia do, with nets stuck up to long poles, and must procure a sufficiency of birds during the summer season. They also wander among the sand ridges immediately after a fall of rain, to hunt the jerboa and talperoo, (see Nat. Hist.,) of which they procure vast supplies; but all these sports are temporary, particularly the latter, as the moment the puddles dry up the natives are forced to retreat and fall back on previous means of subsistence.

With regard to their language, it differed in different localities, though all had words common to each respectively. My friend Mr. Eyre states, that they have not any generic name for anything, as tree, fish, bird; but in this, as far as the fish goes, I think he is mistaken, for the old man who visited our camp before the rains, and who so much raised our hopes, certainly gave them a generic name; for placing his fingers on such fish as he recognised, he distinctly mentioned their specific name, but when he put his fingers on such as he did not recognise, he said "Guia, Guia, Guia," successively after each, evidently intending to include them under the one name. With respect to their religious impressions, if I may so call them, I believe they have none. The only impression they have is of an evil spirit, but however melancholy the fact, it is no less true that the aborigines of Australia have no idea of a superintending Providence.

In conclusion: I have spoken of Mr. Browne and Mr. Piesse throughout my narrative, in terms such as I feel they deserved. I should be sorry to close its pages without also recording the valuable and cheerful assistance I received from Mr. Stuart, whose zeal and spirit were equally conspicuous, and whose labour at the charts did him great credit. To Flood I was indebted for having my horses in a state fit for service, than whom as a person in charge of stock, I could not have had a better; and I cannot but speak well of all the men in their respective capacities, as having always displayed a willingness to bear with me, when ever I called on them to do so, the fatigues and exposure incidental to such a service as that on which I was employed.

Before closing my narrative I would make a few observations on the conduct of such an Expedition as the one the details of which I have just been giving.

It appears to me then that discipline is the first and principal point to be considered on such occasions; unless indeed the leader be implicitly obeyed it is impossible that matters should go on regularly. For this reason it is objectionable to associate any irresponsible person in such an undertaking. When I engaged the men who were to accompany me, I made them sign an agreement, giving me power to diminish or increase the rations, and binding themselves not only to the performance of any particular duty, but to do everything in their power to promote the success of the service in which they were engaged, under the penalty of forfeiture of wages, in whole or part as I should determine. I deemed it absolutely necessary to arm myself with powers with which I could restrain my men even in the Desert, before I left the haunts of civilized man, although I never put these powers in force—and this appears to me to be a necessary precaution on all such occasions. Equally necessary is the establishment of a guard at night, for it is impossible to calculate on the presence of natives—they may be close at hand, when none have been seen or heard during the day. Had Dr. Leichhardt adopted this precaution his camp would not have been surprised, nor would he have lost a valuable companion. Equally necessary is it to keep the stock, whether horses or bullocks, constantly within view. In all situations where I thought it probable they might wander I had them watched all night long. Unless due precaution however is used to ensure their being at hand when wanted, they are sure to wander and give ceaseless trouble.

As regards the consumption of provisions, I had both a weekly and a monthly statement of issues. In addition to this they were weighed monthly and their loss ascertained, and their consumption regulated accordingly, and I must say that I never found that the men were disposed to object to any reasonable reduction I made. I found the sheep I took with me were admirable stock, but I was always aware that an unforeseen accident might deprive me of them, and indeed they called for more watchful care even than the other stock. The men at the Depôt were never without their full allowance of mutton. It was only the parties out on distant and separate services who were reduced to an allowance scarcely sufficient to do their work upon.

The attention of a Leader is no less called to all these minutæ than his eye and judgment to the nature of the country in which he may happen to be. I would observe that in searching for water along the dry channel of a creek, he should watch for the slightest appearance of a creek junction, for water is more frequently found in these lateral branches, however small they may at first appear to be, than in the main creek itself, and I would certainly recommend a close examination of them. The explorer will ever find the gum-tree in the neighbourhood of water, and if he should ever traverse such a country as that into which I went, and should discover creeks as I did losing themselves on plains, he should never despair of recovering their channels again. They invariably terminate in grassy plains, and until he sees such before him he may rest assured that their course continues. Should the traveller be in a country in which water is scarce it will be better for him to stop at any he may find, although early in the day, than to go on in the chance of being without all night—and so entailing fatigue on his men.

I trust that what I have said of the natives renders it unnecessary for me to add anything as to the caution and forbearance required in communicating with them. Kindness gains much on them, and their friendly disposition eases the mind of a load of anxiety—for however confident the Leader may be, it is impossible to divest the minds of the men of apprehension when in the presence of hostile natives. He who shall have perused these pages will have learnt that under whatever difficulties he may be placed, that although his last hope is almost extinguished, he should never despair. I have recorded instances enough of the watchful superintendence of that Providence over me and my party, without whose guidance we should have perished, nor can I more appropriately close these humble sheets, than by such an acknowledgment, and expressing my fervent thanks to Almighty God for the mercies vouchsafed to me during the trying and doubtful service on which I was employed.

PART II.

CHAPTER I.

DUTIES OF AN EXPLORER—GEOGRAPHICAL POSITION OF SOUTH AUSTRALIA—DESCRIPTION OF ITS COAST LINE—SEA MOUTH OF THE MURRAY—ENTERED BY MR. PULLEN—RISK OF THE ATTEMPT—BEACHING—ROSETTA HARBOUR—VICTOR HARBOUR—NEPEAN BAY—KANGAROO ISLAND—KINGSCOTE—CAPT. LEE'S INSTRUCTIONS FOR PORT ADELAIDE—PORT ADELAIDE—REMOVAL TO THE NORTH ARM—HARBOUR MASTER'S REPORT—YORKE'S PENINSULA—PORT LINCOLN—CAPT. LEE'S INSTRUCTIONS—BOSTON ISLAND—BOSTON BAY—COFFIN'S BAY—MR. CAMERON SENT ALONG THE COAST—HIS REPORT—POSITION OF PORT ADELAIDE.

NO MARINER ever shook the reefs from his sails, on the abatement of the storm, under the fury of which his vessel had been labouring, with more grateful feelings than those with which I turn from the dreary and monotonous wastes I have been describing, to the contemplation of fairer and more varied scenes. My weary task has been performed, and however uninteresting my narrative may have proved to the general reader, I would yet hope, that those who shall hereafter enter the field of Australian discovery, will profit from my experience, and he spared many of the inconveniences and sufferings to which I was unavoidably exposed. They may rest assured, that it is only by steady perseverance and unceasing attention, by due precaution and a mild discipline, that they will succeed in such an undertaking as that in which I was engaged. That unless they are fortunate enough to secure such an assistant as I had in Mr. Browne, their single eye must be over every thing, to study the features of the country through which they are passing, to keep their horses and cattle always within view, to prevent disputes in their camp, and to husband their provisions with the utmost care, to ascertain from time to time the quantity they may have on hand, and to regulate their consumption accordingly. Few difficulties present themselves to the explorer in journeying down a river, for that way is smooth before him; it is when he quits its banks, and traverses a country, on the parched surface of which little or no water is to be found, that his trials commence, and he finds himself obliged to undergo that personal toil, which sooner or later will lay him prostrate. Strictly speaking, my work should close here. I am not, however, unmindful of the suggestion I made in my Preface, that a short notice of South Australia at the close of my journal would not be out of place.

In the following pages, therefore, it is proposed to give some account of that province, from whence, as the reader is aware, I took my departure, before commencing my recent, labours. Its circumstances and prospects have, I know, of late, been frequently brought before the public, but, I trust, nevertheless, that my observations will carry something of novelty, if not of interest, and utility with them.

South Australia, then, the youngest of the colonies that have been established round the shores of the Australian Continent, is situate, as its name would imply, upon its southern coast. It extends from the 132nd to the 141st degree of longitude east from Greenwich, and runs up northwards into the interior to the.26th parallel of latitude. The district of Port Phillip bounds it on the east, for which reason, the fixing of the eastern boundary line between those two fine provinces has of late been a point of great interest and importance. Mr. Tyers, an able and intelligent officer, was employed by the Government of New South Wales, primarily to determine the longitude of the mouth of the Glenelg, and from his triangulations and observations it would appear that the 141st meridian falls on the coast about a mile and a half to the eastward of it. Subsequent observations, taken by Captain Stokes, in command of Her Majesty's surveying ship, the Beagle, differ slightly from the result of Mr. Tyers' observations, but they prove beyond doubt, the care and accuracy with which the latter officer carried on his survey. The point, has since, I believe, been finally recognised by the governments of Sydney and Adelaide, and the boundary line been marked to the distance of 123 miles from the coast. The party employed in this useful undertaking, however, was obliged to relinquish it for a time, in consequence of heavy rains; but it is not probable that any dispute will hereafter arise on the question. If the line could have been extended to the Murray river, it would have been as well, but the desert country beyond it is valueless to civilised man. Taking it for granted, then, that the S.E. angle of the province of South Australia has been fixed, we shall in the first instance proceed along its sea line, and notice any thing worthy of observation, before we enter into a detail as to the character of the country itself.

From the mouth of the Glenelg the coast of South Australia trends to the westward as far as Cape Northumberland in long. 140°37' and in lat. 38°;* from Cape Northumberland it turns to the N.N.W., keeping that general direction for more than 100 miles. Between the last mentioned Cape and Cape Morard des Galles in lat. 36½°, there are several bays, two only of which, Rivoli Bay, immediately to the north of Cape Lannes, and Guichen Bay, a little to the south of Cape Bernouilli, have more particularly drawn the attention of the local Government, rendered necessary in

* The reader will be good enough to bear in mind that the longitudes in this work are all east of Greenwich, and that the latitudes are south.

consequence of the rapid settlement of the back country. Recent surveys have enhanced the value of these two bays, and townships have been laid out at each. That at Rivoli bay being called Grey Town, that of Guichen bay Robe Town. At the latter, there is a resident magistrate and a party of mounted police. Many allotments have been sold in both towns, and although the bays offer but little protection to large vessels, they are of great importance to the colonial trade and to the settlers occupying the beautiful and fertile country in the neighbourhood of Mounts Gambier and Shanck. From Cape Morard des Galles, a low dreary and sandy beach extends for five leagues beyond the sea mouth of the Murray, a distance of more than 100 miles. This beach, which varies in breadth from one to three miles, conceals the waters of the Coorong, and the depressed and barren country beyond it is completely hid from view by the bright sand-hills on this long and narrow strip of land.

The sea mouth of the Murray, famous for the tragical events that have occurred near it, and which give a melancholy interest to the spot, is in long. 138°56' and in lat. 35°32'. No one could, I am sure, look on the foaming waters of that wild line of sand-hills through which it has forced a channel, without deep feelings of awe and emotion. Directly open to the Southern Ocean, the swell that rolls into Encounter Bay, is of the heaviest description. The breakers rise to the height of fifteen or eighteen feet before they burst in one unbroken line as far as the eye can see, and as the southerly is the most prevailing wind on that part of the Australian coast, it is only during the summer season, and after several days of northerly wind that the sea subsides, and the roar of breakers ceases for a time. The reader will perhaps bear in mind that the channel of the Goolwa connects Lake Victoria with Encounter Bay, the sea mouth of the Murray being the outlet through which its waters are discharged into the ocean.

The channel of the Goolwa (now called Port Pullen, in compliment to an officer of that name on the marine survey staff of the province, who succeeded, after several disappointments, in taking a small cutter through that narrow passage, and navigating her across the lake into the Murray River, as high as the settlement of Moorundi) is to the westward of the sea mouth as the Coorong is to the eastward.†

But although Mr. Pullen succeeded in getting into the Goolwa, it was only under the most favourable circumstances, nor will the sea mouth of the Murray ever, I fear, be available for navigable purposes. How far it may be practicable to

† The compliment thus paid to Mr. Pullen, who is now employed on the expedition to the North Pole, in search of Sir John Franklin, by Col. Gawler, the then Governor, was well merited, as a reward for the perseverance and patience he had shewn on the occasion—for those only who have been at the spot can form an idea of the disturbed and doubtful character of the place, and the risk there must have been in the attempt to enter such a passage for the first time.

steamers, I would not hazard an opinion, nor is the subject at the present moment one of much importance, for the country to the eastward of the ranges is not yet sufficiently located to call for such a speculation.

The sea mouth of the Murray is about the third of a mile in breadth, and when the river is flooded a strong current runs out of it with such rapidity, that the tide setting in at the same time causes a short and bubbling sea. It took Captain Barker nine minutes and fifty-eight seconds to swim across it on the fatal occasion on which he lost his life—but he was obliged to go somewhat above the outlet, as the stream would otherwise have carried him amidst the breakers. The western shore is very low, but the eastern one is marked by a large sand-hill, now called Barker's Knoll, after that talented and amiable officer. From seaward, nothing but a wild line of sand-hills meets the view, such as few mariners would venture to approach, and through which fewer still could hope to find a passage into the calmer waters of Lake Victoria, so completely hidden is the entrance. It was only by patient watching indeed, that Mr. Pullen seized the opportunity by which he entered the Goolwa. He was not the first, however, who did so, as Captain Gill, the master of a small cutter that was unfortunately wrecked on the strand at some distance to the eastward of the outlet, was the first to come down the Coorong in his boat, in which he ultimately reached Victor Harbour, but he also had to remain three weeks under the sand-hills before he could venture forth. Some years prior to this, however, Sir John Jeffcott, the first judge of South Australia, and Captain Blenkensop, the head of the fishery, both found a watery grave in attempting to pass from the Goolwa into Encounter Bay.

I speak more particularly on the point, however, because, in 1838, during my first visit to the province, I went with a party of hardy seamen, with the intention, if possible, of passing into the Goolwa from seaward. At Encounter Bay, Captain Hart, who had the superintendence of the fishery there, gave me his most experienced steersman, and a strong whale-boat. In this I left Victor harbour for Freeman's Nob, a small rocky point in the very bight of Encounter Bay, where I remained until three A.M. of the next morning, when I started for the outlet under the most favourable auspices. A northerly wind had been blowing of the land for several days, and the sea was so tranquil that I had every hope of success. I had five leagues to pull, and keeping about a mile from the shore, swept rapidly along it. We were still about four miles from the inlet when the sun rose over it, as if encouraging us onwards. On approaching it at low water, I tried in vain to enter. The sea was breaking heavily right across the entrance from one side to the other, and after several ineffectual attempts to run in, I came to an anchor, close to the outer line of breakers, hoping that the sea would subside at high water and that we should then have less difficulty. We had not, however, been in this position more than half an hour, when a heavy southerly swell set in; from a deep blue the water became green, and the wind suddenly flew round to the S.W. Before we could weigh and stand out from the shore, several seas had broken outside of us, and in less than ten minutes the whole coast, to the distance of more than a mile from the shore, was white with foam, and it seemed clear that a gale was coming on. Under these circumstances I determined on returning to the little harbour from which we had started in the morning, but the wind being directly against us, we made very little head. "We shall never get to the Nob," said Mr. Witch, who had the steer oar, to me; "it blows too hard, Sir." "What are we to do, then?" said I. "Why, Sir," he replied, "we must either beach or run out to sea," "We will beach, then," I said; "it is better to try that than to do any thing else." Mr Witch evinced some surprise at my decision, but made no remark. "You had better select your place," I observed, "and be careful to keep the boat's head well on to the seas." "You need not fear me, Sir," said the hardy seaman; "I am accustomed to such work. It looks worse than it really is." The sea, however, was now breaking full a mile and a half from the shore, and in looking towards it I observed a solitary horseman riding slowly along, as if watching our movements. At length Mr. Witch said that he thought we were opposite to a favourable spot, on which I directed him to put the boat's head towards the shore, and to keep her end on as he went in. Round we flew, and in a moment after we were running at railway speed on the top of a heavy wave. "Steady, men," said Mr. Witch: "Steady all," and on we went; but looking round him a moment after—"Back, all. Back, all," he cried. The men did as they were ordered, and the boat's way was stopped. Her stern rose almost perpendicularly over the prow, and the next moment fell into the trough of the sea. The wave, transparent as bottle glass, rushed past us, and topping, as it is called, burst at our very bow, in a broad sheet of foam. "Give way, my lads," was the next order of the watchful steersman, as he again cast his eyes behind him. "Give way, my lads. Give way, all." "Steady, men," he called, as if doubtful of the result of the coming wave. I thought I saw paleness on the face of the rowers, but they pulled regularly and well, and a thundering sound soon told us we had escaped the threatening sea that had come so rapidly up. I do not know if I am doing justice to the occurrence. There was more of apparent than real danger in it, and I myself was less nervous, because I had not long before been accustomed to the heavy surf of Norfolk Island. It was, however, a moment of great excitement. We had literally shot towards the shore, and were now within fifty yards of it, when Mr. Witch said to me, "Take care of yourself, Sir; we shall catch it at last."

I turned round, and saw a large roller close upon us, just on the point of topping—I had scarcely time to stoop and give my back to it when it came upon us, and I never had such a thump in my life. The boat was filled in a moment and we were all thrown out—Mr. Witch, who had been standing, was hurled to a great distance, but the men were up in a moment, the water being about four feet deep, and with admirable dexterity ran her on the beach. I do not remember ever having been in so strong a breeze. The reader may form some idea of it when I assure him that the wind rolled the boat over and over as if she had been as light as a carpenter's chip, and the sand and pebbles came with such violence in our faces, that we were obliged to retreat behind the sand hills until it moderated.

It was my friend Mr. Strangways who had accompanied me from Adelaide, whose figure we had seen on the beach, and he assured me that we seemed to fly as we approached him.

The wind having apparently flown permanently round to the south, and it being hopeless to expect that the sea would subside for many days, I hauled the boat over the sand hills, and launching her in the Goolwa, tried to row through the outlet to sea, but after remaining for eight days, and having my boat four times swamped, I was forced to give up the attempt as I had no time to spare. The distance between my outer and inner points might have been a cable's length. In endeavouring to pass out I shoaled to a quarter less one, having kept the lead constantly going. I abandoned the task therefore under an impression that the outlet was not navigable, yet Mr. Pullen succeeded in taking a small cutter into the Goolwa with perfect safety. I cannot but conclude therefore that it has a shifting bar, and that it will present difficulties to regular navigation that will only be surmounted by a better knowledge of its locality, and in all probability by artificial means.

From Freeman's Nob the coast line turns southwards to Rosetta Head, a bold and prominent conical hill, from the summit of which the whalers look for their game. Under the lea of Rosetta Head there is a small harbour called Rosetta Harbour. It is separated by a rocky island called Granite Island, and a reef that is visible at low water, and connects Granite Island with the main land from Victor Harbour, so called after H.M's ship Victor, when surveying in that quarter. Neither of these harbours however are considered secure, although they are protected from all but south-east winds.

It was in Rosetta Harbour, that during the early settlement of the Colony the South Australian Company's ship South Australian, was driven on shore and lost. The John Pirie, a strongly built schooner, also belonging to the Company, had well nigh shared her fate. This little vessel was lying astern of the Australian when she went ashore, with the reef close astern of her. In this fearful position her anchors began to drag, and her destruction appeared inevitable, when her commander, Captain Martin, determined on attempting to take her over the reef, it being high water at the time. He accordingly cut his cable, set his sails, and ran his vessel on the rocks. Four times she struck and was heaved as often over them, until at length she floated in the deeper water of Victor Harbour, and found her safety under the lea of the very danger from which she expected destruction. It was a bold resolve and deserved the success that attended it. I always feel a pleasure in recording such events, not only from feelings of admiration, but because they are examples for men to follow when placed in equally hazardous circumstances, and shew that firmness and presence of mind are equal to almost every emergency. The anchorage in Victor Harbour is under the lea of Granite Island, but I believe it is foul and rocky, and until both it and Rosetta Harbour shall be better known, the seaman will enter them with caution. Encounter Bay indeed, is not a place into which the stranger should venture, as he would find it extremely difficult to beat out to sea with a contrary wind. Still no doubt vessels may find refuge at these places from strong west and south-west winds, but I have always understood that it is better for a ship encountering a gale at the entrance of Backstairs Passage rather to keep at sea, than seek shelter in any contiguous harbour.

There is room for two or three tolerably sized vessels in Victor Harbour, which is in longitude 188°38'00" and in latitude 35°32', and in certain seasons of the year it may be deemed secure, if it were not liable to other objections, but I have heard it stated by an experienced seaman, one whose intimate knowledge of this part of the coast of South Australia is indisputable, that there is anchorage under the lea of Freeman's Nob, and a small island off it, sufficient for two or three vessels of 250 or 300 tons, altogether preferable to either of those I have mentioned, as being more sheltered, and having better holding ground—but we must not forget that it is deeper in the bay, and there would consequently be a greater difficulty in beating out; but the truth is that the importance and capabilities of these harbours will only be developed as the wants of the colonists render it necessary for them to have ports in this vicinity. When the country to the eastward of the mountains shall be more thickly peopled, and when the rich and fertile valleys of the Inman, the Hindmarsh and Currency Creek, and the available country between the two last, be more generally cultivated, and when the mines at the Reedy Creek and other places are at full work, the want of a harbour at Encounter Bay will be sufficiently apparent.

The principal whale fishery on the coast of South Australia is in Encounter Bay, and has, I believe, of late years proved as advantageous a speculation to those who have carried it on as could be expected; profits are of course dependent on contingencies, as the nature of the season and the number of whales that may visit the coast: but the fishery at Encounter Bay has certainly been as successful as any other on the coast, and would have been more so if the ground had not been intruded upon. As a source of colonial industry, and as a proof of commercial enterprise, I should regret to see this bold and hardy occupation abandoned. See Appendix.

From Rosetta Head the line of coast again trends for a short distance to the west, and forms, together with the opposite shore of Kangaroo Island, the Backstairs Passage, or eastern entrance into St. Vincent's Gulf, of which Cape Jervis is the N.W. point. It is here that the more important navigation of the South Australian seas commences. The line of coast I have already described is not sufficiently known to be approached by the stranger without caution, nevertheless the several bays and harbours I have mentioned may offer better shelter and greater convenience than I am able to point out.

One of the first establishments, if not the very first, of the South Australian Company was on Kangaroo Island, on the shores of Nepean Bay. Here the town of Kingscote was laid out, and some very good houses built, which are now falling to dilapidation and decay, since it has been abandoned by the Company's servants for some years. Nevertheless Kingscote is a very pretty sea-port town, and the harbour is undoubtedly good. The bay is large enough to hold a number of ships, and is secure from all winds, being almost completely land-locked. The water inside moreover is smooth, since the bay is protected by a long spit of sand, whereby the roughness of the outer sea does not affect it, and vessels consequently lie there during heavy weather without any apparent motion. It is to be regretted, that, with such advantages, Kingscote Harbour should have any drawback, but when we have given credit for its capabilities as a harbour, we have done all, and even as a harbour, sailors are divided in opinion, whether or not American River, or a small bay, five miles to the south-east of it, are not to he preferred. In Nepean Bay there is a deficiency of water, which is not the case in either of the last mentioned places. The soil is equally good in the neighbourhood of all three, but Kingscote having been occupied, the ground has been cleared of the dense brush that grew on it in a state of nature, and some of the most productive gardens in the Province are to be found there. It is astonishing what quantities of the finest onions are sent from Kingscote, with other produce, to Adelaide. The island is, however, so generally and so heavily covered with brushwood, that although the soil is good in many places, it has been found impracticable to clear. On the general character of Kangaroo Island, I would observe, that, from the reports of those best acquainted with it, nine-tenths of the surface is covered with dwarf gum-trees, or heavy low brush, that there are no plains of any consequence, no harbours excepting those I have already mentioned—that water is generally scarce, and the best land is most heavily wooded and perfectly impenetrable; but, if it is thus useless and unavailable for pastoral and agricultural purposes, Kingscote, being so short a distance from Adelaide, holds out every inducement as a watering-place to those who, desiring change of air and sea-bathing, would wish to leave the heated neighbourhood of the capital during the summer months. It is a disadvantage to them that there are few places on the shores of St. Vincent's Gulf, on which bathing places could be established, but the change of air at Kingscote would be as great a benefit as sea-bathing itself, for hot winds are not felt there, but a cool and refreshing breeze is almost constantly blowing. As a watering-place therefore, it may, one day or other, be of importance, when the convenience of steam-boats shall render the passage from Adelaide to Kangaroo Island, like a trip across the Channel. But it is to be observed that whatever disadvantages the island may possess, its natural position is of the highest importance, since it lies as a breakwater at the bottom of St. Vincent's Gulf, and prevents the effects of the heavy southerly seas from being felt in it. There is, perhaps, no gulf, whether it is entered by the eastern or western passage, the navigation of which is so easy as that of St. Vincent, and so clear of dangers, that it can only be by the most fortuitous circumstances, or the most culpable neglect, that any accident can befall a ship in its passage up to Adelaide.

Anxious to make this portion of my work as useful as possible, and feeling assured that the remarks I have hitherto made will only lead the seaman to adopt those measures of precaution in approaching any of the harbours and bays I have mentioned, our knowledge of which is still limited, I shall here quote a passage from a small book of Sailing Instructions for South Australia, published some years ago by Captain Lee, an experienced mariner, for the guidance of commanders of vessels bound to Port Adelaide.

I shall only observe that, in running up the Gulf it is extremely difficult to recognise the peak of Mount Lofty; but a pile of stones has been erected upon it, which is easily visible through a good telescope, and that the pilot station spoken of by Captain Lee as being five miles from Glenelg has been abandoned, and the pilots now board ships from the light vessel moored off the bar.

"Vessels from England bound to Port Adelaide, should, after leaving the Cape of Good Hope, run to the eastward in 37° or 38° south latitude, until they arrive in longitude 132° east, when they may haul to the northward, so as to get into latitude 36°25', in longitude 135°30'; then steer to the north-east, and make Kangaroo Island, passing between which and a small island named Althorpe's island, they will enter Investigator's Straits. These Straits form the western entrance to St Vincent's Gulf, and are so free from danger, that it seems almost wonderful how any vessel can get on shore without gross negligence. The only danger that can possibly affect a vessel is the Troubridge Shoal, and this, by a little attention to the lead, may be easily avoided, as on the south side of the shoal the water deepens gradually from four to seventeen or eighteen fathoms. The shores on the side of Kangaroo Island are bold and rocky, whilst on the north side, on Yorke's Peninsula, they are low and sandy. In working up in the night, stand no nearer to the north shore than nine fathoms, or to the southward than twelve fathoms. You will have from sixteen to twenty fathoms in the fair way—fine grey sand, mixed with small pieces of shell. In working up St. Vincent's Gulf, you may stand to the eastward in six fathoms, and towards the Troubridge Shoal in nine fathoms. The prevailing winds are from the south-west to south-east, especially in the summer months, when the sea breeze acts in about nine o'clock. The strength of tide in the Gulf is very irregular, with a strong south-west wind, the flood runs up at the rate of about two miles an hour, whilst with a northerly wind it is scarcely perceptible. The anchorage in Holdfast Bay is hardly safe in the winter months, as it is open to north-west, west, and south-west winds, which, blowing hard, raise a short tumbling sea. The ground is a fine sand, almost covered with weeds, so that when the anchor once starts, the weeds being raked up under the crown, will in a great measure prevent its again holding. In the summer months it may be considered a perfectly safe anchorage, if due caution is exercised in giving the vessel cable in time. The best anchorage for a large vessel is with the summit of Mount Lofty, bearing east in six fathoms. A small vessel will lay better close in, just allowing her depth of water sufficient to ride in.

"The pilot station for Port Adelaide is about five miles north of Holdfast Bay. In running up keep in five fathoms, until abreast of the flag-staff on the beach, when a pilot will come on board. It is always high water in Port Adelaide morning and evening, and consequently low water in the middle of the day. In the present state of the harbour, no vessel drawing more than sixteen feet water ought to go into the port. Several very serious accidents have befallen vessels in this port, for which the harbour itself ought certainly to be held blameless."

"Vessels," he adds, "from Sydney, or from the eastward, bound to Port Adelaide, having arrived at Cape Howe, should shape a course for Hogan's Group in Bass' Straits, when off which, with a northerly wind, the best passage through the Straits is between Redondo and Wilson's Promontory, because should a gale of wind come on from the north-west, as it almost invariably does commence in that quarter, they would have more drift to the south-east than if they passed through near Kent's Group or Sir R. Curtis's Island. It is also a great saving in distance. Having arrived off King's Island, with a north wind, stand well out to the west or south-west, so as to keep well to the southward of Cape Northumberland, as the heavy gales from the north-west seldom last more than forty-eight hours, when they veer to the south-west, and fine weather ensues. Being abreast of Cape Northumberland, a south-west wind will be a favourable wind to proceed to Adelaide. Steer directly for the east end of Kangaroo Island, which you may pass at a distance of one mile; and if the wind is from the south or south-east, you may then steer across Backstairs Passage to Cape Jarvis; having arrived off which, proceed as directed before: should the wind be strong from south-west or west-south-west, keep Kangaroo Island close on board until abreast of Cape Jarvis, when you will have the Gulf open. Should it be night time or thick weather, and you have sighted Cape Willoughby at the entrance after passing that Cape, steer north-west fifteen miles, and you may lay to or run up north-east by east under snug sail until daylight. There are four rocks at the entrance of this passage, called the Pages; with a beating wind, you may pass on either side of them, but with a leading wind there is no necessity to approach them at all, as it is best to pass close round Cape Willoughby. Should the wind be so strong that a vessel could not carry sufficient canvas to fetch through the passage, it would be better for a stranger to stand out to the southward, rather than attempt to run into Encounter Bay. The anchorage in Encounter Bay is close round Granite Island, where a vessel may lay sheltered from all winds, save from south-east. There are several good anchorages where a vessel may run to, should she be caught in a gale of wind in Bass' Straits: one behind Wilson's Promontory, the corner inlet of Flinders; another in Western Port; two under King's Island, besides several on the Van Diemen's Land side, as Circular Head, George Town, Preservation Island, &c., the whole of which may be attained by a proper consideration of the chart; but it is always better, provided a vessel has sufficient sea room, to keep at sea than to run for an anchorage, as the sea will seldom hurt a good ship properly managed, and she is always ready to take advantage of any change that may take place.

"Should a gale of wind come on when a vessel is far to the westward of King's Island, she may run for Portland Bay. In going in, you pass to the eastward of the St. Lawrence Islands, and haul directly in for the land west-north-west; keep along the south shore of the bay, at a distance of one mile, until you see the flag-staff at Mr. Henty's; bring that to bear west, and you will have six fathoms water about three-quarters of a mile from shore."

From Cape Jarvis the coastline tends to the north along the eastern shore of St. Vincent's Gulf. The scenery, as you turn the point, is extremely diversified. Dark cliffs and small sandy bays, with grassy slopes almost to the water's edge, succeed each other, backed by moderate hills, sparingly covered with trees, and broken into numerous valleys. Thus you pass Yankelilla, Rapid Bay, and Aldingis; but from Brighton the shore becomes low and sandy, and is backed by sand hummocks, that conceal the nearer country from the view, and enable you to see the tops of the Mount Lofty Range at a distance of from eleven to twelve miles.

Port Adelaide, a bar harbour, is about nine miles from Glenelg, and situate on the eastern bank of a large creek, penetrating the mangrove swamp by which the shore of the Gulf is thereabouts fringed. This creek is from ten to eleven miles in length. Its course for about two miles after you cross the bar is nearly east and west, but at that distance it turns to the south, and runs parallel to the coast; and there is an advantage in the direction it thus takes, that would not be apparent to the reader unless explained. It is, that, as the land breeze blows off the shore in the evening, and the sea breeze sets in in the morning vessels can leave the harbour, or run up to it as they are inward or outward bound.

The landing-place of the early settlers was too high up the creek, and was not only the cause of great inconvenience to the shipping, but of severe loss in stores and baggage to the settlers; but at the close of the year 1839, Mr. McLaren, the then manager of the South Australian Company commenced and finished a road across the swamp to a section of land belonging to his employers, that was situated much lower down the creek, and on which the present Port now stands. The road, which is two miles in length, cost the Company £12,000. It has, however, been transferred to the local Government, in exchange for 12,000 acres of land, that were considered equivalent to the sum it cost.

The removal of the Port to this place was undoubtedly a great public benefit; and whatever perspective advantages might have influenced Mr. McLaren on the occasion, he merited all due praise for having undertaken such a work at a time when the Government itself was unable to do so. Both the wharf and the warehouse belonging to the Company are very creditable buildings, as is the Custom House and the line of sheds erected by the Government; but the wharf attached to them is defective, and liable to injury, from the chafing of tide between the piers, which are not placed so as to prevent its action. Mr. Phillips' iron store is also one of a substantial description; but there was not, when I left the province, another building of any material value at the Port. Numerous wooden houses existed in the shape of inns, stables, etc.; but the best of these were unfortunately burnt down by a fire a few days before I embarked for Europe. Whether it is that a misgiving on the minds of the public as to the permanency of the Port has been the cause of, and prevented the erection of more substantial and better houses at Port Adelaide, it is difficult to say; but any one might have foreseen, that as the colony progressed, and its commerce increased, the Port would necessarily have to be moved to some part of the creek where there was deeper and broader water, for the convenience of the shipping. I felt assured, indeed, that the removal of the Port would take place sooner than was generally supposed. The following extract from the South Australian Gazette of the 4th of December last, will prove that I judged truly:—

"NEW ROAD TO THE NORTH ARM.—This road was commenced last Tuesday week; and at the rate at which the work is progressing, will be completed (except as regards the subsequent metalling and ballasting) within four months from the present time. The line adopted is the one which was proposed by Mr. Lindsay in

1840, as requiring less outlay in the original construction than either of the other lines proposed. Taking Adelaide as the starting point, the course will be either along the present Port Road between Hindmarsh and Bowden as far as section No. 407, thence along the cross track between that section and section No. 419 (preliminary), as far as the south-east corner of Mr. Mildred's section, No. 421; then in a straight line through the last named section and Mr. Gilles's, No. 2072, after leaving which it passes through an opening in the sand-hills, and then winds along the highest ground between the creeks, leaving the South Australian Company's road about a mile on the left, till it joins the main road or street running through section G. at the North Arm; or through North Adelaide and along the road at the back of Bowden, parallel with the main Port Road as far as Mr. Torrens' residence, to the south-east corner of Mr. Mildred's section, thence through that section as before. The soil of the so-termed swamp, or rather marsh, is of the most favourable description for embanking and draining operations, consisting at the part of the line where the work has been commenced, of a good loam for the first spit, and then clay to the depth of eighteen inches or two feet, resting upon a stratum composed for the most part of shells of numberless shapes and sizes, which extends to the bottoms of the drains (four feet), being the level of high water at spring tides, and at about the same above the low-water level. The shelly stratum continues below the bottoms of the drains to an uncertain depth. From the commencement of the "Swamp" to the Great Square or public reserve at the junction of the North Arm with the main channel of the Creek, the distance along the line of road is 4,800 yards, or nearly two miles and three-quarters. The breadth of the road between the ditches will be 114 feet, or between three and four times the breadth of the Company's road."

If there is anything more justly a subject of congratulation to the Province than another, it is the commencement of the work thus notified. The road is now, in all probability, finished, and that part of the creek rendered available where these permanent improvements may be made, without the fear of any future change; and when the shores of the North Arm shall be lined by wharfs, and the more elevated portions of Torrens' Island shall be covered with houses, few harbours will be able to boast of more picturesque beauty. There was something dreary in sailing up the creek with its dense and dark mangroves on either side, and no other object visible beyond them save the distant mountains; but the approach to the new Port will not fail to excite those pleasurable feelings in the heart of the stranger which give a colouring to every other object.

The removal of the port to the proposed locality will bring it within three miles of the bar, and will be of incalculable advantage to the shipping, since there will no longer be any delay in their putting to sea. The following letter, addressed by Captain Lipson, the Harbour-master, to the Colonial Secretary, in reference to the improvements that have been effected at the bar, will best explain its present state, and the description of vessels it will admit into the Port.

"Port, 6th July, 1847.

"SIR,—In answer to your letter of this day's date, requesting that I would report to you, for the information of the Legislative Council, what beneficial effects have been produced by the use of the mud barge in deepening the bar at the entrance of Port Adelaide, since the commencement of its operation, in the year 1845, up to the present date, also what additional depth of water, if any, has been obtained by the work alluded to.

"I have the honour to state, that at the commencement of the colony, her Majesty's storeship 'Buffalo' was brought out by the then governor, Captain Hindmarsh, to be detained here nine months for the protection and convenience of the colonists. It was, therefore, much wished to have her inside the bar; but after attending and carefully watching successive spring-tides, it was given up as impracticable, she drawing fifteen feet. The Governor then appointed a board to examine the bar, consisting of the masters of the 'Buffalo,' 'John Renwick,' and another, who, in their report, stated as their opinion, that no vessel above 300 tons ought to be brought into the harbour; however, last week two vessels exceeding 600 tons have been brought up to the wharf. But the most beneficial effect is now felt from a ship being able to cross the outer bar so much sooner on the tide than before, thereby having sufficient time to take her round the bar, and, if moderate, to beat up and anchor at the North Arm the same tide. Ships may now be brought in on the springs in winter, drawing seventeen or eighteen feet, as the time of high water is in the day, and the wind generally fair to beat in, but not so in going out, from the difficulty of reaching the bar at the time required, and the tide leaving so quickly after the ebb is made great care is required; and I find it unsafe to allow any vessel to load deeper than 15 or 16·6 inches at most. With a tug, there would be less difficulty and danger in loading to 18 feet than there now is to 15.

"There is now three feet more water on the bar than there was previous to its being deepened, and if the work be continued next summer, to enlarge a cut which has been made, there will he five feet.

"I have the honour to be, Sir, your obedient servant,

"Thomas Lipson, Harbour Master.

"The Honourable Colonial Secretary."

It is not clear to me, however, that the admission of larger class shipping into the Port will be of any great advantage. I am led to believe that ships smaller tonnage than those drawing 16 to 17 feet, have been found to be most convenient for the ordinary purposes of commerce. However, it is evident, that if Captain Lipson continues the same praiseworthy exertions he has hitherto used, he will deepen the bar for vessels of any tonnage. Under existing circumstances, it may be as well to state that any ship arriving off the bar when there is not sufficient water on it for them to enter the port, will find good anchorage all round the lightship, particularly a little to the westward of it. The whole Gulf, indeed, from this point, may be considered as a safe and extensive roadstead. As regards Port Adelaide itself, I cannot imagine a securer or a more convenient harbour. Without having any broad expanse of water, it is of sufficient width for vessels to lie there in perfect safety, whether as regards the wind or the anchorage.

The head of St. Vincent's Gulf is in latitude 34½°. Between that point and Port Adelaide, the shore is either lined by mangroves, or is low and sandy. There are, nevertheless, several inlets similar to, but much smaller than Port Adelaide, and other commodious anchorages for small craft along it. The principal of these is the inlet connected with the Gawler, of which I shall hereafter speak. York Peninsula forms the western shore of St. Vincent's Gulf, and separates it from that of Spencer. It is a long, low tongue of land—Cape Spencer, its southern extremity, being in 35°17', and in long. 136°52'. Though embracing a considerable area, the character of the Peninsula is unfavourable to the growth of nutritive herbage; the surface soil is a species of calcareous limestone, the rock formation of a tertiary description, although, at the lower extremity, granite and trap rock are known to exist. The surface of the country is undulating, covered in many places by scrub, and the trees being very short-lived, the whole is matted with dead timber, and difficult of access. A deficiency of water renders York Peninsula still more unfavourable for location; nevertheless, several sections of land have been purchased on that part which is immediately opposite to Port Adelaide, and it is said that indications of copper have been found there, a fact I should be inclined to doubt. In 1840, a company applied for a special survey on the shores of the Peninsula to the southward of Point Pearce, and gave the name of Victoria Harbour to the locality; but the survey was subsequently abandoned in consequence of the unfavourable character of the interior, from the great deficiency of water.

If we except the results of a survey made by the late Lieut-Governor, Colonel Robe, of the upper part of Spencer's Gulf, during which, as is the case in the same part of the neighbouring gulf, his Excellency found convenient bays and inlets, but little is known of the eastern shore of that splendid gulf, beyond this point. Double the size of St. Vincent's Gulf, it runs up to the 32½ parallel, and was at one time or other very probably connected with Lake Torrens. The higher part is backed by a range of mountains, the more prominent of which were named by Captain

Flinders—Mount Remarkable, Mount Browne, and Mount Arden. On the first of these there were so many indications of copper, that a special survey of 20,000 acres was taken by a company for the purpose of working any lodes that might be found. The country round about Mount Remarkable is stated to be exceedingly picturesque and good; so that independently of any value it may possess as a mineral survey, it possesses both agricultural and pastoral advantages. After passing the Mount Remarkable Range, however, the country falls off in character. A dreary region extends round the head of the Gulf, and, it is to be feared, to a much greater distance. The description given by Mr. Eyre, and the reports of those who have endeavoured to penetrate to the westward of Lake Torrens both agree as to the sterile and inhospitable character of the remote interior. Little improvement takes place in it on following down the western shore of the Gul& Several individuals, indeed, have perished in endeavouring to take stock round the head of the Gulf to Port Lincoln, either from the want of water, or from having wandered and lost themselves amidst the low brush with which it is covered. The whole of the country, indeed, lying to the westward of Spencer's Gulf is, as far as I have been able to ascertain, of very inferior description. There are, it is true, isolated patches of good land, and a limited run for sheep, but the character of the country corresponds but little with the noble feature for which Spencer's Gulf is so justly celebrated. In reference to this magnificent basin, Captain Lee, from whom I have already quoted, observes—

"The harbour of Port Lincoln, including Boston Bay, is situated near the extremity of the Peninsula, which forms the west side of Spencer's Gulf in the Province of South Australia, and from its great extent, and the number of its safe anchorages, is capable of containing the largest fleets, and as a depôt, is not, perhaps to be surpassed by any port in the world. Vessels from England, bound to Port Lincoln, should run along in about 35°20' south latitude, until they arrive in 135°20' east longitude, when they may haul up to the north-east, and make Cape Catastrophe. After arriving near the Cape, they may then shape a course to pass between it and Williams' Island. There are strong tide ripplings here, which, to a stranger, would present the appearance of reefs; but as the channel is perfectly clear, no danger need be apprehended. Having passed through the channel, should night be approaching, it would be advisable for a stranger to keep the main land aboard, leaving another Island (Smith's Island), on the starboard hand, and bring up in Memory Cove, a perfectly safe anchorage, in about five fathoms. and wait for daylight. Proceeding then along shore to the northward, he will arrive at Taylor's Island, which may be passed on either side; after which he may run along shore at a distance of one mile, until he arrives at Cape Donnington. This Cape may be known by its having a small islet laying about half a mile from the point. Rounding this islet, at half a cable's length, in about nine fathoms' water, and hauling to the westward, he will open the magnificent harbour of Port Lincoln, stretching to the south-west as far as the eye can reach. Should the wind be fresh from the south or south-west, it would be better if bound to Boston Bay, to beat up between Boston Island and the promontory of Cape Donnington. The shores are steep on both sides, so that a vessel may stand close in on either tack. Should the wind be so strong as to prevent a vessel beating in, she may run up under easy sail to a bay on the north-east end of Boston Island, and bring up in seven fathoms opposite a white sandy beach, three-quarters of a mile off shore. There is also excellent anchorage at the entrance to Spalding Cove, bringing the western point of the promontory of Cape Donnington to bear north by east, and the northernmost of Bicker's Island west by north, you will lay in seven fathoms, muddy bottom. Having arrived at Bicker's Island and bound for Boston Bay, stand directly over to the westward, passing the mouth end of Boston Island, until you open the bay, when you may choose a berth according to circumstances, and in any depth from ten to four fathoms.

"The positions of the various points and islands are so correctly laid down on Flinders' chart, that the skilful navigator will at once know his exact situation by cross-bearings.

"The anchorage in Port Lincoln itself is not so safe as in Boston Bay, and more difficult of access, especially in the winter months, when the winds are strong from the south-west, and in the summer months it is quite open to the north-east. In working up, a vessel may stand close in to the eastern shore, and to within half a mile of the western, but should not attempt to pass between the two Bicker's Islands, as there is a reef running from the northernmost island nearly across to the other.

"Vessels from Adelaide, bound to Boston Bay, after arriving at Althorpe's Island, should shape a course so as to pass between the Gambier Islands and Thistle's Islands. There is a small island bearing west five miles from the south end of Wedge Island, the largest of the Gambier group, which is not laid down in Flinders, which should be left on the starboard hand. Bring the highest part of Thistle's Island to bear west, distant about six miles, and in twenty-two fathoms water, and a north-west half-west course will carry you through midway between the Horse-shoe Reef and the rocks which lay off the north-west end of Thistle's Island, and in the direct track for Cape Donnington. The passage between the reefs is about three miles wide, and ought not to be attempted in the night, as the tides set directly across the channel. There is very good anchorage on the north-east side of Thistle's Island, well sheltered three fourths of the year. Bring the rocks before-mentioned to bear north-north-west, and two remarkable sand hills south by west, and you will lay in five fathoms, one mile off shore-north end Thistle's Island west by south. Should the wind be so strong from south-west or west-south-west, so that a vessel from the eastward cannot carry sail sufficient to fetch up to Cape Donnington, or under Thistle's Island, it would be advisable to bear up for Hardwick Bay; passing to the eastward of Wedge Island, come no nearer to the shore of York's Peninsula than two miles, until you arrive within five miles of Corny Point, when you may haul in for that point, rounding it a distance of half a mile, you may bring up in five fathoms, one mile from shore: Corny Point bearing west. Vessels from Sydney, bound to Port Lincoln, may pass through Backstairs Passage, and proceed according to the foregoing directions, or by keeping well to the southward, pass outside Kangaroo Island, until they arrive in longitude 136°E, when they may shape a course either to pass between Gambiers and Thistle's Islands, or else for Cape Catastrophe, taking care to give the Neptune Islands a wide berth, and then proceed according to either of the foregoing directions."

To this extract which refers exclusively to the navigation of Spencer's Gulf, I may add, that Boston Island lies immediately opposite to the bay, and that there are two channels of entrance round the island, through which vessels of the largest size can pass with any wind or in any weather, for the harbour is so sheltered by the headlands forming the entrance, that the swell of the sea is broken before reaching it.

The high ground which almost surrounds Boston Bay, protects it in like manner from the winds, more especially those coming from the west and south-west, in which directions some of the hills attain the height of several hundred feet.

The depth of water in the central parts of the Bay is about twelve fathoms, varying from five to seven at the distance of less than a quarter of a mile from the shore all round; whilst at Boston Point, where the town of Boston has been laid out, there is a depth of two, three, and four fathoms, at about a boat's length from the land. The bottom consists in some places of mud, in others of shells and sand, so that the anchorage is safe.

The tide sometimes rises seven feet, but that is considered a high tide, the ordinary rise not being more than five; this depends, however, on the outward state of the Gulf, and the quarter from which the wind may happen to be blowing.

In the summer season, the land and sea breezes blow very regularly, for three weeks or a month at a time. They are then succeeded by strong winds from the south-west, that last for three or four days, and are sometimes very violent. In winter these interruptions to the usual calm state of the weather are more frequent, but the harbour is little influenced by them; taking it altogether, indeed, as a harbour, it is unquestionably as safe and commodious as any in the world, and it is deeply to be regretted, that its position, of which I shall have to speak, and the nature of the country behind it, should be any drawbacks to its becoming one of the most important ports on the Australian Continent.

In the vicinity of Port Lincoln, the land is of very varied character. To the west and south-west it is poor and scrubby, covered with a diminutive growth of she oak (Casuarinæ) or dwarf gum-trees (Eucalypti), or it is wholly destitute of timber; but along the line of hills, stretching to the north, at a short distance from the shores of the Gulf, there is an improvement in the soil. The pasture is well adapted for sheep, and there are isolated valleys in which the soil is very good and fit for cultivation; but this kind of country only occupies a narrow strip of about ten miles, and although tracts of available land have been found in the interior, and it has been ascertained that water is not deficient, it must still, I fear, be considered as a very inferior district. As regards Port Lincoln itself, the inhabitants procure their water from a spring, on the sea-shore, which is covered by every tide. This spring does not appear to undergo any sensible diminution, even in the height of summer, and is stated to be so copious, that it would yield a most abundant supply.

It has been reported, that strong indications of the presence of copper have been found in the neighbourhood of Port Lincoln, and this report may be correct. The discovery of mines there, would at once raise the harbour to importance, and make it the resort of shipping. Mines might be worked at Port Lincoln with more advantage perhaps to the province, than where they have been already in operation, for it admits of great doubt whether the benefit from the distribution of wealth from mining speculations, makes up for the interference of such speculations with other branches of industry. Unless some local advantage, of the kind to which I have alluded, should give this noble harbour an impulse however, it would appear to have but little prospect of becoming a place of importance, for although Spencer's Gulf penetrates so deep into the northern interior, the country is altogether unprofitable, and although there is depth of water sufficient for the largest ships to the very head of the Gulf, yet, as far as our present knowledge extends, it is not probable that it will be the outlet of any export produce. It is to be remembered, however, that if there should be minerals in any abundance found on the Mount Remarkable special survey—the ore must necessarily be shipped, from some one of the little harbours examined by the Lieutenant-Governor during his survey of that part of Spencer's Gulf—In such case, Port Lincoln will be brought more immediately into notice.

From Port Lincoln, the shore of the Gulf still trends to the south, as far as Cape Catastrophe, in lat. 35°. It then turns with an irregular outline to the N.N.W., and several bays succeed each other. The first of these is Sleaford Bay, sometimes occupied as a whaling station, but of no other importance. Coffin's Bay, almost immediately behind Port Lincoln, is rather an inlet than a bay, and runs so far into the interior, as to approach Boston Bay, to within 16 miles. Coffin's Bay is exceedingly wide, and objectionable for many reasons, but as it is a whaling station of some importance, and visited by numerous whalers, I shall quote Captain Lee's remarks upon it, and give his directions for going to it.

"This is a very large bay, perfectly secure from all winds, save from north to east, but unfortunately a great portion of it is rendered useless by the shallowness of the water. The best anchorage is with Point Sir Isaac, bearing north-north-west, about one mile and a half from the western shore in four or five fathoms. In working in with a southerly wind, you may stand to the eastward until you bring the above point to bear south-west by west, after which it would be better to make short tacks along the western shore. You must be careful to keep the lead going, as the water shoals from five and four fathoms to one and a half at a single cast. This bay seems well adapted for a fishing station. The inner part of the bay extends a long way back into the country, at least thirty miles from Point Sir Isaac, and contains two or three secure harbours and excellent anchorages, a new chart of which is in course of publication.

"Vessels from Sydney bound to Coffin's Bay, should proceed as if bound to Port Lincoln until arrived off the Neptune Islands, when they should steer for Perforated Island, having passed which, steer for Point Whidbey, giving it a berth of at least two miles. In running along shore from Point Whidbey to Point Sir Isaac, come no nearer the shore than two miles, until you get the latter point to bear east-south-east as the rocks lay a long way from the shore. Having arrived at Point Sir Isaac proceed as directed before.

"Althorp's Island is of moderate height, situated at the entrance of Investigator's Straits; may be passed close to on the south aide. Several other islands and reefs lay between it and York's Peninsula, rendering that passage highly dangerous.

"Wedge Island, one of the Gambier Group, may easily be known by its wedge-like form, sloping from south-east to north-west. There are two peaked rocks off the south-east end, one mile off shore, also a small island, bearing west five miles from the south end, not laid down in Flinders' charts.

"Thistle's Island, is low at each end but high in the middle, it lays in a north-west and south-east direction. There are some rocks which lay off the northern point about three miles, which being connected with the island itself, forms a good anchorage behind, secure from all but north and east winds, another good place for a fishing party. See Port Lincoln directions.

"Neptune's Islands are low, three in number, and having numerous rocks and reefs amongst them; ought not to be approached too closely, there being generally a strong swell from the south-west, the sea breaks over them with great violence.

"Liguanea Island is of moderate elevation, and may be passed on the south side at a distance of two miles.

"Perforated Island, as its name imports, may be known by its having a hole through it near the north end and close to the top of the island, it may be passed close on any side. Four Hummocks may easily be known from their appearance answering to their name.

"Greenly Island, this is a peaked island, rather high, and may be seen ten leagues off. There is another island laying south and by west, seven miles, not laid down in Flinders' and two other reefs between them, rendering the passage unsafe.

"Proceeding along shore to the northward you will fall in with Flinders' Island. This is a large island, covered with wood, with plenty of fresh water, possessing a secure anchorage on the northern side, and is admirably adapted for a whaling station. In going on from the southward, keep outside the top Gall's Island, and steer directly for the north-east point, rounding which, you will open the anchorage, and as there is no danger, but may be seen, you may choose a berth according to circumstances.

"Waldegrave's Island, close to the main land, has good anchorage on the northern side, secure from south-east and south-west winds.

"The shore, from Waldegrave's Island to Point Weyland is low and sandy. There is a large body of water running in a direction parallel to the coast, all the way from Point Weyland to the northward of Cape Radstock, having an entrance at both points. It appears as if the action of the sea from the south-west, had broken through the coast range and filled up the valley immediately behind. Indeed the whole coast from Kangaroo Island to as far to the north-west as has been visited by the author, bears evident marks of the encroachments of the sea. In some places marked down as small islands in Flinders', there are now only reefs, other places which were formerly points of land, are transformed to islands."

In the year 1840, I was instructed by the then Governor of South Australia, to send an officer of the survey in a small vessel, with a supply of provisions for Mr. Eyre, who was at that time supposed to have reached Fowler's Bay, during the first of his expeditions; I accordingly selected Mr. John Cannan, in whose zeal and ability I had every confidence. This officer left Port Adelaide the 9th September, 1840, with instructions from me, in addition to the immediate object he had in view, to survey such parts of the coast along which he was about to sail, as had only been partially examined by Captain Flinders. Unfortunately it was during the winter time, and the task I had assigned him would, I knew, be attended with considerable risk in beating along that dangerous and stormy coast. Mr. Cannan arrived at Streaky Bay on the 27th September, but was disappointed in finding Mr. Eyre, or a letter he had buried for him under Cape Bauer, he therefore proceeded to the examination of the coast, as I had instructed him to do; and the following extract from his report will not only enable the reader to judge how he performed that service, but will give him the best information as to the character of the several bays and inlets he examined.

"I send you a chart of Streaky, Smoky, and Denial Bays, by which you will be better able to judge of the capabilities of the harbours they contain, than by any description I can give. I may mention however, that the entrance to Smoky Bay, between the shoals of St. Peter's and Eyre's islands, is dangerous, for with any swell on the sea breaks right across. In the inlet, on the west side of Denial Bay, there is a salt water creek with two fathoms of water; and adjoining some high sand-hills, among which we found fresh water by digging. Our vessel being the first, I believe, that ever entered Smoky Bay, on finding an island at its southern end, I named it after that enterprising traveller Mr. Eyre. I also found an island and reef not laid down by Flinders, to the southern of St. Francis Islands. There is also an island 10 miles west of the rocky group of Whidbey's Isles, and about 12 miles from Greenly's Isles. The captain of a French whaler also informed me, that a sunken rock lays 6 miles N.W., off Point Sir Isaac, on which the sea breaks in heavy weather.

"The desert country surrounding these bays has been sufficiently explored, and so correctly described by Mr. Eyre, as not to require to be mentioned. The absence of any rise that can be called a hill, from Mount Greenly to Mount Barren, the eternal limestone cliff, the scarcity of water and grass, surely prove this coast to be the most miserable in the world, whilst the harbours are as good as could be wished for, and it must be owing to the deficiency of charts, that whalers do not frequent these bays, for there are generally two or three French or American vessels in the neighbourhood during the season. I found no bones or carcasses of whales in Streaky, Denial, or Smoky Bays, but the shores of Fowler's and Coffin's Bays, I found strewed with their remains. In the latter place, Captain Rossiter, of the Mississippi shewed me his chart, and told me there was no shelter for a vessel on this side of the Bight, except at Fowler's Bay, and that was indifferent. The great extent of smooth water at Denial and Streaky Bays, and a well of water on St. Peter's, dug by a sealer who lived on it many months, afford more advantages for fishing, and more especially to a shore party, than are to be found any where else in the Province.

"From the general flatness of the country, it may be presumed that its character does not alter for a great distance inland. I observed nothing in the formation of the island, differing from the mainland, and I may mention that the rocks of the isles of St. Francis presented the same appearance as the Murray Cliffs."

It will appear from the above, that Mr. Cannan did not proceed farther to the westward than Fowler's Bay, and that he did not therefore prolong his survey to the western limits of the Colony, by a distance of about five leagues, since the 132° meridian falls on that coast a little to the westward of Cape Adieu, and between 12 and 15 leagues from the bottom of the Great Australian Bight.

Although some of the bays and harbours I have described in running along its coast, are not so good as might be desired, yet it is evident that, as a maritime country, South Australia is particularly favoured, not only in having anchorage of the safest description, but also in possessing two or three known harbours, capable of containing ships in any number or any size, and as safe and capacious as any in the world. Looking indeed at Port Adelaide, one cannot but admire its appropriate and convenient position. Had such a harbour not existed there, the produce of that fertile portion of the Province would hardly have been available to the inhabitants in the shape of exports, so difficult would it have been to have found another harbour of equal security, or of equal size, for the commercial wants of the settlers. Added to this, it has the double advantage of being close to the capital, being so easy of access, and in so central a position, as to he able to communicate with the neighbouring colonies with the greatest ease.

It will be remembered that I stated in the former part of my work, that the remarkable wall forming the Great Australian Bight, was thrown up simultaneously with the great fossil bed of the Murray.

As the principal object of the Expedition into Central Australia was to ascertain the past and present structure of the Continent, I have been led to allude to the subject again, in consequence of two or three remarks in Mr. Cannan's letter, which has been quoted above, bearing strongly upon it, and corroborative of the hypothesis I have entertained as proving a striking uniformity in the rock formation of those two localities. To those remarks I would beg to call the attention of my readers. They will be found at the commencement and termination of the last paragraph.

CHAPTER II.

PLAINS OF ADELAIDE—BRIDGES OVER THE TORRENS—SITE OF ADELAIDE—GOVERNMENT HOUSE BUILDINGS AND CHURCHES—SCHOOLS—POLICE—ROADS—THE GAWLER—BAROSSA RANGE—THE MURRAY BELT—MOORUNDI—NATIVES ON THE MURRAY—DISTANT STOCK STATIONS—MOUNT GAMBIER DISTRICT—ITS RICHNESS—ASCENT TO MOUNT LOFTY—MOUNT BARKER DISTRICT—SCENE IN HINDMARSH VALLEY—PROPORTION OF SOIL IN THE PROVINCE—PASTORAL AND AGRICULTURAL—PORT LINCOLN—CLIMATE OF SOUTH AUSTRALIA—RANGE OF THE THERMOMETER—SALUBRITY.

HAVING, in the preceding chapter, run along the coast of South Australia, and noticed such parts as have been sufficiently examined to justify our observations, it remains for me to give an account of its interior features, of its climate, soil, mineral, and other sources of wealth, and lastly of its fitness as a colony for the peculiar habits of an English population.

The city of Adelaide, the capital of South Australia, stands on the eastern shore of St. Vincent's Gulf, and is about six miles from the coast. Any one landing either at the old or new port, and proceeding to the capital for the first time, would perhaps be disappointed at the description of country through which he would pass. It consists indeed of extensive level plains, over the eastern extremity of which the Mount Lofty Range is visible. They are bounded southwards by a line of trees, marking the course of the river Torrens across them, but extend northwards for many miles without any visible termination. Their monotony however, is, at the present date, in some measure broken by belts of wood, and the numerous cottages that have been built upon them, with their adjoining corn-fields, have changed their aspect, and removed the appearance of loneliness which they first exhibited. Still neither the gloomy swamp over which the stranger has in the first instance to travel, on landing at the Port—or the character of the plains themselves, are calculated to raise his anticipations, as to the beauty or fertility of the interior. The first town through which he will pass after leaving the Port, is Albert Town, which has been laid out on the first available ground near the swamp. When I left the colony in May last, several tolerable buildings had been erected in Albert Town, but it was nevertheless a wretched looking and straggling place, and will never perhaps advance beyond its present state.

On his nearer approach to the capital the traveller will pass between the villages of Bowden and Hindmarsh, in both of which he will observe numerous kilns of bricks. He will then enter on the Park Lands, by which North and South Adelaide are separated from each other. On this land the scene at once changes, and he will find himself riding through an open forest, shading rich, alluvial, and grassy flats; and, strictly speaking, will then be traversing the Valley of the Torrens. In May, 1847, there were four bridges over that little river. The Company's bridge a little above the city. The Frome bridge, a light wooden structure, built by the sappers and miners, under the direction of Captain Frome, the Surveyor-General, after whom it was called. The City bridge, constructed of stone, but then incomplete, and a rude wooden bridge between Adelaide and Hindmarsh, erected by an innkeeper, with a view of drawing the traffic from the Port past his door. The City bridge, which was undertaken by contract, promised to grace the approach to Adelaide, and was intended to be the principal bridge to connect the north and south portions of the city, as well as to form the chief line to the Port and to the north. The occurrence of an unusual flood, however, in the latter part of the year 1847 deprived the good citizens of Adelaide of these necessary means of communication with the country on the right bank of the Torrens, by the injury it did to them. The Company's bridge suffered less than any other, but was so shaken as to be impassable for several days. Aware, as I am, of the general character of the Australian streams, and seeing no reason why the Torrens should differ from others, taking into consideration, too, the reports of the natives as to the height to which the river had been known to rise in former years, and the fact that no rain had fallen since the establishment of the Colony to cause any very great or sudden flood, it appeared to me, that the place selected for the City bridge was too low. Ordinary floods so completely change the channel of the river, and make such devastation in its bed, that it is hardly to be recognised when the water subsides, so that unless the banks are high, and the soil of which they may be composed stiff enough to resist the impetuosity of the stream, I fear no bridge across the Torrens will be permanently safe.

The position and ground chosen by the first Surveyor-General of South Australia, as the site of its future capital is a remarkable instance of the quick intelligence of that officer. For although he had but little time to make his selection, a more intimate knowledge of the coast has proved that no more eligible point could have been found. Fault has, I am aware, been found with Colonel Light in this matter, but without just grounds, I think, for in no other locality could the same quantity of water have been found, or the same facility offered for the construction of those reservoirs and other works so necessary to the health and comfort of a large metropolis. A principal objection raised to the situation of Adelaide is its distance from the Port, but that we must remember is a disadvantage common to many other large and mercantile cities. The Surveyor-General seems to have been fully aware of the responsible duty that devolved upon him, and to have acted with great judgment. Port Lincoln, indeed, is a splendid harbour, one with which Port

Adelaide, as far as size goes, cannot be compared, but having said this nothing farther can be advanced in its favour, for it is not only deficient in its supply of water, but the contiguous country is far from rich, whereas Adelaide is backed by one of undoubted fertility.

Established where it is, the city of Adelaide stands on the summit of the first elevated ground, between the coast and the mountain ranges.

It is separated, as the reader will have learnt, by the valley of the Torrens, and occupies the northern and southern slopes and brows of the hills on either side. The view to the westward from the more elevated parts of the city commands the whole of the plains of Adelaide, and St. Vincent's Gulf; to the eastward, it extends over the rich and dark wooded valley of the river, the lighter wooded country at the base of the Mount Lofty Range, and the peaks and elevations of that beautiful mountain chain.

South Adelaide is on flat ground and twice the size of the northern part of the town. It has also been more extensively built upon, and is the established commercial division of the city. The Government House and all the public buildings and offices are in South Adelaide, and the streets in the vicinity of the North Terrace, have assumed a regularity and uniformity greater than any street in North Adelaide. Hindley and Rundle streets, indeed, would do no discredit to any secondary town in England. Every shop and store that is now built is for a substantial and ornamental character, and those general improvements are being made which are the best proofs of increasing prosperity and opulence.

There is scarcely any article of European produce that cannot be obtained in Adelaide, at a very little advance on home prices, nor is it necessary, or indeed advisable that Emigrants should overload themselves in going out to any of the Australian Colonies. Experience, the best monitor, leads me to give this advice,

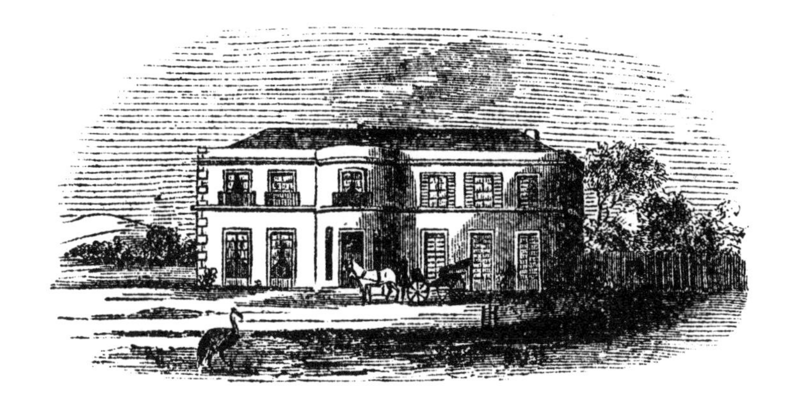

which, however, I am bound to say, I did not adopt when I went out to New South Wales; but the consequence was, that I purchased a great many things with which I could have dispensed, and that I should have found the money they cost much more useful than they proved.

King William Street divides Hindley from Rundle Street, and is immediately opposite to the gate of Government House, which is built on a portion of the Park lands, and is like a country gentleman's house in England. It stands in an enclosure of about eight or ten acres; the grounds are neatly kept, and there is a shrubbery rapidly growing up around the House.

The Public Offices are at the corner of King William Street and Victoria Square, facing into the latter. The building is somewhat low, but a creditable edifice, to appearance at all events, although not large enough for the wants of the public service.

I am not aware that there is any other public building worthy of particular notice, if I except the gaol, which is a substantial erection occupying the north-west angle of the Park land, but is too low in its situation to be seen to advantage at any distance. Like Government House, it was built with a view to future addition, but fortunately for the colony, Government House is the first which seems to call for completion.

The number of Episcopalian Churches in Adelaide is limited to two, Trinity Church and St. John's. The former was originally built of wood, and may be said to be coeval with the colony itself. It has of late however been wholly built of stone, and under the active and praiseworthy exertions of Mr. Farrell, the colonial chaplain, an excellent and commodious school-room has been attached to it.

Trinity Church stands on the North Terrace, and is a prominent object as you ascend from the Park lands. St. John's is situated on the East Terrace at a greater distance, but it has a commanding view of the Mount Lofty Range, and the intervening plains. Perhaps considering that the city has not extended much in the direction of East Terrace, it may be little too far for public convenience, but this is a question that admits of doubt. It is a neat and unostentatious brick building, at which the Rev. Mr. Woodcock performs service, whose exertions amongst the natives in the West Indies have stamped him both as a Christian and a philanthropist. The two churches are calculated to hold about 1,000 sittings, and the average attendance is about 900.

It may appear to the reader that the number of churches in Adelaide, where there is a population of between 8,000 and 10,000 souls, is not sufficient, as is the case. Ere this however, a third church, to be called "Christ's Church," will have been erected in North Adelaide, where such a place of worship was much required. £500 had been subscribed for the purpose in December last, and it was confidently anticipated that the further contributions of the colonists would enable the committee to commence and finish it. The arrival of the Bishop on the 24th of the above month, of which accounts have been received had given great satisfaction, and his Lordship was to begin his useful ministry on the following day (Christmas Day), by preaching at Trinity Church.

However few the Episcopalian churches in the capital of South Australia, we cannot accuse the Dissenters of a similar want of places of public worship, of which there are 9, the whole number throughout the province being 31; whilst the number of churches is 6. The Congregational chapels are calculated to accommodate 4,700 communicants, the average attendance being about 2,300, and are, generally speaking, good looking and ornamental buildings, and do no discredit to those who superintended their erection, and approved the places.

There is a Roman Catholic Bishop of South Australia, but he had, during the latter period of my residence in the province, been absent in Europe. The Catholic Church stands on the West Terrace, and is, perhaps, in one of the most healthy situations that could have been chosen. There is an excellent school attached to the church, which is equally open to all denominations of Christians, and is, I have understood, more numerously attended than any other in the capital. The total number of Sunday-schools in the province, in 1841, was 26, at which 617 boys and 582 girls attended. The average number of Sunday and other schools in 1845 was 55, at which 780 males and 670 female children attended.

In the year 1846, when His Excellency Colonel Robe laid the estimates on the table of the Legislative Council, its attention was drawn to the state of education and religion in the province, and after a long discussion on the subject, a grant of 2s. per head was voted to the different sects in aid of religion and education. It was left to the ministers of the Protestant Church, and to the proper officers of the other persuasions to appropriate the sum received by each, according to the last census, as they deemed best, for the promotion of one or the other of the above purposes, with the sole condition that they should render an account yearly to the Council of the manner in which the several sums had been appropriated. Yet this provision, which without interfering in the slightest degree with any religious sect, gave to the heads of each the greater power of doing good, caused very great dissatisfaction. All I can say is, that it was an instance of liberal and enlightened views of government, of which the Council of South Australia in having set the example ought to be proud.

The Legislative Council of New South Wales has since, I believe, followed its example, and I sincerely trust the good that is anticipated, will result from this proof on the part of both Governments to raise the moral and social character of the people.

In addition to the schools already noticed, there is a school for the natives on the Park lands. At this school there were in 1847, thirty-five boys and twenty-nine girls. The establishment being entirely under the superintendence of the Government, is kept in the very best order; the apartments are neat and clean, the master is patient and indulgent, and if we could hope for any improvement in the moral and social habits of the aborigines, it would be under circumstances so promising, but as I propose, in another place, to make some observations on the natives generally, it may not be necessary for me to add to the above remarks, at the present moment.

Of other public buildings not under the immediate control of the Government, the Bank of South Australia is certainly the first. It stands on the North Terrace and is a prominent and pleasing object from whatever point of view it is seen. There are, however, several other very creditable buildings in different parts of the city.

Had the city of Adelaide been laid out in the first instance on a smaller plan, it would now have been a compact and well-built town, but unfortunately it was planned on too large a scale, and it will necessarily have a straggling appearance for many years to come.

North and South Adelaide are, as I have already stated, separated from each other by the valley of the Torrens, than which nothing can be prettier. Its grassy flats are shaded by beautiful and umbrageous trees, and the scenery is such as one could not have expected in an unimproved state. The valley of the Torrens is a portion of the Park lands which run round the city to the breadth of half a mile. Nothing could have been more judicious than the appropriation of this open space for the amusement and convenience of the public, and for the establishment of those museums and institutions which tend so much to direct the taste, and promote the scientific improvement of a people.

Beyond the Park lands, the preliminary sections, of 134 acres each, extend to a certain distance—many of which have been laid out into smaller sections, and the city is surrounded by numerous villages, few of which add to its appearance. This certainly may be said of Thebarton, Hindmarsh, Bowden, and several other villages, but those of Richmond, and Kensington, embosomed in trees, and picturesque in scenery, bear a strong resemblance to the quiet and secluded villages of England.

In Hindmarsh, Mr. Ridley, whose mechanical genius has been of such public utility, and whose enterprise is so well known, has established his steam flour-mill, which is the largest in the province. In addition to this, the South Australian Company has a steam-mill at the tipper bridge; there are several of a smaller size in the city, and the total number of flour-mills in the Colony, including wind and water mills is twenty-two.

This general description of the capital of South Australia will perhaps suffice to shew its rapid growth during the eleven short years since the first wooden dwelling was erected upon its site.

It may be necessary for me to state that its peace and order are preserved by a body of police, whose vigilance and activity are as creditable to them as their own good conduct and cleanliness of appearance; and whilst the returns of the supreme court, and the general infrequency of crime, prove the moral character of the working classes generally, the fewness of convictions for crimes of deeper shade amongst that class of the population from whose habit of idleness and drinking we should naturally look for a greater amount of crime, as undoubtedly proves the vigilance of the police. From the return of convictions before Mr. Cooper the Judge, it is clear that the majority of those who have been brought before him are men who have already suffered for former breach of the laws, and who, having escaped from the neighbouring Colonies, have vainly endeavoured to break themselves of former evil habits. The eyes of' the police are however so steadily kept on such men, that they have little chance of escaping detection if they commit themselves, and they consequently level their aim at those who encourage them in vice, and who, in reality, are little better than themselves in morals, as knowing that, in many instances, they will not dare to bring them to punishment.

There are five principal roads leading from Adelaide; three into the interior, and two to the coast. Of the three first, one leads to the north. through Gawler Town, one as the Great Eastern Road leads to Mount Barker and the Murray, and the third running southwards, crosses the range to Encounter Bay. Of the roads leading to the coast, the one goes to the Port, the other to Glenelg. In endeavouring to give a description of the country, and enabling the reader to judge of it, I would propose to take him along each of these roads, and to point out the character and changes of the country on either side, for the one is peculiar and the others are diversified. My desire is to present such a view of the colony to the minds of my readers, as shall enable them to estimate its advantages and disadvantages. I would speak of both with equal impartiality and decision. The grounds of attachment I entertain for this colony rest not on any private stake I have in its pastoral or mineral interests, and I hope the reader will believe that my feelings towards it are such as would only lead me to speak as it really and truly should be spoken of. There is no country, however fair, that has not some drawback or other. There are no hopes, however promising, that may not be blighted; no prospects, however encouraging, that may not wither. Unfitness for the new field of enterprise on which a man may enter unpropitious seasons, the designs of others, or unforeseen misfortunes; one or more of these may combine to bring about results very opposite from those we had anticipated. I would not therefore take upon myself the responsibility of giving advice, but enter upon a general description of the province of South Australia as a tourist, whose curiosity had led him to make inquiries into the capabilities of the country through which he had travelled, and who could therefore speak to other matters, besides the description of landscape or the smoothness of a road.

If we take our departure from Adelaide by the great Northern Road, we shall have to travel 25 miles over the plains, keeping the Mount Lofty Range at greater and less distances on our right, the plains extending in varying breadth to the westward, ere we can pull up at Calton's Hotel in Gawler Town, where, nevertheless, we should find every necessary both for ourselves and our horses.

That township, the first and most promising on the Northern Road, is, as I have stated, 25 miles from Adelaide; and occupies the angle formed by the junction of the Little Para and the Gawler Rivers; the one coming from south-east, and the other from north-north-east; the traveller approaching from the south therefore, would have to cross the first of these little streams before he can enter the town.

Still, in its infancy, Gawler Town will eventually be a place of considerable importance. Through it all the traffic of the north must necessarily pass, and here, it appears to me, will be the great markets for the sale or purchase of stock. From its junction with the Little Para, the Gawler flows to the westward to the shores of St. Vincent's Gulf. It has extensive and well wooded flats of deep alluvial soil along its banks, flanked by the plains of Adelaide—the river line of trees running across them, only with a broader belt of wood, just as the line of trees near Adelaide indicates the course of that river. If I except these features, and two or three open box-tree forests at no great distance from Albert Town, the plains are almost destitute of timber, and being very level, give an idea of extent they do not really possess, being succeeded by pine forests and low scrub to the north from Gawler Town.

The Gawler discharges itself into a deep channel or inlet, which, like the creek at Port Adelaide, has mangrove swamps on either side; still the inlet is capable of great improvement, and the anchorage at its mouth, so high up the gulf is safe, and if it were only for the shipment of goods, for transhipment at Port Adelaide, Port Gawler as it is called, would be of no mean utility, but it is probable that ships might take in cargo at once, in which case it would be to the interest of the northern settlers to establish a port there. Captain Allen and Mr. Ellis, two of the most independent settlers in the province, are the possessors of the land on both sides the Gawler, and I feel confident it is a property that will greatly increase in value. The alluvial flats along this little stream, are richer and more extensive than those of the Torrens, and they seem to me to be calculated for the production of many things that would be less successfully cultivated in any other part of the province. Apart, however, from any advantages Gawler Town may derive from the facilities of water communication, it will necessarily be in direct communication with Port Adelaide, as soon as a road is made between them. At present the drays conveying the ore and other exports are obliged to keep the great northern line to within a few miles of the city, before they turn off almost at a right angle to the Port; but there can be no doubt as to the formation of a direct line of communication with the Port from Gawler Town, if not of the establishment of a railway, ere many years shall elapse, for not only are the principal stock stations of the province, but the more valuable mines to the north of this town.

Up to this point the traveller does not quit the plains of Adelaide, the Mount Lofty Range being to the eastward of him and the plains, bounded by the mangrove swamps extending towards St. Vincent's Gulf. Generally speaking, for their extent the soil is not good, but there are patches of alluvial soil, the deposits of creeks falling from the hills, that are rich and fertile. Yet, notwithstanding the quality of the soil, a great portion of the Adelaide plains have been purchased and are under cultivation. There is a great deficiency of surface water upon them, but it is procurable by digging wells; and Mr. Ellis I believe has rendered those parts of them contiguous to the Gawler available as sheep stations, by sinking wells for the convenience of his men and stock; neither can there be a doubt but that many other apparently unavailable parts of the province might be rendered available by the adoption of similar means, or by the construction of tanks in favourable situations.

This is a point it is impossible to urge too much on the attention of the Australian stock holder. There is generally speaking a deficiency of water in those Colonies, and large tracts of country favourable to stock are unoccupied in consequence, but the present liberal conditions on which leases of Crown lands are granted will make it worth the sheep farmer's while to make those improvements which shall so conduce to his prosperity and comfort.

In proof of this, I would observe that I had several capacious tanks on my property at Varroville, near Sydney, for which I was indebted to Mr. Wells the former proprietor, and not only did they enable me to retain a large quantity of stock on my farm, when during a season of unmitigated drought my neighbours were obliged to drive their cattle to distant parts of the Colony—but I allowed several poor families to draw their supplies from, and to water some of their cattle at my reservoirs.

Beyond Gawler Town the country changes in character and appearance, whether you continue the northern road across the river, or turn more to the eastward, you leave the monotonous plain on which you have journeyed behind, and speedily advance into an undulating hilly country, lightly wooded withal, and containing many very rich, if not beautiful valleys. The Barossa Range and the districts round it are exceedingly pretty. Here, at Bethany, the Germans who have fled from the religious persecution to which they were exposed in their own country have settled, and given the names of several places in their Fatherland to the features around them. The Keizerstuhl rises the highest point in the Barossa Range, the outline of which is really beautiful, and the Rhine that issues from its deep and secluded valleys flows northwards through their lands.

In this neighbourhood Mr. Angas has a valuable property, as also the South Australian Company. Angas Park is a place of great picturesque beauty, and is capable of being made as ornamental as any nobleman's estate in England. The direct road to the Murray River passes through Angas Park, but a more northerly course leads the traveller past the first of those valuable properties to which South Australia is mainly indebted for her present prosperous state. I mean the copper mines of Kapunda, the property of Captain Bagot, who, with Mr. Francis Dutton, became the discoverer and purchaser of the ground on which the principal lode has been ascertained to exist. There has been a large quantity of mineral land sold round this valuable locality, but although indications of copper are everywhere to be seen, no quantity sufficiently great to justify working had I believe been found up to the time I left the Colony. As however I shall have to give a more detailed account of the mines of South Australia, it may not be necessary for me to speak of them at length in this place.

Captain Bagot is anxious to establish a township in the vicinity of Kapunda, and he will no doubt succeed, the very concourse of people round such a place being favourable to his views.

Beyond this point to the north the coast range of Mount Lofty, which thus far preserves a northerly direction, throws off a chain to the westward of that point, but the main range still continues to run up into the interior on its original bearing, rather increasing than decreasing in height. Upon it, the Razor Back Mount Brian, to the south of which is the great Burra Burra mine, and the Black Rock Hill, rise to the height of 2,922, 3,012 and 2,750 respectively. On the more western branch of the chain, Mount Remarkable, Mount Brown, and Mount Arden, so named by Captain Flinders, form the principal features. This chain has been traced by Mr. Eyre to Mount Hopeless, in lat. 29½, and has been found by him to terminate in the basin of Lake Torrens. The main range on the contrary has only been followed up to lat. 32°10', beyond which point it cannot extend to any great distance, as if it did, I should necessarily have seen something of it during my recent expedition. It is a remarkable fact that the further the northern ranges have been followed up, the more denuded of trees they have become. Immense tracts of land, through portions of which the Wakefield flows, rich in soil and abundant in pasture, have scarcely a tree upon them. The scenery round Mount Remarkable on the contrary is bold and picturesque, and much diversified by woodland.

Here again the indications of copper were so abundant, that 20,000 acres were taken as a special survey a short time before I left the Colony. The occupation of this land will necessarily extend the boundaries of location, but up to the period when the survey was taken, Mr. White, formerly a resident at Port Lincoln, was the most distant stockholder to the north.

Proceeding eastward from Angas Park, the road to the Murray river leads through a hilly country of an inferior description, portions only of it being occupied as sheep stations. From the brow of the last of these hills, the eye wanders over the dark and gloomy sea of scrub, known as the Murray belt, through which the traveller has to pass before he gains the bank of the river or the station at Moorundi. He descends direct upon the level plain over which he has to go, and after passing some pretty scenery on the banks of a creek close to which the road runs and crossing an open interval, he enters the belt, through which it will take him four hours to penetrate. This singular feature is a broad line of wood, composed in the lower part of Eucalyptus dumosa, a straggling tree, growing to an inconsiderable height, rising at once from the ground with many slender stems, and affording but an imperfect shade. About the latitude of 34° the character of the Murray belt changes—it becomes denser and more diversified. Pine trees on sandy ridges, Acacia, Hakea, Exocarpi, and many other shrubs form a thick wood, through which it is difficult to keep a correct course. Occasionally a low brush extends to the cliffs overlooking the valley of the Murray, but it may be said, that there is an open space varying in breadth from half a-mile to three miles between the Murray belt and the river. It is a flat table land about 250 or 300 feet above the level of the sea, the substratum being of the tertiary fossil formation. The surface is a mixture of red sand and clay, mixed with calcareous limestone in small rounded nodules. The very nature of this soil is heating, and the consequence is that it has little herbage at any one time. There is however a succession of vegetation, especially during the spring months, which, from the fact of the cattle being particularly fond of it, must I should imagine be both sweet and nutritious.

Any one who has ever been on the banks of the Murray will admit that it is a noble river. The description I have already given supersedes the necessity of my dwelling on it here. In another place I shall have to speak of it, not in a commercial point of view, but as a line of communication between two distant colonies, and the important part it has acted in the advancement of the province of South Australia. As a commercial river, I fear it will not be of practical utility. To prove this, it may be necessary for me to observe that the Murray runs for more than five degrees of latitude through a desert. That it is tortuous in its course, and is in many places encumbered with timber, and its depth entirely depends on the seasons. The difficulties, therefore, that present themselves to the navigation of the central Murray are such as to preclude the hope of its ever being made available for such a purpose, even admitting that its banks were located at every available point. Moorundi, the property of Mr. Eyre, the present Lieutenant-Governor of New Zealand, is ninety miles from Adelaide, and twenty-six from the N.W. bend of the Murray. It is part of a special survey of four thousand acres taken by Mr. Eyre and Mr. Gilles on the banks of the river, and in consequence of its appropriate position, was selected by Captain Grey, the then Governor of South Australia, as a station for a Resident Magistrate and Protector of the Aborigines, to fill both which appointments he nominated Mr. Eyre. There can be no doubt, either as to the foresight which dictated the establishment of this post on the banks of the Murray, or the selection of Mr. Eyre as the Resident. At the time this measure was decided on, the feelings of the natives on the river were hostile to the settlers. The repeated collisions between them and the Overlanders had kindled a deep spirit of revenge in their breasts, and although they suffered severely in every contest, they would not allow any party with stock to pass along the line of the river without attempting to stop their progress; and there can be no doubt but that, in this frame of mind, they would have attacked the station next the river if they had been left to themselves, and with their stealthy habits and daring, would have been no mean enemy on the boundaries of location. The character and spirit of these people is entirely misunderstood and undervalued by the learned in England, and the degraded position in the scale of the human species into which they have been put, has, I feel assured, been in consequence of the little intercourse that had taken place between the first navigators and the aborigines of the Australian Continent. I have seen them under every variety of circumstances—have come suddenly upon them in a state of uncontrolled freedom—have passed tribe after tribe under the protection of envoys—have visited them in their huts—have mixed with them in their camps, and have seen them in their intercourse with Europeans, and I am, in candour, obliged to confess that the most unfavourable light in which I have seen them, has been when mixed up with Europeans.

That the natives of the interior have made frequent attacks on the stations of the settlers I have no doubt; very likely, in some instances, they have done so without any direct provocation, but we must not forget their position or the consequences of the extension of boundaries of location to the aborigines themselves. The more ground our flocks and herds occupy, the more circumscribed become the haunts of the savage. Not only is this the inevitable consequence, but he sees the intruder running down his game with dogs of unequalled strength and swiftness, and deplores the destruction of his means of subsistence. The cattle tread down the herbs which at one season of the year constituted his food. The gun, with its sharp report, drives the wild fowl from the creeks, and the unhappy aborigine is driven to despair. He has no country on which to fall back. The next tribe will not permit him to occupy their territory. In such a state what is he to do? Is it a matter of surprise that in the confidence of numbers he should seek to drive those who have intruded on him back again, and endeavour to recover possession of his lost domain? It might be that the parties concerned were not conscious of the injury they were inflicting, but even that fact would not lessen the fancied right of the native to repossess himself of his lost territory. Yet on the other hand we cannot condemn resistance on the part of the white man; for it would be unjust to overlook the fearful position in which they are placed, and the terrible appearance of a party of savages working themselves up to the perpetration of indiscriminate slaughter. No doubt many parties have gone to take up stations in the interior, with the honest intention of keeping on good terms with the natives, and who in accordance with such resolution have treated them with hospitality and consideration; but, it unfortunately happens that a prolonged intercourse with the Europeans weakens and at length destroys those feelings of awe and uncertainty with which they were at first regarded. The natives find that they are men like themselves, and that their intrusion is an injury, and they perhaps become the aggressors in provoking hostilities. In such a case resistance becomes a matter of personal defence, and however much such collisions may be regretted, the parties concerned can hardly be brought to account; but, it more frequently happens, that the men who are sent to form out-stations beyond the boundaries of location, are men of bold and unscrupulous dispositions, used to crime, accustomed to danger, and reckless as to whether they quarrel, or keep on terms with the natives who visit them. Thrown to such a distance in the wild, in some measure out of the pale of the law, without any of the opposite sex to restrain their passions, the encouragement these men give to their sable friends, is only for the gratification of their passions. The seizure of some of their women, and the refusal to give them up, provokes hostility and rouses resentment, but those who scruple not at the commission of one act of violence, most assuredly will not hesitate at another. Such cases are generally marked by some circumstances that betray its character, and naturally rouse the indignation of the Government. If the only consequence was the punishment of the guilty, we should rejoice in such retributive justice; but, unfortunately and too frequently, it happens, that the station belongs to a stockholder, who, both from feelings of interest and humanity, has treated the natives with every consideration, and discountenanced any ill-treatment of them on the part of his servants, but whose property is nevertheless sacrificed by their misconduct.

I have been unintentionally led into this subject, in the course of my remarks on the policy of Captain Grey, in establishing the post at Moorundi. The consequences have been equally beneficial to the settlers and aborigines. The eastern out-stations of the province have been unmolested, and parties with stock have passed down the Murray in perfect safety. If any act of violence or robbery has been committed by the natives, the perpetrators have been delivered up by the natives themselves, who have learnt that it is their interest to refrain from such acts; and instead of the Murray being the scene of conflict and slaughter, its whole line is now occupied by stock-stations, and tranquillity everywhere prevails.

About fifteen miles below Moorundi is Wellington, where a ferry has been established across the Murray, that township being on the direct road from Adelaide to Mount Gambier, and Rivoli Bay. A little below Wellington, Lake Victoria receives the waters of the Murray, which eventually mingle with those of the ocean, through the sea mouth.

The country immediately to the eastward of the Murray affords, in some places, a scanty supply of grass for sheep, but, generally speaking, it is similar in its soil and rock formation, and consequently in its productions to the scrubby country to the westward. The line of granite I have mentioned, in the former part of my work, as traversing or crossing the Murray below Wellington, continues through the scrub, large blocks being frequent amongst the brushes on a somewhat lower level than the tertiary fossil limestone in its neighbourhood. Round these blocks of granite the soil is considerably better, and there is a coating of grass upon it, as far as the ground consists of the decomposed rock.

About sixty miles to the E.S.E. of Wellington is the Tatiara country, once celebrated for the ferocity and cannibalism of its inhabitants, but now occupied by the settlers, who have of late crossed the Murray in considerable numbers to form stations there. The distance from Wellington to the district of Mount Gambier, said to be the fairest portion of South Australia, whether as regards its climate or its soil, is more than 200 miles. The first portion of the road, to almost the above distance, is through a perfect desert, in which, excepting during the rainy season, water is scarcely to be found, so that the journey is not performed without its privation. After passing Lake Albert the traveller has to journey at no great distance from the Coorong over a low country, once covered by the waters of the ocean, the noise of whose billows he hears through the silence of the night. The first elevation he reaches is a continuation of the great fossil bed through which the volcanic hills, where he will ultimately arrive, have been forced up. Mount Gambier, the principal of these, is about 40 miles from the Glenelg, and 50 from Rivoli Bay. The country from either of these points is low for many miles, but well grassed, of the richest soil, and in many places abundantly timbered. Mount Gambier is scarcely visible until you almost reach its base—nor even then is its outward appearance different from other hills. On reaching its summit, however, you find yourself on the brink of a crater, standing indeed on a precipice, with a small sheet of water of about half-a-mile in circumference, two hundred feet below you; the water of which is as blue as indigo, and seems to be very deep; no bottom indeed has been found at 50 fathoms. The ground round the base of Mount Gambier is very open, and you may ride your horse along it unchecked for many miles. At the lower parts, and at some distance from it, the ground is moist, and many caverns have been found in which water of the very purest kind exists, no doubt deposited in the natural reservoirs by percolation from the higher ground. The whole formation of the district, these capacious caverns, and the numerous and extensive tea-tree swamps along the coast, plainly demonstrate that they are supplied by gradual filtration, or find their way through the interstices, or cells of the lava to the lower levels.

It is generally admitted that the greater part of the land in the neighbourhood of Mount Gambier is equal to the richest soil, whether of Van Diemen's Land or of Port Phillip, the general character indeed of this district, and the fact of its being so much farther to the south than Adelaide, its perpetual verdure and moister climate would lead to the supposition that it is capable of producing grain of the very finest quality, and there can, I think, be but little doubt that it will rival the sister colonies in its agricultural productions, and considering the nature of the soil is similar to that round the volcanic peaks in the Mediterranean, it will also produce wine of a superior description.

Settlers both from the province of South Australia and neighbouring colonies have vied with each other in securing stations in this fertile, but remote district, and it would appear from the number of allotments that have been purchased in the townships which have been established on the coast that settlers are fast flocking to it.

From what has been stated it would seem that the district of Mount Gambier is adapted rather for agricultural than pastoral pursuits, and that it is consequently favourable for occupation by a rural population. Tea-tree swamps (melaleuca) are a feature, I believe, peculiar to South Australia, and generally indicate the presence of springs, and always of moisture. The soil is of the very richest quality, and there is, perhaps, no ground in the world that is more suitable for gardens, and as these swamps are both numerous and extensive in the lower country, behind Rivoli and Guichen Bays, this portion of the province promises equally fair for the growth of those European fruits which are less advantageously cultivated in the more northern parts of the province.

Returning to Adelaide, and proceeding from thence to the eastward, along the great eastern or Mount Barker line, we cross, in the first instance, the remaining portion of the plains lying between the city and the hills, to the base of which the distance is about three miles, the whole is laid out in farms, and is extensively and carefully cultivated. As you approach the hills, the country becomes lightly wooded and undulating, affording numerous sites for villas, on which many have already been erected, both by settlers and the more opulent tradesmen. Individuals indeed, residing in England, can form but a faint idea of the comforts and conveniences they enjoy, at such a distance from their native country. Being at sufficient elevation to catch the sea breeze, which passes over the plains of Adelaide, without being felt, they have almost the advantage of living near the sea coast, and the cool winds that sweep down the valleys behind them, and constitute the land breeze, ensure to them cool and refreshing evenings, when those dwelling at a lower elevation are oppressed by heat. On the first rise of the mountains is the Glen Osmond Lead Mine, which will be noticed hereafter. The Mount Barker district being more numerously settled than most other parts of the province, and being one of its most important and fertile districts, more labour has been expended on the road leading into it, than on any other in the colony. From the level of the Glen Osmond Mine, it winds up a romantic valley, with steep hills of rounded form, generally covered with grass, and studded lightly with trees on either side, nor is it, until you attain the summit of the Mount Lofty range, that any change takes place in the character of the hills or the vegetation, you then find yourself travelling through a dense forest of stringy barks, the finest of which have been levelled to the ground, with the axe, for the purpose of being sawn into planks for building, or split into rails for fencing. From Crafer's Inn, situated under the peak of Mount Lofty, the road to Mount Barker passes through a barren country for some miles, and crosses several steep valleys, in the centre of which there are rippling streams; the summit of the ranges still continues to be thickly wooded, the ground underneath being covered with shrubs and flowers of numberless kinds and varied beauty. In illustration of this, I may observe, that the first time I crossed the Mount Lofty range, I amused myself pulling the different kinds of flowers as I rode along, and on counting them when I reached

Adelaide for the purpose of arranging them in a book, found that I had no less than ninety-three varieties. The majority of these, however, consisted of papilionaceous plants, and several beautiful varieties of Orchideæ. On descending to a lower level, after crossing the Onkaparinga, the scenery and the country at once change, you find yourself upon rich alluvial flats, flanked by barren rocky hills, the air during the spring being perfumed by the scent of the Tetratheca, a beautiful hill flower, at that time in splendid blossom, and growing in profusion on the tops of the hills, mingled with the Chyranthera, with its light blue blossoms; both these plants it has always appeared, are well adapted for the edges of borders, but there are not many plants in Australia that would be fit for such a purpose.

It does not appear necessary, in a work like this, to trouble the reader with an account of every village or of every valley in the districts through which I lead him; my object is to give a general and faithful description of the country only, reserving the power of drawing attention to any thing I may deem worthy of notice. Taking the district of Mount Barker therefore in its full range, I would observe, that it is one of the finest agricultural districts in the province. It abounds in very many beautiful alluvial valleys, which, when I first crossed, had grass that rose above the horses middles as they walked through it, and looked luxuriant beyond description. These valleys are limited both in length and breadth, but are level and clear; their soil is a rich alluvial deposit, and the plough can be driven from one end to the other without meeting a single obstacle to check its progress. Independently of these valleys, there are other portions of good grazing land in the Mount Barker district, but there are, nevertheless, very many stony ranges that are entirely useless even to stock. The Mount Barker district may be said to extend from the village of Nairne to Strathalbyn, on the River Angas, the latter place being 15 miles from the shores of Lake Victoria. Within the range of this district, there are also the villages of Hahansdorf and Macclesfield, the former being a German village, at no great distance from Mount Barker. Immediately to the north of the village of Nairne is Mount Torrens, the river of that name has several branches to the north-east of it as high up as Mount Gould. The first of the Company's special surveys, and perhaps some of the finest soil in the province is in this locality. The surveys on the sources and tributaries of the Torrens are splendid properties, and the Company may well consider them as amongst the most valuable of its acquisitions; beyond the heads of the Torrens the country is more hilly and less available. There are, nevertheless, isolated spots sufficiently large for the most comfortable homesteads. From this point, a west-south-west course will soon lead the traveller into the plains of Adelaide, and at less than 10 miles after entering upon them, he will again find himself in the metropolis. Again departing from it for the southern parts of the province, he will keep the Mount Lofty range upon his left, and will really find some difficulty in passing the numberless fences which now enclose the plains. The land indeed in this line of road is more fenced than in any other direction, a reason for this may be that the road runs nearer the base of the hills, and the land is consequently better than that on the lower ground. Many very excellent farms are to be found on the banks of the Sturt and the Onkaparinga, on the latter of which the village of Noorlunga has been established, at the point where the road crosses it. The Sturt has a tortuous course, somewhat to the northward of west, and falls into the gulf at Glenelg, after spreading over the flats behind the sand-hills at that place. The direction of the road is parallel to that of the ranges, or nearly south-south-west as far as the village of Noorlunga, when it turns more to the eastward of south, for Willunga, which is 28 miles distant from Adelaide. The banks of the Onkaparinga, above the crossing place, are extremely inaccessible, insomuch that stock can hardly be driven down to water for many miles above that point. The hills however are rounded in form, grassy, and clear of trees, consequently well adapted for grazing purposes. It was at Noorlunga, which is not more than two miles from the gulf, and can be approached in boats, as high as the bridge there, that Captain Barker first landed on the South Australian shore. The country between it and Willunga is generally good, portions of it are sandy and scrubby, but Morphett's Vale is a rich and extensive piece of land, and I can well remember before it was settled seeing several large stacks of hay that had been cut, as it then lay in a state of nature. Willunga is close under the foot of the hills, which here, trending to the south-south-west, meet the coast line extremity of the Southern Aldinga plains. Close to this point is a hill, called Mount Terrible, almost of a conical shape, over the very summit of this, in the early stages of the colony, the road led to Encounter Bay; and I shall not forget the surprise I experienced, when going to that place, on finding I could not by any possibility avoid this formidable obstacle. On the other side of Mount Terrible the country is very scrubby for some miles, until, all at once, you burst upon the narrow, but beautiful valley of Mypunga. This beautiful valley, which had scarcely been trodden by the European when I first encamped upon it, was then covered with orchideous plants of every colour, amidst a profusion of the richest vegetation. A sweet rippling stream passed within five yards of my tent-door, and found its way to the Gulf about a mile below me to the west. It was on the occasion of my going to the sea mouth of the Murray, that I first stopped at this spot. Amongst the boat's crew I had brought with me from Adelaide a young lad, of not more than twenty-one, who had, for some weeks before, been leading a very hard life. At Mypunga he was seized with delirium tremens, and became so exceedingly outrageous, that I was obliged to have his feet and hands tied. In the morning he was still as frantic as ever, but the policeman, under whose charge I had placed him, having imprudently loosened the cord from his ankles, he suddenly started upon his feet, and gaining the scrub, through which we had descended into the valley, with incredible swiftness, secreted himself amongst it. Nor could we, by the utmost efforts during that and the succeeding day, discover his hiding place. I was accompanied by a man of the name of Foley, a bushranger of great notoriety, who had been captured by the Adelaide police, and was sent with my party in the hope that his knowledge of the coast would be of use to me, but neither could he discover the unfortunate runaway, who, there is no doubt, subsequently perished. Beyond Mypunga, to the south, are the valleys of Yankalilla and Rapid Bay, but very little, if in any respect inferior to the first mentioned place. The country between them is, however, extremely hilly, and contains some beautifully romantic spots of ground. The rock formation of this part of the ranges is very diversified; the upper part of Rapid valley is a fine grey limestone; a little to the southward veins both of copper and lead have been discovered, and I have good reason for supposing that quicksilver will one day or other be found in this part of the province. At Willunga there is a small stream, which issues from a valley close behind the township, and appears in former times to have laid many hundred acres of the flats below under water. Their soil is composed of the very richest alluvial deposit, and has produced some of the finest crops of wheat in the province. Aldinga plains lie to the south-west of Willunga, and are sufficiently extensive to feed numerous sheep, but unavailable in consequence of the deficiency of water upon them, and are an instance of a large tract of land lying in an unprofitable state, which might, with little trouble and expense, by sinking wells in different parts, be rendered extremely valuable. On ascending the hills above Willunga, in following up the southern line of road to Encounter Bay, it leads for several miles through a stringy-bark forest, and brings the traveller upon the great sandy basin, between Willunga and Currency Creek. This gloomy and sterile feature bears a strong contrast to the rich and fertile valleys I have described, and is really a most remarkable formation in the geology of the province. At an elevation of between 600 and 700 feet this basin is surrounded on all sides by rugged stony hills, excepting to the south and south-east, in which direction it falls into the valley of the Hindmarsh and Currency Creek respectively. Mount Magnificent, Mount Compass, and Mount Jagged, rise in isolated groups in different parts of the basin, the soil of which is pure sand, its surface is undulating, and in many parts covered with stunted banksias, through which it is difficult to force one's way in riding along. The Finniss rises behind Mount Magnificent, and is joined by a smaller branch from Mount Compass, as it flows from the eastward. At about 25 miles from Willunga the traveller descends into the valley of Currency Creek, and finds the change from the barren tract over which he has been riding as sudden as when he entered upon it from the rich flats of Willunga, The valley of Currency Creek is not, however, the same as those I have already described in other parts of the colony; it is prettily wooded and grassy, but continues narrow for some distance after you have entered it; a small running stream, with a rocky bed, occupying the centre of the valley, which ultimately escapes from the hills by a kind of gorge, and discharges itself into an arm of the Goolwa. The extent of good land in Currency Creek is not very great, and is bounded both to the north and south by barren scrub. Due south, at the distance from 15 to 18 miles, is Encounter Bay, the country intervening between the two points to the shores of the Goolwa is very level, the soil is light but rich, and there appeared to me to be many thousand acres that were adapted for agricultural purposes, better adapted indeed than the richer soils. Whether that view be correct or not, the valleys of the Inman and Hindmarsh immediately behind Encounter Bay would fully make up for the want of agricultural land in this part of the province. Hindmarsh valley is not of any great extent, but the soil is good, and its scenery in my humble opinion surpasses any other I remember in South Australia. I shall never, indeed, forget the beautiful effect of sunset, on a fine bold mountain at the head of it, called the Black Hill. The glowing orb was fast descending behind it to the west, and the Black Hill was cast into deep shade, whilst the sun's rays shooting down two valleys on either side gave the grass the appearance of young wheat. The extent of arable land in the valley of the Inman is very considerable, but in point of scenery bears no comparison with the first. I do not know whether I have made it sufficiently clear that there is a high range at the back of the coast hereabouts. If not, I would observe that it runs uninterruptedly from Mount Lofty to Cape Jarvis. Opposite to Encounter Bay it occupies nearly the centre of the promontory, and consequently forms a division of the eastern and western waters, there being a considerable breadth of barren stringy-bark forest between the heads of the opposite valleys, here as on the higher parts of the ranges near Mount Lofty, from the ascent of the great eastern road to the valley of the Onkerparinga.

It is a remarkable fact, but one that I believe I have already adverted to, that the farther north, towards the valley of the Wakefield, the more denuded of timber the country becomes, until at last not a tree of any kind can be seen. These extensive and open downs are, nevertheless, well grassed, and covered with a profusion of orchideous plants. Whether, however, there is any salt present in the soil, to check the growth of the trees, it is impossible to say. Undoubtedly many of the ponds in the Wakefield, as well as other parts of the province are brackish, but the same denuded state of the country exists not any where else. These districts are far too valuable to be overlooked, and are therefore extensively occupied by cattle and sheep. My most worthy friend, Mr. Charles Campbell, and my companion Mr. John Browne, and his brother, both occupy the most distant stations to the north.

Mr. Campbell has one of the finest cattle runs in the province, and my comrade, I believe, is perfectly satisfied with his run. The condition of their cattle and sheep would at all events lead to the conclusion, that neither suffer from the nature of the water they drink or the pasture on which they feed.

As regards the general appearance of the wooded portion of the province, I would remark, that excepting on the tops of the ranges where the stringy-bark grows; in the pine forests, and where there are belts of scrub on barren or sandy ground, its character is that of open forest without the slightest undergrowth save grass. The trees are more or less numerous according to the locality, as well as more or less umbrageous, a character they generally have on river flats, but the habit of the eucalyptus is, generally speaking, straggling in its branches. In many places the trees are so sparingly, and I had almost said judiciously distributed as to resemble the park lands attached to a gentleman's residence in England, and it only wants the edifice to complete the comparison.

The proportion of good to bad land in the province has generally been considered as divisible into three parts; that is to say, land entirely unavailable—land adapted for pastoral purposes only, and land of a superior quality. On due consideration, I am afraid this is not a correct estimate, but that unavailable country greatly preponderates over the other two. If, in truth, keeping the distant interior entirely out of view, and confining our observations to those portions of the colony into which the settlers have pushed in search for runs, we look to the great extent of unavailable country between the Murray and the Mount Gambier district, along the line of the Murray belt, and the extensive tracts at the head of the Gulfs, we shall find that South Australia, from the very nature of its formation, has an undue proportion of waste land. Those parts, however, which I have mentioned as being unavailable, were once covered by the sea, and could hardly be expected to be other than we now see them, and it may, therefore, be questioned how far they ought to he put into the scale. In this view of the matter, and taking the hilly country only into account, the proportion of unavailable and of pastoral land maybe nearly equal; but that of the better description will still, I think, fall short of the other two. Taking South Australia in its length and breadth, the quantity of available land is, beyond doubt, very limited, but I regard it as exceedingly good, and believe that its capabilities have by no means, been ascertained. I feel satisfied, indeed, that necessity will prove, not only, that the present pastoral districts are capable of maintaining a much greater number of stock upon them than they have hitherto borne, but that the province is also capable of bearing a very great amount of population; that it is peculiarly fitted for a rural peasantry, and that its agricultural products will be sufficient to support masses of the population employed either in its mining or manufactures. In this view of the subject it would appear that

Providence has adapted the land to meet its new destinies, and that nothing we can say, either in praise or censure of its natural capabilities, will have the effect of concealing either the one or the other, as time shall glide on.

On the better soils the average crop of wheat is rather over than under twenty-five bushels to the acre. In many localities, and more especially when the ground is first cropped, it exceeds forty; and on some lands, once my own, in the Reed Beds, at the termination of the Torrens's river, five acres, which I sold to Mr. Sparshott, averaged fifty-two bushels to the acre. The Reed Beds may be said to be on the plains of Adelaide, and their very nature will account to the reader for the richness of their soil; but the soil of the plains is not generally good, excepting in such places where torrents descending from the hills have spread over portions, and covered them with an alluvial deposit to a greater or less depth. The average crop of wheat on the plains does not exceed twelve or fifteen bushels to the acre, and depends on the time when the hot winds may set in. Barley on the light sandy soil of the plains is much heavier than wheat.

In the description I have thus endeavoured to give of South Australia, I have omitted any mention of the district of Port Lincoln, chiefly because sufficient was not known of it when I sailed for England to justify my hazarding any remark. Recent advices from the colony state that a practicable line of route from Adelaide has been discovered along the western shore of Spencer's Gulf, and therefore, the disasters that overtook early explorers in that quarter, are not likely again to occur. It is farther said, that the number of sheep now depastured on the lands behind Port Lincoln, amounts to 70,000—a proof of the utility, if not the richness of the country—as far, however, as I am aware, the soil must be considered of an inferior description—in other respects, the Port has advantages that will always render it an agreeable, if not altogether a desirable residence. It appears to be gradually improving, but the amount of its population is still low, not more than sixty. It is frequented by American and other whalers, but the duties collected add little to the revenues of the province. Port Lincoln, however, could hardly now be abandoned, since there are considerable interests at stake there. It has been stated that copper has been found in the interior, and I see no reason why it should not exist in the mountain formation of the Gawler Range, in such case an impulse will be given to the whole district, that would even change its prospects, and increase the mercantile operations of the province.

It does not appear to be the disposition of the English settlers to try experiments on the growth of intertropical productions. It must be admitted, however, that there are not many places in South Australia where they could be cultivated with advantage; for although both the plains of Adelaide and the valley of the Murray are warm in summer, the frosts, which are sufficient to blight potatoes, would necessarily injure, if they did not destroy, perennials, whilst in the hills the cold is adverse to any plants the growth of a tropical climate, if we except those which, as annuals, come to maturity in the course of a summer; but the true reason why the growth of extraneous productions is neglected in South Australia, is the expense consequent on the state of the labour market—for no doubt many pursuits might be followed there that would be remunerative. It is exceedingly difficult, however, to lead the pursuits of a community out of their ordinary course, and it is only where direct advantages are to be gained, that the spirit of enterprise and speculation breaks forth.

The climate of South Australia is admirably adapted for the growth of fruit trees of the hardier tropical kinds, for although the tenderer kinds grow there also, they do not arrive at perfection. The loquat, the guava, the orange, and the banana, are of slow growth, but the vine, the fig, the pomegranate, and others, flourish beyond description, as do English fruit trees of every kind. It is to be observed, that the climate of the plains of Adelaide and that of the hills are distinct. I have been in considerable heat in the former at noon, and on the hills have been in frost in the evening. The forest trees of Europe will grow in the ranges, but on the plains they languish; in the ranges also the gooseberry and the currant bear well, but in the gardens on the plains they are admitted only to say you have such fruits; the pomegranate will not mature in the open air, but melons of all kinds are weeds. Yet, such trees as are congenial to the climate arrive at maturity with incredible rapidity, and bear in the greatest abundance. The show of grapes in Mr. Stephenson's garden in North Adelaide, and the show of apples and plums in Mr. Anstey's garden on the hills are fine beyond description, and could not be surpassed in any part of the world—it may readily be imagined, therefore, that the intermediate fruit trees, such as the peach, the nectarine, the pear, the cherry, the greengage, and others, are of the most vigorous habits. All of them, indeed, are standards, and the wood they make during one season, is the best proof that can be given of their congeniality to the soil and climate of the province.

There are in South Australia two periods of the year which are equally deceptive to the stranger. The one is when the country is burnt up and suffering under the effects of summer heat—when the earth is almost herbless, and the ground swarms with grasshoppers—when a dry heat prevails in a calm still air. The other when vegetation is springing up under the early rains and every thing is green. Arriving at Adelaide during the first period, the stranger would hardly believe that the country, at any other season of the year, would be so clothed with herbage and look so fresh; arriving at the other, he would equally doubt the possibility of the vegetable kingdom being laid so completely prostrate, or that the country could assume so withered and parched an appearance; but these changes are common to every country under a similar latitude, and it would be unjust to set them down to its prejudice, or advantage.

The following mean of heat at 2 P.M. throughout the year, will give the reader a correct idea of the range of the thermometer. I have taken 2 P.M. as being the hottest period of the day, and, therefore, nearest the truth.

January	85	106½	70
February	9	94	71
March	77	103½	68½
April	67½	85	55½
May	62	76	53
June	58	67	49
July	55	60	49
August	59	68	52
September	61	72½	55½
October	68½	94½	55
November	74	94	59
December	83	100	68

The west and south-west winds are the most prevalent, blowing for 130 or 140 days in the year. During the summer months the land and sea breezes prevail along the coast, but in the interior the wind generally commences at E.N.E., and going round with the sun settles at west in the afternoon.

I need not point out to the reader, that the above table only shews the mean of the thermometer during a certain hour of the day; the temperature during the night must necessarily be much lower; the coolness of the night, indeed, generally speaking, makes up for the mid-day heat. There are some days of the year when hot winds prevails, which are certainly very disagreeable, if not trying. Their occurrence, however, is not frequent, and will be easily accounted for from natural causes. They sometimes continue for three or more days, during which time clouds of dust fill the air, and whirlwinds cross the plains, but the dryness of the Australian atmosphere considerably influences the feelings on such occasions, and certainly produces a different effect upon the system from that which would be produced at a much lower temperature in a more humid climate; for, no doubt, it is to the united effects of heat and moisture, where they more or less exist, that the healthiness or unhealthiness of a country may be ascribed. In such countries, generally speaking, either teaming vapours, or malaria from dense woods or swamps naturally tries the constitution, but to its extreme dryness, and the absence of all vegetable decay, it appears to me that the general salubrity of South-east Australia is to be attributed. So rarified, indeed, is the atmosphere, that it causes an elasticity of spirits unknown in a heavier temperature. So the hot winds, of which I have been speaking, are not felt in the degree we should be led to suppose. Like the air the spirits are buoyant and light, and it is for its disagreeableness at the time, not any after effects that a hot wind is to be dreaded. It is hot, and that is all you can say; you have a reluctance to move, and may not rest so well as usual; but the spirits are in no way affected; nor indeed, in the ordinary transactions of business does a hot wind make the slightest difference. If there are three or four months of warm weather, there are eight or nine months of the year, during which the weather is splendid. Nothing can exceed the autumn, winter, and spring of that transparent region, where the firmament is as bright as it would appear from the summit of Mount Blanc. In the middle of winter you enjoy a fire, the evenings are cold, and occasionally the nights are frosty. It is then necessary to put on warmer clothing, and a good surtout, buttoned across the breast, is neither an uncomfortable nor unimportant addition. Having said thus much of the general salubrity of the climate of South Australia, I would observe, in reference to what may be said against it, that the changes of temperature are sudden and unexpected, the thermometer rising or falling 50° in an hour or two. Whether it is owing to the properties I have ascribed, that the climate of this place as also of Sydney should be fatal to consumptive habits, I do not know, but in both places I have understood that such is the case, and in both I have had reason to regret instances. It has been said that influenza prevailed last year in Adelaide to a great extent, and that it carried off a great many children and elderly persons. An epidemic, similar in its symptoms, may have prevailed there, and been severe in its progress, but it hardly seems probable that the epidemic of this country should have been conveyed through constant change of air, the best cure for such a disease, to so distant a part of the world. With all its salubrity, indeed, I believe it may be said, that South Australia is subject to the more unimportant maladies like other countries, but that there are no indigenous disorders of a dangerous kind, and that it is a country which may strictly be called one of the healthiest in the world, and will, in all probability, continue so, as long as it shall be kept clear of European diseases.

Having thus endeavoured to give a description of the general character and climate of this limited but certainly beautiful portion of the Australian continent, without encumbering my description with any remark on the principal and particular sources of wealth it possesses, which not being usual, could not, or rather would not, have been considered applicable. I hope the object I have had in view will be sufficiently clear to the reader. I have endeavoured to point out with an impartial pen, the real capabilities of the province, and the nature of those productions which are most congenial to her soil. Without undue praise on the one hand, or unjust depreciation on the other, it has been my desire to present a faithful picture of her to my readers, and I hope it will appear from what I have said, as is really and truly the case, that both in climate and other respects it is a country peculiarly adapted to the pursuits and habits of my countrymen. That its climate so far approaches that of England, as to be subject to light and partial frosts, which render it unfit for the cultivation of tropical productions, but make it essentially an agricultural country, capable of yielding as fine cereal grain as any country in the world, of whatever kind it may be—that at the same time the greater mildness of the climate makes it favourable to the growth of a variety of fruits and vegetables, independently of European fruit trees and culinary herbs, which put it in the power of the settler to secure the enjoyment of greater luxuries and comforts, than he could possibly expect to have done in his own country, except at a great expense, and that as far as the two great desiderata go, on which I have been dwelling, it is a country to which an Englishman may migrate with the most cheerful anticipations.

CHAPTER III.

SEASONS—CAUSE WHY SOUTH AUSTRALIA HAS FINE GRAIN—EXTENT OF CULTIVATION—AMOUNT OF STOCK—THE BURRA-BURRA MINE—ITS MAGNITUDE—ABUNDANCE OF MINERALS—ABSENCE OF COAL—SMELTING ORE—IMMENSE PROFITS OF THE BURRA-BURRA—EFFECT OF THE MINES ON THE LABOUR MARKET—RELUCTANCE OF THE LOWER ORDERS TO EMIGRATE—DIFFERENCE BETWEEN CANADA AND AUSTRALIA—THE AUSTRALIAN COLONIES—STATE OF SOCIETY—THE MIDDLE CLASSES—THE SQUATTERS—THE GERMANS—THE NATIVES—AUTHOR'S INTERVIEWS WITH THEM—INSTANCES OF JUST FEELING—THEIR BAD QUALITIES—PERSONAL APPEARANCE—YOUNG SETTLERS ON THE MURRAY—CONCLUSION.

IT WAS my object in the last chapter, to confine my observations strictly to the agricultural and pastoral capabilities of the province of South Australia, which I thought I could not better do than by describing the nature of its climate and soil, for on these depend the producing powers of every country. In speaking of the climate, however, I merely adverted to its temperature, leaving its seasons out of question for the time, intending to close my remarks on these heads, by a short review of the state of the agricultural and pastoral interests of the colony at the present date.

It will be borne in mind that the seasons of Australia are the reverse of our own; that when in England the ground is covered with snow, there the sun is hottest, and that when summer heats are ripening our fruits, in Australia it is the coldest season of the year, December, January, February, and March being the summer months; June, July, August, and September the winter ones. An experience of ten years has shewn that the seasons of South Australia are exceedingly regular, that the rains set in within a few days of the same period each successive year, and that during the winter the ground gets abundantly saturated. This regularity of season may be attributed to the almost insular position of the promontory of Cape Jarvis, and may be said to be almost local, in elucidation of which, I may refer to what I have stated in the former part of my work, of the state of the weather in the valley of the Murray when the expedition was proceeding up its banks in the month of August, 1844. For some time before there had been heavy rains in the hills, and it was with some difficulty the drays crossed them. During our stay at Moorundi, the ranges were covered with heavy clouds, and the mountain streams were so swollen as to stop one of my messengers; but the sky over the valley of the Murray was as clear as crystal, morning mists it is true curled up at early dawn from the bosom of its waters, but they were soon dissipated, and a sharp frosty night was succeeded by a day of surpassing beauty.

The regularity, however, both in its commencement and in the quantity of moisture that falls during the rainy season in the colony, enables the agriculturist to calculate with certainty upon it, and the only anxiety of the farmer is to get his grain into the ground sufficiently early, if possible, to escape the first hot winds. In a region, portions of which are subject, it must be confessed, to long continued drought, this is no inconsiderable advantage, although South Australia is not singular in this respect, for the rainy seasons in the Port Phillip districts are, I believe, equally regular and more abundant, whilst the climate of Van Diemen's Land almost approaches to that of England; neither, indeed, fairly speaking, is South Australia more favoured than those of her immediate neighbours in the quality of her soil. Van Diemen's Land is the granary of the southern seas, and there is unquestionably a very great proportion of the very best soil in the Port Phillip district. Nevertheless that of South Australia has yielded a finer and a heavier grain than has ever been produced in those colonies, but the reason of this is, that with a naturally rich soil to work upon, the agriculturists of South Australia have spared no pains in cultivating their lands, but there can be no doubt that with equal care and attention both the Vandemonians and the settlers of Port Phillip would produce an equally fine sample. The farmers of South Australia have enhanced the value of their colony by their energy and skill in cultivating it, and can boast of having sent the finest sample of wheat to England that has ever been exhibited in her market.

South Australia, in its length and breadth, contains about 300,000 square miles, or in round numbers more than 190,000,000 acres. The limits of location, however, do not exceed 4,000 miles, or 7,000,000 acres. In this area, however, a great portion of desert country is included, or such, at least, as at the present moment is considered so. Of the more available land, 470,000 acres have been purchased, but the extent of country occupied by sheep and cattle stations is not known.

It may be necessary here to observe, that the returns of the land under cultivation last year were published after I left the colony; but the comparison between the two previous years will shew the increase and decrease of the different grains, sufficiently to establish the progress of agricultural pursuits in the colony. In the year 1845, the number of acres of wheat sown was 18,848. In 1846 it was 26,135. Of barley, there were in the former year 4,342 acres, in the latter only 3,490. Of oats, there were 1,485 in the first year, which, in 1846, increased to 1,963. It would thus appear, that the increase of cultivated land in the course of one year amounted to between 6,000 and 7,000 acres, and that more than 400 agriculturists were added to the list of landed proprietors. The necessary consequence of such extensive farming operations is that the produce far exceeds the wants of the settlers, and that there is a considerable surplus for exportation; the price of the best flour being from £12 to £13 per ton, whereas for a short period in 1839 it was £120!!!

Whilst the agriculturists have been so earnest in the development of the productive powers of the colony, another class of its inhabitants were paying equal attention to its pastoral interests. The establishment of stock stations over its surface followed its occupation, and a mild climate and nutritive herbage equally contributed to the increase of cattle and sheep that had been introduced. In 1844 the number of sheep assessed was 355,700, in the following year that number had increased to 480,669, or an addition of 120,000. At the present moment there cannot be far short of a million of sheep in the province, with an increase of 200,000 annually, at a moderate computation. The number of other kinds of stock in the possession of the settlers, at the close of last year, was as follows:—of cattle, 70,000; 30,000 having been imported during the two previous years from New South Wales. The number of horses was estimated at 5,000, and of other smaller stock, as pigs and goats, there were supposed to be more than 20,000.

It is impossible to contemplate such a prosperous state of things in a colony that has only just completed the eleventh year of its existence, without feeling satisfied that some unusually favourable circumstances had brought it about. Had South Australia been as distant from the older colonies on the continent as Swan River, the amount of stock she would have possessed in an equal length of time, could not have amounted to a tenth of what they now number. It is to the discovery of the Darling and the Murray that South Australia owes the superabundance of her flocks and herds, and in that superabundance the full and complete establishment of her pastoral interests. I stated in the course of my preliminary observations on the progress of Australian discovery, that when I was toiling down those rivers, with wide spread deserts on either side of me, I had little idea for what purposes my footsteps had been directed into the interior of the Australian Continent. If I ever entertained even a distant hope that the hilly country from which I turned back at the termination of the Murray, after having floated on its broad waters for eighty-eight days, might ever be occupied, I certainly never hoped that the discoveries I was then making would one day or other prove of advantage to many a friend, and that I was marking the way for thousands of herds and flocks, the surplus stock of New South Wales, to pass into the province of South Australia.

If then such consequences have resulted from enterprises, apparently of almost as hopeless a character as the one from which I have so recently returned, why, I would ask, should I despair, as to its one day or other being instrumental in benefiting my countrymen. There may yet be that in the womb of time which shall repay me for all I suffered in the performance of that dreary task—when I shall have it in my power to say, that I so far led the way across the continent as to make the remainder of easy attainment, and under the guidance and blessing of Providence have been mainly instrumental in establishing a line of communication between its northern and southern coasts. I see no reason why I should despair that such may one day be the case. The road to the point which may be termed my farthest north is clear before the explorer. That point gained, less probably than 200 miles—a week's journey with horses less jaded than mine unfortunately were, and with strength less reduced—would place him beyond the limits of that fearful desert, and crown his labours with success. I believe that I could, on my old route, make the north coast of Australia, to the westward of the Gulf of Carpentaria, before any party from Moreton Bay. If it is asked what practical good I should expect to result from such an undertaking, I would observe, that nothing would sooner tend to establish an intercourse with the inhabitants of the Malay archipelago, than the barter of cattle and sheep, that in truth there is no knowing what the ultimate results would be. The Malays who visit the northern coasts of Australia to collect the sea slug, have little inducement to keep up an intercourse with our settlements in Torres Straits, but there can be no doubt of their readiness to enter into commercial intercourse with us, which, if Torres Straits are to be navigated by steamers, would be doubly important.

When the stock from New South Wales was first brought down the Murray, the journey occupied from three to four months. Latterly it did not take half that time. In less than fifty days, from the Murray, on his way to the north, the stock-holder would find that he had passed the centre, and an equal number of days from that point would, it appears to me, take him to his journey's end. This, however, would depend on the nature of the country beyond where it is at present known, and the nature of the season during which it was undertaken, but experience alone, as in the instance of the journey down the Murray, would be the best guide and the best instructor.

In the early part of the year 1840, I had occasion to address a number of the colonists at the conclusion of a public entertainment and availed myself of the opportunity to state that whatever prospects of success the pastoral capabilities of the province appeared to hold out, I felt assured it was to the mountains, the colonists would have to look for their future wealth, for that no one who pretended to the eye of a geologist could cross them as I had done, without the conviction that they abounded in mineral veins. There is something, in truth, in the outline and form of the Mount Lofty chain that betrays its character. Rounded spurs, of very peculiar form, having deep valleys on either side, come down from the main range, the general outline of which bears a strong resemblance to that of the Ural chain.

In the year 1843, the first discovery of copper was made, but even this was scarcely sufficient to rouse the colonists to a full sense of its importance, and it was only by degrees, as other mines were successively discovered, that the spirit of speculation burst forth, and the energies of the settlers were turned for a time from their legitimate channels. A short time before this, their circumstances had been reduced to the lowest ebb. There was no sale for agricultural produce, no demand for labour, the goods in the shops of the tradesmen remained unsold, and the most painful sacrifices of property were daily made at the auction mart. The amount of distress indeed was very great and severe, but such a state of things was naturally to be expected from the change that had taken place in the monetary affairs of the province. It was a change however which few anticipated, and for which few therefore were prepared.

It is a painful task to advert to past scenes of difficulty and distress, such at least I feel it to be, more especially where there is no immediate object to be gained by a reference to them; let me therefore turn from any inquiry into the causes which plunged South Australia into difficulties that threatened to overwhelm her, to those which raised her from them.

Notwithstanding the spirit and firmness with which the colonists bore their reverses, there could not but be a gloom over the community where everything seemed to be on the brink of ruin. Men's minds became depressed when they saw no relief in the present, and no hope in the future. But Time, with a rapid wing, brought about changes that appear permanently to have altered the circumstances of the colony, and to have placed it at once as one of the most flourishing of the British possessions. The first circumstance, I have understood, which partially cheered the drooping spirits of the settlers, was a slight rise in the price of wool, in the year I have mentioned. The discovery of the mines following soon upon this, the sun of prosperity burst at once upon the province, and gladdened every heart. From this period, mine after mine of copper and lead continued to be discovered. Every valley and hill-top was searched for hidden treasures, and the whole energies of the colonists seemed to be turned to this new source of wealth. I was absent in the interior when the Burra Burra mine was secured, but the excitement it created had not subsided when I reached Adelaide.

I do not know whether the presence of mineral veins is indicated in other countries as in South Australia by means of surface deposits. The opinion I formed that ores would be discovered in the Mount Lofty ranges did not rest upon the discovery of any such deposit myself, but on the peculiar form of the hills, which appeared to me to have settled into their present state from one of extreme fusion. The direction of the ranges being from north to south, these deposits lie also in the same direction. Those of iron are greater than those of copper, and it is impossible to describe the appearance of the huge clean masses of which they are composed. They look indeed like immense blocks, that had only just passed from the forge.

The deposits at the Burra Burra amounted, I believe, to some thousand tons, and led to the impression that where so great a quantity of surface ore existed, but little would be found beneath. In working this gigantic mine, however, it has proved otherwise. I was informed by one of the shareholders just before I left the colony, that it took three hours and three-quarters to go through the shafts and galleries of the mine. Some of the latter are cut through solid blocks of ore, which glitter like gold where the hammer or chisel has struck the rock, as you pass with a candle along them.

It would be out of place in me, nor indeed would it interest my readers, were I to enter into a statistical account of the profits of the Burra Burra mine. A general notice will convey every necessary information on that head, and enable the public to judge as well of its value and importance as if I entered into minuter details. It will give the reader some idea of the scene of bustle and activity the Burra mine and road must present, and the very great amount of labour it requires.

The quantity of ore sent weekly from the mine to the port is from 430 to 450 tons, employing from 150 to 160 drays, and more than double that number of men. The total quantity of ore received at the port in December last was 10,000 tons, the average value of which at £20 per ton, amounts to £200,000, and the price of shares, originally of £5, had, by last advices, reached £160.

Considering the gigantic scale of the Burra Burra mine, it was supposed that few other mines would be found in the colony that would at all approach it, that indeed, it had been the principal deposit, and that whatever indications other mines might give, they would soon cease in working, or produce so little as to be valueless. I confess that such was my own opinion—surprised at the immense size of this magnificent mine, I hardly thought it possible that in mountains, after all of limited range, mines of great value would still be found, and that discoveries of new mines were frequently taking place, and that too in situations where no such feature would be supposed to exist. On York's Peninsula for instance, immediately across St. Vincent's Gulf, opposite to Port Adelaide, and directly on the sea shore, there are two sections, on which copper ore is abundant. The position of this mine can at once be determined by the reader, on a reference to the map. The land is very low, and the rock formation, tertiary fossil, but the various and anomalous positions in which copper is found in South Australia, baffles all ordinary calculations—as likely to exist in the valley, as on the hill—at the sea side as well as inland: there is not a locality in which it may not be looked for and found.

The whole of the mountain chain indeed, is a mass of ore from one end to the other, and it is impossible to say what quantity, or how many of the richer metals will ultimately be found in a country through which the baser metals are, without doubt, so abundantly diffused. The quantity of gold hitherto discovered has not been important, but it is reasonable to suppose, that where a small quantity has been found, large deposits must be at no great distance. This gold however, like the baser metals of South Australia, is very pure, there being few component parts mixed with it.

From the various examinations of the hills that have at different times been made, it would appear that precious stones, as well as metals, exist amongst them. Almost every stone, the diamond excepted, has already been discovered. The ruby, the amethyst, and the emerald, with beryl and others, so that the riches of this peculiar portion of the Australian continent may truly be said to be in their development only.

With such prospects before it, there can be but little doubt that the wealth of South Australia will, one day or other, be very great, neither can there be any doubt but that the discovery of the mines at the critical period, made a complete revolution in the affairs of that colony, and suddenly raised it from a state of extreme depression to one of independence, even as an individual is raised to affluence, from comparative poverty by the receipt of an unlooked-for legacy. The effect, however, which the discovery had on its present prospects, and the effect it must have on the future destinies of that colony, can hardly, it appears to me, be placed to the credit of any ordinary process of colonization. It has rather been in the shape of an unexpected auxiliary, that this immense and valuable supply of ore has been brought to bear upon its fortunes, for the condition to which the colony was reduced at one time, was such, that it would have taken many years to have acquired the appearance of returning prosperity, but the discovery of the mines was like the coming up of a rear-guard, to turn the tide of battle, when the main army had apparently been all but defeated. The assistance the colony received was complete and decisive, and has seemingly placed her beyond the hazard of failure or reverse: but, admitting the state of depression to which it was reduced, and the length of time it would have taken to bring about a healthy change, I yet believe, that the favourable position of the province as regards its connection with the other colonies, the character of its climate and soil, and the energies of its inhabitants, would have ensured its ultimate success. Before the depression in 1841, South Australia had become a pastoral country, in consequence of the number both of cattle and sheep that had been imported. In 1838, the city of Adelaide had scarcely been laid out, no portion of it had yet been sold, when flocks and herds were on their way to the new market, and from that period, even to the present, there has been no cessation to their ingress—first of all, as I have stated, the Murray, and then the Darling, became the high roads along which the superfluous stock of Port Phillip and New South Wales were driven to browse on South Australian pastures, and to increase the quantity and value of her exports.

However low therefore the price of wool might have kept, the natural increase of stock would still have gone on, and if we may judge from the unflinching energies of the agricultural portion of the community, their efforts to develop the productive powers of the soil, would rather have been stimulated than depressed by the misfortunes with which they were visited. I do them nothing more than justice when I assure the reader, that settlers in the province from the neighbouring colonies, could not help expressing their surprise at the state of cultivation, or their admiration of the unconquerable perseverance, that could have brought about so forward and creditable a state of things.

I have already stated that the general outline and form of the Mount Lofty chain, bears a strong resemblance to the outline and form of the Ural mountains. But it is of trifling elevation, running longitudinally from north to south, with a breadth of from 15 to 20 miles. The metalliferous veins crop out on the surface of the ground, preserving the same longitudinal directions as the ranges themselves, and the rock in which the ores are imbedded, generally speaking, is a compact slate. As the Mount Lofty ranges extend northwards, so does the Barrier or Stanley range, over which the recent expedition crossed on leaving the Darling; no copper ores were found amongst those hills, but an abundance of the finest ore of iron, running, as the out-croppings of the copper ores, from north to south, and occurring in depressed as well as elevated situations, the rock formation being very similar to that of the more western ranges.

If we are to judge from these facts, it is very evident that strong igneous action has influenced the whole, nor can I help thinking, from general appearances, that the continent of Australia has been subjected to a long subterranean process, by which it has been elevated to its present altitude, and it appears to me that that action, though considerably weakened, is still going on. The occurrence of two slight shocks of earthquake felt at Adelaide, since the establishment of the colony, would further strengthen this opinion.

The copper ores of South Australia fetch a higher price at the Swansea sales than those from any other part of the world, not only because they are intrinsically rich, but because they are generally composed of carbonates, which are necessary to facilitate the smelting of the ores of sulphuret of copper from Cuba and other places. The necessity for sending the ores from Adelaide to some foreign port to undergo the process of smelting, will probably exist for a considerable length of time; until such time, indeed, as the electric process shall be found to answer on a sufficiently large scale to be profitable, or, until smelting works are established; but, the great difficulty to be apprehended in carrying on such operations would be the want of fuel, which scarce even at the present moment, would soon be more so— for there is not sufficient wood in the vicinity of any of the mines to keep up the supply for such a consumption as that which would be required; besides which, the cartage of the wood, and the expenses attending its preparation for the furnace, would materially diminish any profits arising from the smelting of the ores. In such a view of the case I cannot but think that the establishment of works at the mines will be found to be as unprofitable to their proprietors as to the smelter, and that such works will only be remunerative when carried on under more favourable circumstances—for it would appear that coal is the only mineral South Australia does not possess, and I am apprehensive that no bed of it will ever be found in the colony. I have ever thought the geological formation of the country unfavourable to the presence of coal, but, still, it is said to exist as a submarine formation close to Aldingi Bay. The discovery of this mineral in the province would immediately give to it, within itself, the means of the most unbounded wealth, and would undoubtedly fill up the measure of its prosperity to the brim.

By a late report of the Directors of the Burra Burra mine, it would appear, that they had made several successful attempts to smelt the ore, but, that the cost, having exceeded that of cartage to the port, and freight, the process has been abandoned. Parties, however, had offered to enter into an engagement to smelt the whole of the ore from the mine at about Swansea prices; notwithstanding the unfavourable circumstances under which such smelting would necessarily be carried on.

As I understand the nature of this arrangement, the ore will be smelted at the mine, and the remuneration to the smelter will be between fifty and sixty shillings per ton perhaps, by way of "return charges," or we will say between sixty and seventy shillings, which is a sum exactly equal to the cartage of the ore to the port. If then the Directors abandoned their intentions, because they found they could not smelt at so low a sum as the price of cartage and freight, how will the contractor make it pay under more unfavourable circumstances? No doubt, if he should find it remunerative, the share holders of the Burra Burra would find it still more so, and it would be the interest of the proprietors of the larger mines to enter into similar engagements; but, on a due consideration of this important subject, I am led to believe that to make smelting works successful in South Australia, Companies must purchase the ore, and carry it off to localities suitable for the operation. Such an arrangement would still considerably increase the profits to the proprietors of the mine, nor would there be any difficulty in determining the value of the ore, by processes similar to those adopted at Swansea, by which the interests of both parties are equally protected.

In the South Australian Register of the 27th of November of last year, it is stated that a Mr. Hunt, one of the auctioneers in Sydney, offered for sale thirteen tons of pure copper ore of colonial manufacture, from ore the produce of the Burra Burra, in ingots weighing 80lbs. each; the ore having been smelted by Mr. James at Mr.

Smith's foundry at Newtown. This copper was however bought in at £80, the limit being £85 per ton.

It will give the reader some idea of the character of this prodigious mine, and of the profits arising from it, to know, that during the four months preceding the 23rd October, 1847, the directors declared and paid three dividends, amounting to 200 per cent. on the subscribed capital, and that the credits of the Association on the 30th September were £104,694 4s 8d. The Burra Burra mine however is not the only one of importance. Several others have of late been discovered, and South Australia may be said to be a thriving country in every sense of the word, and one in which those profitable interests will rapidly increase.

We have hitherto been speaking of the mines of South Australia as the sources of wealth, and as the sudden, if not the remote cause of the prosperity of that province. It now becomes our duty to consider how far the discovery of the mines has benefited or interfered with the other branches of industry and sources of wealth; and as regards both these, it must be admitted that their discovery has had an injurious effect. The high rate of wages given by the proprietors of mines, not only to the miners, but to all whom they employ, draws the labourers from every other occupation to engage with them. The consequence has been a general want of labourers throughout the whole colony, still more severely felt by reason of the previous want of labour in the labour market. Every man who could obtain sufficient money to purchase a dray and team of bullocks, hurried to the mines for a load of ore to take to the port, and disdained any ordinary employment when by carting ore he could earn £6 or £7 in a fortnight. The labourer was quite right in going where he received the best remuneration for his services; but the consequences were in many instances fatal to their former employers. Many farmers were unable to put in seed or to cultivate their land; many, after having done so, were unable to gather it, and had it not been for the use of Mr. Ridley's machine, the loss in the crops would have been severely felt. Not only did the farmers suffer, but the stock-holders, and the colonists generally. The want of hands, indeed, was felt by all classes of the community, since the natural consequence of the high wages given by the mining proprietors to the men they employed, tended still more to depress the labour market, and to increase the demand upon it by leading many of the more frugal labourers to purchase land with the money they were enabled to save. As landed proprietors they not only withdrew their labour from the market, but in their turn became employers; but I feel called upon to say at the same time, that equal distress was felt in the neighbouring colonies for working hands, where no mines had been discovered, and where they could not therefore possibly have interfered.

From what has been said of the province of South Australia, and setting its mines entirely out of the question, the description that has been given of its pastoral and agricultural capabilities, of its climate, and of the prospects of success which present them selves to the intending emigrant, it will naturally be inferred that the impression I have intended to convey is, that, as a colony, it is most peculiarly adapted for a British population, whether rural or other. The state of the colony is now such, that the way of the emigrant in landing is straight before him, for with honesty, sobriety, and industry, he cannot lose it. When I stated, in a former part of my work, that I would not take upon myself to give advice, which if followed, and not successfully, might subject me to the reproach of any one, I referred to those who have similar means of acquiring information to myself, and whose stakes, being considerable, make the responsibility of giving advice the greater. With the lower orders—the working classes—the case is different. They have not the means of acquiring information on these matters, and it becomes the duty of those who can promote their welfare to do so. I am quite aware that there are many of my poor countrymen who would gladly seek a better home than they possess at this moment, but who, clinging to the spot where they were born, disheartened at the thought of abandoning their hearth, and bound by early recollections to their native country, cannot make up their minds to turn their backs on the companions of their youth, and the haunts of their childhood.

Such a feeling undoubtedly claims our sympathy and respect. It is that very feeling—the love of Home—the belief that they can no where be happier, which has been the strength of England, and has given her sons the heart to love, and the spirit to defend her. But the period however, when those feelings were so strong, has passed away—more general ones have taken their place, and the circum. stances of the times have so changed, that neither hearth nor home have the same attractions; a restlessness pervades the community, and a desire to escape from those scenes, and that spot which they or their forefathers once thought the most hallowed upon earth. But two circumstances have militated against the migration of the rural population in this country, to the Australian colonies, at all events.

The one has been an apprehension as to the length and nature of the voyage; the other the expense, more especially to a family man. Had it not been for these causes, the Australian colonies would not have had to complain of the want of labour. The truth is, that the ignorance which prevails in the inland counties as to any matters connected with foreign parts, and the little means the labouring classes possess of defraying their own expenses, has kept them, except in a few instances, from seeking to go to that distant part of the world, which assuredly holds out to them the brightest prospect, and is most like their own home. They may however rest satisfied that the voyage to Australia is as safe as that to New York, that it is far more pleasant as regards the weather, and that little or no sickness has ever thinned the number of those who have embarked for the Australian colonies. The expense of the voyage is certainly greater than that of a passage to the Canadas, or to the United States, but it is to be hoped that the means of transport will soon be at their command. I would only in this place offer the remarks I conscientiously think the case requires, as one who, having witnessed the happiness of thousands in the land of which he is speaking, would gladly be instrumental in opening the way for thousands more of his countrymen to the same happy destiny. Having been both to Canada and the Australian colonies, if I were asked which of the two I preferred, I should undoubtedly say the latter. I do not desire to disparage the Canadas by this assertion, for I know that they have advantages in their soil and in the magnificence of their rivers beyond comparison, but Australia, on the other hand, has advantages over our transatlantic possessions, such as her increased distance from England, cannot counterbalance. Her climate, in the first place, is surpassing fine. There the emigrant is spared the trouble of providing against the severities of a Canadian winter. That season passes over his head almost without his knowledge, and the ground, instead of being a broad sheet of snow, is covered with vegetation. Her lands, unencumbered by dense forests, are clear and open to the plough, or are so lightly wooded as to resemble a park, rather than a wild and untouched scene of nature. Instead of having to toil with the saw and the axe to clear his ground before he can cultivate it, and instead of consuming a year's provisions before he can expect any return, he can there run the plough from one end to the other of his enclosures, without meeting a stone or a root to turn its point, and at once reap the produce of the soil. These surely are advantages of no ordinary kind, and, if the expense of a voyage to the Australian colonies is greater than that to America, I cannot but think that the contingent expenses to which the Canadian or Union emigrant is put, before he can consider himself as finally settled down, must necessarily exceed those of the Australian.

As before observed, the aspect of South Australia, and indeed of many parts of the neighbouring colonies, is essentially English. There, as in England, you see the white-washed cottage, and its little garden stocked with fruit trees of every kind, its outward show of cleanliness telling that peace and comfort are within. To sever oneself from our kindred, and to abandon the dwelling of our fathers, is a sacrifice of no imaginary magnitude, whether we are rich or poor, and the prospects of reward should be bright indeed to compensate for it. I conclude that it has been to combat the reluctance in the lower orders to leave their homes, that inducements too highly coloured in many instances, have been held out to them, the consequence of which has been that many, whose expectations were excited, suffered proportionate disappointment at the outset of their career as emigrants. Convinced of the injurious tendency of such a practice, and regarding it as a culpable and cruel mockery of misfortunes, which, having been unavoidable, claim our best sympathies, I should not have said so much as I have done on this important subject, had I not felt justified in so doing. The reader may rest assured that to the sober, the honest, and the industrious, the certainty of success in South Australia is beyond all doubt. An individual with these qualities may experience disappointment on landing, but he must recollect that this is always a period of anxiety, and the circumstances in which he first finds himself placed, may not come up to his expectations; his useful qualities and regular habits cannot be immediately known, and we seldom alter our condition, even for the better, without some trouble or vexation.

I have, in the course of my remarks, in my recommendation of the Australian colonies as being favourable to the views of emigrants, given a preference to South Australia. I have done so because I am better acquainted with its condition than with that of either of the other settlements. Of it I have spoken as to what I know; but of the others, to a great extent, from hearsay. The character however of those colonies needs no recommendation from me. As far as its pastoral and agricultural capabilities go, I believe Port Phillip to be as fine a district as any in the world. The advantages indeed of the Australian colonies must be nearly equal, from the fact that the pursuits of their respective inhabitants are so nearly the same. Local circumstances may give some parts of the continent a preference over others, but, as points of emigration there is little choice. The southern portions are not subject to the withering droughts to which parts of the eastern coast are liable, and may be preferred on that account but still there are districts in New South Wales as unexceptionable as any in Port Phillip or South Australia.

It now remains to make some observations on the present state of society in the last-mentioned colony; for it appears to me, that in order to give a correct picture of it, some notice on that head is required. I think too, I am the more called upon to do so, because many very mistaken notions are held of it. As in most of Her Majesty's possessions, so in South Australia, the Government officers form a prominent, and I may say, distinct class. Colonel Robe, the late Governor of the province, made Government House the seat of the most unmeasured hospitality, which he exercised beyond the point to which there was any public call upon him. His table was covered with every delicacy the season could afford, his wines were of the very best, and there was a quiet but effective manner about him, which gained universal esteem. As a soldier, he was exceedingly particular in the order and appearance of his establishment, nor was there anything wanting to complete the comfort of it. The number of the colonists who assembled round him occasionally, was from 50 to 60; on more public festive occasions they exceeded 300, and I may add, that on both, the scene differed not in the slightest degree from that of similar parties in this country, save that there was less of formality in the interchange of friendly communications between the visitors. Except also in giving a tone to society, and setting an irreproachable example to the community, the officers of the Government are exceedingly retired, their salaries are too limited to enable them to follow the example of their chief.

They live quietly, and as gentlemen, are ever happy to see their friends, but public parties are seldom given by any of them. Prudence indeed calls upon them to refrain from those displays, which they cannot reasonably afford, and the consequence was, that a warmer intimacy existed in their quiet intercourse with each other, than could have sprung from more formal entertainments.

The truth is, the salaries of the Government officers, bear no proportion to the means of the majority of the settlers, who have risen into affluence from a combination of circumstances, that have been unprecedented in the history of colonization. There are few private individuals in the province, who have not, at one time or other, benefited by some speculation, but I am not aware that any one of the Government officers have any private interests in the colony, if I except the possession of a section or two of land, on which they have built and reside, nor do I know that any of them have allowed a spirit of speculation to interfere with public duties.

Amongst the leading or upper classes of society, there are many very estimable persons. I do not mention names, but my recollection will bear me back to the many happy days I have spent with them, and certainly any one not desiring an extended circle of acquaintance could no where, whether amongst gentlemen or the ladies, find individuals more worthy of his regard or friendship than in the still limited society of South Australia.

Many of the tradesmen having succeeded in business, or acquired an independence from their interests in the mines, have retired, and live in suburban residences, which they have built in well selected situations, and with considerable taste. Attached to the customs of Home, many of the citizens of Adelaide possess carriages of one kind or another, and are fond of devoting their Sunday evenings to visiting places in the neighbourhood. As regards the lower classes, I do not think there is in any of Her Majesty's possessions, a greater amount of mechanical genius and enterprise than amongst the mechanics of South Australia. I speak confidently on this head, since I have had very many points referred to me, which have long satisfied me of this fact.

There are many societies in South Australia, of which the lower orders are members, all of them tending to promote social interests. The order of Odd Fellows is prominent amongst these, and spreads a feeling throughout all classes which cannot fail of doing good, for the charities of this order are extensive, and it supports a well-attended school. Taking then the lower orders of the province in the aggregate, they may be said to be thoroughly English, both in their habits and principles.

In speaking of the upper classes I did not notice a portion of them included under the denomination of the "Squatters." It is a name that grates harshly on the ear, but it conceals much that is good behind it; they in truth are the stockholders of the province, those in whom its greatest interests would have been vested if the mines had not been discovered. Generally speaking, the squatters are young men who, rather than be a burthen on their families, have sought their fortunes in distant lands, and carried out with them almost to the Antipodes the finest principles and feelings of their forefathers. With hearts as warm as the climate in which they live, with a spirit to meet any danger, and an energy to carry them through any reverse of fortune, frank, generous, and hospitable, the squatters of the Australian colonies are undoubtedly at the head of their respective communities, and will in after days form the landed, as they do now the pastoral interests, from whom every thing will be expected that is usually required of an English country gentleman. Circumstanced as they are at the present moment, most of them leading a solitary life in the bush, and separated by such distances from each other as almost to preclude the possibility of intercourse, they are thus cut off as it were from society, which tends to give them feelings that are certainly prejudicial to their future social happiness, but I would fain hope that the time is coming round when these gentlemen will see that they have it very much in their own power to shorten the duration of many of the sacrifices they are now called upon to make, and that they will look to higher and to more important duties than those which at present engage their attention.

The views taken by the late Sir George Gipps of the state of society in the distant interior of New South Wales is perfectly correct, nor can there be any doubt but that it entails evils on the stock holders themselves which, on an abstract view of the question, I cannot help thinking they have it in their power to lessen, or entirely to remove, when an influx of population shall take place; but, however regular their establishments may be, they cannot, as single men, have the same influence over those whom they employ, or the settlers around them, as if they were married; for it is certainly true, that the presence of females puts a restraint on the most vicious, and that wherever they are, especially in a responsible character, they must do good. I do not know anything, indeed, that would more conduce to the moral improvement of the settlers, and people around them, than that squatters should permanently fix themselves, and embrace that state in which they can alone expect their homes to have real attractions. That they will ultimately settle down to this state there cannot, I think, be a doubt, and however repugnant it may be to them at the present moment to rent lands, on the occupation of which any conditions of purchase is imposed, I feel assured that many of the squatters will hereafter have cause to thank the Secretary of State for having anticipated their future wants, and enabled them to secure permanent and valuable interests on such easy terms. Nothing, it appears to me, can be more convincing in proof of the real anxiety of Earl Grey for the well being of the Australian provinces than the late regulations for the occupation of crown lands.

I believe I am right in stating that every word of those regulations was penned by Earl Grey himself, and certainly, apart from local prejudices, I am sure a disinterested person would admit the care and thought they evince, and how calculated they are to promote the best interests of the squatters, and the future social and moral improvement of the people under their influence. There seems to me to run throughout the whole of these regulations an earnest desire to place the stockholder on a sure footing, and to remove all causes of anxiety arising from the precarious tenure upon which they formerly held property.

There is another division of the population of South Australia I have hitherto omitted to mention, I mean the German emigrants. They now number more than 2,000, and therefore form no inconsiderable portion of the population of the province. These people have spread over various districts, but still live in communities, having built five or six villages.

The Germans of South Australia are quiet and inoffensive, frugal and industrious. They mix very little with the settlers, and, regarded as a portion of the community, are perhaps too exclusive, as not taking a due share in the common labour, or rendering their assistance on occasions when the united strength of the working classes is required to secure a general good—as the gathering in of the harvest, or such similar occasions. Their religious observances are superintended by different pastors, all of them very respectable persons. The oldest of these is Mr. Kavel, to whom the Germans look with great confidence, and hold in deserved esteem. Many of the Germans have been naturalized, and have acquired considerable property in various parts of the province, but very few have taken to business, or reside in Adelaide as shopkeepers. The women bring their market or farm produce into the city on their backs, generally at an early hour of the morning, and the loads some of them carry are no trifle. Here, however, as in their native country, the women work hard, and certainly bear their fair proportion of labour. The houses of the Germans are on the models of those of their native country, and are so different in appearance from the general style, as to form really picturesque objects. There is nowhere about Adelaide a prettier ride than through the village of Klemzig, on the right bank of the Torrens, that having been the first of the German settlements. The easy and unmolested circumstances of these people should make them happy, and lead them to rejoice that in flying from persecution at home they were guided to such a country as that in which they now dwell, and I have no doubt that as a moral and religious people, they are thankful for their good fortune, and duly appreciate the blessings of Providence.

My anxiety to raise the character of the natives of Australia, in the eyes of the civilized world, and to exhibit them in a more favourable light than that in which they are at present regarded, induces me, before I close these volumes, to adduce a few instances of just and correct feeling evinced by them towards myself, which ought, I think, to have this effect and to satisfy the unprejudiced mind that their general ideas of right and wrong are far from being erroneous, and that, whatever their customs may be, they should not, as a people, occupy so low a place in the scale of human society, as that which has been assigned to them. I am quite aware that there have been individual instances of brutality amongst them, that can hardly be palliated even in savage life—that they have disgusting customs—that they are revengeful and addicted to theft. Still I would say they have redeeming qualities; for the first, I would fain believe that the horrors of which they have been guilty, are local; for the last, I do not see that they are worse than other uncivilized races. Treachery and cunning are inherent in the breast of every savage. I question, indeed, if they are not considered by them as cardinal virtues; but, admitting the Australian native to have the most unbridled passions, instances can be adduced of their regard for truth and honesty, that ought to weigh in any general estimate we may form of their character. No European living, not even Mr. Eyre, has seen so many of the Aborigines of the Australian continent as myself; and that, too, under circumstances when strife might have been expected; and no man certainly has had less reason to complain of them. If my party has ever been menaced by these people, if we have ever had their spears raised in hundreds against us, it has been because they have been taken by surprise, and have acted under the influence of fear. If I had rushed on these poor people, I should have received their weapons, and have been obliged to raise my arm against them, but, by giving them time to recover from their surprise, allowing them to go through their wonted ceremonies, and, by pacific demonstrations, hostile collisions have been avoided. If I had desired a conflict, the inclination might have been indulged without the fear of censure, but I saw no credit, no honour to be gained by such a course, and I therefore refrained. I can look back to my intercourse with the Australian aborigines, under a consciousness that I never injured one of them, and that the cause of humanity has not suffered at my hands;—but, I am travelling out of my proper course, and beg the reader to excuse me, it is for him, I allow, not for me, to draw such conclusions.

I have said, that I thought I could adduce instances of a regard for justice and honesty that would weigh in favour of the Australian native. As one instance, let me ask, if anything could have been more just, than the feeling which prompted the native to return the blanket one of his tribe had stolen from the camp on the banks of the Castlereagh, as detailed in my former work, vol. I. page 141. The man who restored the lost property was apprehensive of danger, from the fact of his having come armed, and from his guarded and menacing attitude when the soldier approached to ascertain what he wanted. Had he been the father of the thief, we could only have said that it was a singular proof of honest pride by a single individual, but such was not the case, the whole tribe participated in the same feeling, for we learnt from them, that the thief had been punished and expelled their camp. Could anything have been more noble than the conduct of the native, who remained neuter, and separated himself from them, when the tribes attempted to surprise my camp on the Murrumbidgee, because I had made him presents as I went down that river, vol. II. page 212. On the other hand, could anything have been more just than the punishment inflicted on the boy who stole my servant Davenport's blanket at Fort Grey? as mentioned in the present work; or the decision of the two sons of the Boocolo of Williorara, as regarded the conveyance of our letter-bag to Lake Victoria? Here are broad instances of honesty that would do credit to any civilized nation. Surely men, who can so feel, should not be put lowest in the scale of the human race? It is true that all attempts to improve the social condition of the Australian native has failed, but where is the savage nation with which we have succeeded better? The natives of New Zealand will perhaps be the only instance, in modern times, of a barbarous race surviving the introduction of civilization amongst them. Without venturing to compare the natives of Australia, to a people so much superior, I would only claim for them a due share of consideration. All I can say is that they have submitted to our occupation of their country with a forbearance that commands our best sympathies.

It will be borne in mind, that I have not here spoken of their personal appearance. That that generally is against them, cannot be doubted. If there is any truth in phrenology, they must have their share of the brutal passions. The whole appearance of the cranium indeed, would lead to the conclusion that they possess few of the intellectual faculties; but, in a savage state, these are seldom called forth. They are, nevertheless, capable of strong attachment, are indulgent parents, and certainly evince a kindly feeling towards their relations, are improvident and generous, having no thought for the morrow. On the other hand, they are revengeful and crafty, and treat their wives with much harshness, imposing on them the burthen of almost everything: that man being considered the richest who has the greatest number, because he can sit in his hut, and send them out to procure food.

I think it is agreed on all hands that the natives of Australia are sprung from the same parent stock. Their personal appearance and customs, if not their dialects, shew this. From what race they originally sprang it is more difficult to determine, for there is not one of the great families into which the human race has been divided, with which they may properly be classed. With such features as they generally possess, in the flattened nose, thick lip, and overhanging brow, one can hardly fancy that they would be good looking, but I certainly have seen very good looking men amongst them—I may say tribes, indeed, on the Darling for instance, and on the Murrumbidgee, (see page 53, vol. II. of my last work.) The men on Cooper's Creek were fine rather than handsome. Generally speaking, the natives have beautiful teeth, and their eye, though deep sunk, is full of fire. Although their muscular development is bad, they must have a very remarkable strength of sinew, or they could not otherwise raise themselves, as they do, on so slender a footing in climbing up the trees, and in many other occupations. I have read in several authors that the natives of Australia have woolly hair. This is a mistake; their hair is as fine and as curly as that of an European, but its natural beauty is destroyed by filth and neglect. Nothing can prove its strength more than the growth of their beards, which project from their chins, and are exceedingly stiff.

In many places the natives have but a scanty and precarious subsistence, which may in some measure account for the paucity of their numbers in some localities. In many parts of the country in which I have been I feel satisfied they can seldom procure animal food, as they would not otherwise resort to the use of some things which no time could, I should imagine, make palateable. Their dexterity at the chase is very great, although in hunting the kangaroo they become so nervous that they frequently miss their mark. I have seen them sink under water and bring up a fish writhing on the short spear they use on such occasions, which they have struck either in the forehead, or under the lateral fin, with unerring precision. Still some of our people come pretty close to them in many of their exercises of the chase, and the young settlers on the Murray very often put them to the blush. At the head of them is Mr. Scott, Mr. Eyre's companion, who has now succeeded him in the post at Moorundi. There is not a native on the river so expert in throwing the spear, in taking kangaroo or fish, or in the canoe, as he is. His spear is thrown with deadly precision, and he has so mixed with the natives, that he may be said to be one of themselves, having the most unbounded influence over them, and speaking their language as fluently as themselves. Mr. Scott is at the same time very firm and decided, and is exceedingly respected by the settlers on the Murray. Under such circumstances it is to be hoped he will emulate Mr. Eyre and effect much good among his sable friends. Their devotion and attachment to him is very remarkable, and every native on the Murray knows "Merrili," as he is called.

One great cause of the deaths amongst the Aborigines is their liability to pulmonary diseases from being constantly in the water. They are much annoyed by rain, nor will any thing induce them to stir during wet weather, but they sit shivering in their huts even in the height of summer. There is no people in the world so unprovided against inclemency or extremes of weather as they are. They have literally nothing to cover them, to protect them from the summer heat or the winter's cold; nor would any charity be greater than to supply these poor people with clothing. A few blankets, a few Guernsey shirts, and woollen trowsers, would be to them a boon of the first importance, and I would that my voice in their favour could induce the many who are humane and charitable here to devote a small portion of that which they bestow in works and purposes of charity to think of these children of the desert. It is only by accustoming them to comforts, and to implements which they cannot afterwards do without, to supersede as it were their former customs, that we can hope to draw them towards civilized man and civilization; for what inducement has the savage with his wild freedom and uncontrolled will, to submit to restraint, unless he reap some advantage?

The yearly and monthly distribution of blankets and of flour to the natives at Moorundi is duly appreciated. They now possess many things which they prefer to their own implements. The fish-hooks they procure from the Europeans are valued by them beyond measure, since they prevent the necessity of their being constantly in the water, and you now see the river, at the proper season, lined by black anglers, and the quantity of fish they take is really astonishing, and those too of the finest kinds. I once saw Mr. Scott secure a Murray cod, floating on the top of the water, that weighed 72lbs. This beautiful and excellent fish is figured in Mitchell's first work. It is a species of perch, and is very abundant, as well as several others of its own genus, that are richer but smaller; the general size of the cod varying from 15lbs. to 25lbs.

The manners and customs of the natives have been so well and so faithfully recorded by Mr. Eyre that I need not dwell on them here. My views have been philanthropic, my object, to explain the manner in which I have succeeded in communicating with such of them as had never before seen Europeans, in order to ensure to the explorer, if possible, the peaceable results I myself have experienced. There are occasions when collisions with the natives are unavoidable, but I speak as to general intercourse. I feel assured no man can perform his duty as an explorer, who is under constant apprehension of hostility from the people through whose country he is passing.

The province of South Australia could never at any time have been thickly inhabited. There are some numerous tribes on the sea-coast at the head of the Gulfs and in Encounter Bay, as well as on the Murray River, but with the exception of a few scattered families on the northern hills, and in the scrub, the mountain ranges are, and it appears to me have been, almost uninhabited. There are no old or recent signs of natives having frequented the hills, no marks of tomahawks on the trees, or of digging on the flats. The Mount Lofty ranges, indeed, are singularly deficient of animal life, and seem to be incapable of affording much subsistence to the savage, however luxuriant and beneficial the harvest they now yield.

The Adelaide tribe is not numerous; they occupy a portion of the Park lands, called the native location, and every encouragement has been given them to establish themselves in comfort on it, but they prefer their wild roving habits to any fixed pursuit. Nevertheless, they are variously employed by the townspeople, in carrying burthens, in cutting up wood, in drawing water, and similar occupations; and, independently of any assistance they may receive from the Government, earn an immense quantity of food from the citizens. The natives properly belonging to the Adelaide tribe are all more or less clothed, nor are they permitted by the police to appear otherwise, and as far as their connection with the settlers goes, they are fast falling into habits of order, and understand that they cannot do any thing improper with impunity.

The Murray tribe, as well as the tribes from the south, frequently visit their friends near the capital, and on such occasions some scene of violence or dispute generally ensues. Frequently the abduction of a lubra, or of an unmarried female of another tribe, brings about a quarrel, and on such occasions some angry fighting is sure to follow; and so long as that custom remains, there is little hope of improvement amongst them. The subject of ameliorating their condition is, however, one of great difficulty, because it cannot be done without violating those principles of freedom and independence on which it is so objectionable to infringe; but when a great ultimate good is to be obtained, I cannot myself see any objection to those restraints, and that interference which should bring it about. There is nowhere, not even in Sydney, more attention paid to the native population than in South Australia, and if they stand a chance of improvement it is there. Whilst every kindness is shewn to the adult portion, the children are under the direct care of the Government. There is, as I have elsewhere stated, a school, at which from thirty to forty boys and girls attend. Nothing can be more regular or more comfortable than this institution. The children are kindly treated, and very much encouraged, and really to go into it as a visitor, one would be disposed to encourage the most sanguine expectations of success. As far as the elementary principles of education go, the native children are far from deficient. They read, write, and cypher as well as European children of their own age, and, generally speaking, are quiet and well behaved; but it is to be regretted that, as far as our experience goes, they can advance no farther; when their reason is taxed, they fail, and consequently appear to be destitute of those finer qualifications and principles on which both moral feeling and social order are based. It is however questionable with me whether this is not too severe a construction to put on their intellect, and whether, if the effect of ancient habits were counteracted, we should find the same mental defect.

At present, the native children have free intercourse with their parents, and with their tribe. The imaginations of the boys are inflamed by seeing all that passes in a native camp, and they long for that moment, when, like their countrymen, they will be free to go where they please, and to join in the hunt or the fray. The girls are told that they are betrothed, and that, at a certain age, they must join their tribe. The voice of Nature is stronger even than that of Reason. Why therefore should we be surprised at the desertion of the children from the native schools? But it will be asked—What is to be done? The question, as I have said, is involved in difficulty, because, in my humble opinion, the only remedy involves a violation, for a time at all events, of the natural affections, by obliging a complete separation of the child from its parents; but, I must confess, I do not think that any good will result from the utmost perseverance of philanthropy, until such is the case, that is, until the children are kept in such total ignorance of their forefathers, as to look upon them as Europeans do, with astonishment and sympathy. It may be argued that this experiment would require too great a sacrifice of feeling, but I doubt this. Besides which, it is a question whether it is not our duty to do that which shall conduce most to the benefit of posterity. The injury, admitting it to be so, can only be inflicted on the present generation, the benefit would be felt to all futurity. I have not, I hope, a disposition for the character of an inhuman man, and certainly have not written thus much without due consideration of the subject, but my own experience tells me we are often obliged to adopt a line of conduct we would willingly avoid to ensure a public good.

It will not then, I trust, be thought that I have ventured to intrude this opinion on the public, with any other views than those which true philanthropy dictates. I am really and sincerely interested in the fate of the Australian Aborigine, and throw out these suggestions, derived from long and deep practical experience, in the ardent hope that they may help to produce the permanent happiness of an inoffensive and harmless race.

MR. KENNEDY'S SURVEY OF THE RIVER VICTORIA

WHILST I was endeavouring to penetrate into the heart of the Australian Continent, there were two other Expeditions of Discovery engaged in exploring the country to the eastward of me. Dr. Leichhardt, an account of whose successful and enterprising journey from Moreton Bay to Port Essington is already before the public, was keeping the high lands at no great distance from the coast, and Sir Thomas Mitchell, the Surveyor-General of New South Wales, was traversing the more depressed interior, between my own and Dr. Leichhardt's tracks. The distance at which Dr. Leichhardt passed the extreme westerly point gained by me was 600 geographical miles, and his distance from my extreme easterly one was 420 miles; Sir Thomas Mitchell's distance from my extreme west, being about 380 miles, and that from my last position, (on Cooper's Creek), about 260. He had been traversing a country of great richness and fertility, a country, indeed, such as he had never before seen, and in a despatch addressed to the Governor of New South Wales, thus describes it and the river he discovered on the occasion:—

"On ascending the range early next morning, I saw open downs and plains with a line of river in the midst, the whole extending to the N.N.W., as far as the horizon. Following down the little stream from the valley in which I had passed the night, I soon reached the open country, and during ten successive days I pursued the course of that river, through the same sort of country, each day as far as my horse could carry me, and in the same direction again approaching the Tropic of Capricorn. In some parts the river formed splendid reaches, as broad and important as the river Murray; in others it spread into four or five branches, some of them several miles apart. But the whole country is better watered than any part of Australia I have seen, by numerous tributaries arising in the downs.

"The soil consists of rich clay, and the hollows give birth to numerous water-courses, in most of which water was abundant. I found at length that I might travel in any direction, and find water at hand, without having to seek the river, except when I wished to ascertain its general course, and observe its character. The grass consists of Panicum and several new sorts, one of which springs green from the old stem. The plains were verdant indeed, the luxuriant pasturage surpassed in quality, as it did in extent, any thing I had ever seen. The Myall-tree and salt bush, (Acacia pendula and salsolæ), so essential to a good run, are also there. New birds and new plants marked this out as an essentially different region from any I had previously explored; and although I could not follow the river throughout its long course at that advanced season, I was convinced that its estuary was in the Gulf of Carpentaria; at all events the country is open and well watered for a direct route thereto. That the river is the most important of Australia, increasing as it does by successive tributaries, and not a mere product of distant ranges, admits of no dispute; and the downs and plains of Central Australia, through which it flows, seem sufficient to supply the whole world with animal food. The natives are few and inoffensive. I happened to surprise one tribe at a lagoon, who did not seem to be averse that such strangers were in that country; our number being small, they seemed inclined to follow us. I crossed the river at the lowest point I reached, in a great southerly bend in long. 144°34' east, lat. 24°14' south, and from rising ground beyond the left bank, I could trace its downward course far to the northward. I saw no Callitris (Pine of the colonists) in all that country, but a range, shewing sandstone cliffs appeared to the southward, in long. 145° and lat. 24°30' south. The country to the northward of the river, is, upon the whole, the best, yet, in riding ninety miles due east from where I crossed the southern bend, I found plenty of water, and excellent grass, a red gravel there approaches the river, throwing it off to the northward. Ranges extending N.N.W. were occasionally visible from the country to the northward."

Sir Thomas Mitchell's position at his extreme west was more than 460 miles from the nearest part of the Gulf of Carpentaria; he was in a low country, and on the banks of a river which had ceased to flow. Whatever the local appearances might have been, which led the Surveyor-General to conclude that it would reach the northern coast, I do not know, but notwithstanding the favourable report he made of it, I never for a moment anticipated that this river would do so; I felt assured, indeed, that however promising it might be, it would either enter the Stony Desert or be found to turn southward, and be lost amongst marshes and lagoons. The appearance of Cooper's Creek might have justified my most sanguine expectations, but I was too well aware of the character of Australian rivers, and had seen too much of the country into which they fall, to trust them beyond the range of sight. My natural course on the discovery of Cooper's Creek would have been to have traced it downwards, but I was not unmindful that I should keep it between myself and the track on which Mr. Browne and I had last returned from the north-west interior, in pursuing the northerly course I intended, and I consequently felt satisfied, after a little consideration, that if it continued northerly, I should strike it again; if not, that it would either spread over the Stony Desert, or fall short of it altogether.

On making this discovery, therefore, my hopes were centered in its upward, not its downward course, for judging that in crossing the Stony Desert, I had crossed the lowest part of the interior, my anticipations of finding any important river in the central regions of Australia were destroyed. My endeavour had been, not only to examine the country through which I was immediately passing, but to deduce from it, what might be its more extended features, and to put together such facts as I reasonably could, to elucidate the past and present state of the continent. In the course of my investigations, I saw grounds for believing that the fall of the interior was from north to south and from east to west. However much the more northerly streams might hold to the northward and westward, whilst in the hilly country, I felt assured, that as soon as they gained the depressed interior, they would double round to the southward, and thus disappoint the explorer. Sir Thomas Mitchell himself tells us, that every river he traced on his recent journey, excepting the Victoria, disappointed him, by turning to that point and entering a sandy country. It is evident, indeed, upon the face of Sir Thomas Mitchell's journal, that there are no mountains in that part of the interior, in which the basins of the Victoria must lie, or from which a river could emanate, of such a character, as to lead even the most sanguine to expect, that after having ceased to flow, it would continue onwards for another 460 miles through such a country. From the favourable nature of the Surveyor-General's report, however, it was deemed a point of great importance to ascertain the further course of the river, and Mr. Kennedy, a young and intelligent officer, who had accompanied Sir Thomas Mitchell into the interior, was ordered on this interesting service. Before I make any observations, however, on the result of his investigations, I shall give the following extract from is letter to the Colonial Secretary, on his return from the interior.

"Having reached the lowest point of the Victoria attained by the Surveyor-General, I was directed to pursue the river, and determine the course thereof as accurately as my light equipment, and consequent rapid progress, might permit. Accordingly, on the 13th of August we moved down the river, and at 41 miles crossed over to its proper right bank; the Victoria is there bounded on the south by a low sand-stone ridge, covered with brigalow; and on the north by fine grassy plains, with here and there clumps of the silver leaf brigalow; at seven miles we passed a fine deep reach, below which the river is divided into three channels, and inclines more to the southward; at thirteen miles we encamped upon the centre channel; the three were about half a mile apart, the southern one under the ridge being the deepest; we found water in each, but I believe it to be only permanent in the southernmost, which contains a deep reach, one mile below our encampment, in latitude 24°17'34"S; an intelligent native, whom we met there with his family on our return, gave me the name of the river, which they call Barcoo. I also obtained from him several useful words, which he seemed to take a pleasure in giving, and which I entered in my journal.

"Between the parallels of 24°17' and 24°53' the river preserves a generally very direct course to the south-south-west, and maintains an unvaried character, although the supply of water greatly decreases below the latitude of 24°25'. It is divided into three principal channels, and several minor watercourses, which traverse a flat country, lightly timbered by a species of flooded box; this flat is confined on either side by low sand-stone ridges, thickly covered with an acacia scrub. In latitude 24°50' we had some difficulty in finding a sufficiency for our own consumption, but after searching the numerous channels, the deep (though dry) lagoons and lakes formed there by the river, we at length encamped at a small water-hole in latitude 24°52'55" and longitude 144°11'26".

"Being aware that the principal view of the Government in sending me to trace the Victoria, was the discovery of a practical route to the Gulf of Carpentaria, I then began to fear that I should be unable, with my small stock of provisions, to accomplish the two objects of my Expedition. My instructions confined me to the river, which had now preserved almost without deviation a south-south-west course for nearly a hundred miles; the only method which occurred to me, by the adoption of which I might still hope to perform all that was desired, was to trace the river with two men as far as latitude 26°, which the maintenance of its general course would have enabled me to do in two days, and then to hasten back to my party, to conduct them to the extreme northern point attained by the Victoria, and endeavour to prolong the direct route carried that far, from Sydney towards the Gulf of Carpentaria, by Sir Thomas Mitchell.

"With this intention I left the camp on the 20th of August, and at twelve miles found several channels united, forming a fine reach, below which the river takes it turn to the west-south-west, receiving the waters of rather a large creek from the eastward, in latitude 25°3'0". In latitude 25°7', the river having again inclined to the southward, impinges upon the point of a low range on its left, by the influence of which it is turned in one well watered channel to the west and west by north, for nearly thirty miles; in that course the reaches are nearly connected, varying in breadth from 80 to 120 yards; firm plains of a poor white soil extend on either side of the river; they were rather bare of pasture, but they are evidently in some seasons less deficient of grass. In latitude 25°9'30", and longitude about 143°16', a considerable river joins the Victoria from the north-east, which I would submit may be named the 'Thomson,' in honour of E. Deas Thomson, Esquire, the Honourable the Colonial Secretary. It was on one of the five reaches in the westerly course of the Victoria that I passed the second night; the river there measured 120 yards across, and seemed to have a great depth; the rocks and small islets which here and there occurred in its channel giving it the semblance of a lasting and most important river; this unexpected change, however, both in its appearance and course caused me to return immediately to my camp for the purpose of conducting my party down such a river whithersoever it should flow.

"On the 25th August, we resumed our journey down that portion of the Victoria above described, and made the river mentioned from north-east three miles above its junction; following it down we found an unbroken sheet of water in its channel, averaging fifty yards in breadth; we forded it at the junction, and continued to move down the Victoria, keeping all the channels, into which it had again divided, on my left. At about one mile the river there turns to the south-south-west and south, spreading over a depressed and barren waste, void of trees or vegetation of any kind, its level surface being only broken by small doones of red sand, resembling islands upon the dry bed of an inland sea, which, I am convinced, at no distant period did exist there.

"On the 1st September, we encamped upon a long, though narrow, reach in the most western channel, at which point a low sandstone ridge, strewed with boulders, and covered with an acacia scrub, closes upon the river. This position is important, as a small supply of' grass will, I think, in most seasons, be found on the bank of the river, when not a blade, perhaps, may be seen within many miles above or below: my camp, which I marked K IV was in latitude 25°24'22" longitude 142°51'. Beyond camp IV the ridge recedes, and the soil becomes more broken and crumbling; our horses struggled with difficulty over this ground to my camp, at a small water-hole, in latitude 25°43'44", where I found it necessary to lighten some of their loads by having buried 400lbs. flour, and 70lbs. sugar, still retaining a sufficient supply to carry us to Captain Sturt's farthest, on Cooper's Creek, to the eastward, (to which point I was convinced this river would lead me) and from thence back to the settled districts of New South Wales; which was all I could then hope to accomplish. At about sixteen miles further, the ground becoming worse, so that our horses were continually falling into the fissures up to their hocks, I was compelled to leave 270lbs. more of flour and sugar at my camp of the 4th September, in latitude 25°51', at another small water-hole, found in the bed of a very dry and insignificant channel; here a barren sandstone range again impedes the river in its southerly course, and throws it off to the westward, thus causing many of its channels to unite and form a reach of water in latitude 25°54'; this, the lowest reach we attained, I did not discover until my return. Having found a sufficient supply in a channel more to the westward, in latitude 25°55', and longitude, by account, 142°23', the river, having rounded the point of the range which obstructs it, resumes its southerly course, spreading in countless channels over a surface bearing flood marks six and ten feet above its present level; this vast expanse is only bounded to the eastward by the barren range alluded to, which, ending abruptly, runs parallel with the river at a distance varying from four to seven miles. On the 7th September, I encamped upon a small water-hole in 26°0'13" in the midst of a desert not producing a morsel of vegetation; yet so long as we could find water, transient as it was, I continued to push on with the hope of reaching, sooner or later, some grassy spot, whereon by a halt I might refresh the horses; however, that hope was destroyed at the close of the next day, for although I had commenced an early search for water when travelling to the southward, with numerous channels on either side of me, I was compelled at length to encamp in latitude 26°13'9" and longitude, by account, 142°20', on the bank of a deep channel, without either water or food for our wearied horses. The following morning, taking one man and Harry with me, we made a close search down the most promising watercourses and lagoons, but upon riding down even the deepest of them, we invariably found them break off into several insignificant channels, which again subdivided, and in a short distance dissipated the waters, derived from what had appeared the dry bed of a large river, on the absorbing plain; returning in disappointment to the camp, I sent my lightest man and Harry on other horses to look into the channels still unexamined, but they also returned unsuccessful. We had seen late fires of the natives at which they had passed the night without water, and tracked them on their path from lagoon to lagoon in search of it; we also found that they had encamped on some of the deepest channels in succession, quitting each as it had become dry, having previously made holes to drain off the last moisture. My horses were by this time literally starving, and all we could give them was the rotten straw and weeds which had covered some deserted huts of the natives. Seeing, then, that it would be the certain loss of many, and consequently an unjustifiable risk of my party to attempt to push farther into a country where the aborigines themselves were at a loss to find water, I felt it my imperative duty to at once abandon it. I would here beg to remark, that although unsuccessful in my attempt to follow it that far, from the appearance of the country, and long-continued direction of the river's course, I think there can exist but little doubt that the 'Victoria' is identical with Cooper's Creek, of Captain Sturt; that creek was abandoned by its discoverer in latitude 27°46', longitude 141°51', coming from the north-east, and as the native informed him, 'in many small channels forming a large one;' the lowest camp of mine on the Victoria was in latitude 26°13'9" longitude 142°20'; the river in several channels trending due south, and the lowest point of the range which bounds that flat country to the eastward, bearing south 25° east; Captain Sturt also states that the ground near the creek was so blistered and light that it was unfit to ride on; but that before he turned, he had satisfied himself that there was no apparent sign of water to the eastward.

"Having marked a tree EK 1847, we commenced our return journey along the track at two p.m. of the 9th of September; at eight miles I allowed one of the horses to be shot; for being an old invalid, and unable to travel further, he must have starved if left alive. At thirteen miles we reached the water. Some while after dark the following day we made our next camp; but it was with much difficulty that my private horse and two or three others were brought to water, one being almost carried by three men the latter part of the day. Upon discovering the reach, in latitude 25°54', near the range, and finding a little grass in the channel about the water, I gave the horses two days' rest. My camp on the reach is marked K III it is in latitude 25°55'37", longitude, by account, 142°24'; the variation of the compass 8° east; water boiled at 214° the temperature of the air being 64°. On the 14th September we proceeded on our journey, and reached the firm plains beyond the desert. On the 22nd, having halted a day, we again moved on, and arrived within five miles of the carts; on the 7th October, leaving my party on the south channel, I rode to the spot, and found them still safe, although a native had been examining the ground that very morning. Lest he should have gone to collect others to assist him in his researches, I brought my party forward the same evening, had the carts dug out during the night, and at sunrise proceeded to our position of the 4th August on the south channel."

From the above account, which is equally clear and distinct, it would appear, that, just below where the river Alice joins the Victoria, the latter river had already commenced its south-west course, and that the last thirty miles down which the Surveyor-General traced this river was a part of the general south-west course, which it afterwards maintained to the termination of Mr. Kennedy's route, and consequently the latter traveller never had an opportunity of approaching so near the Gulf of Carpentaria as the Surveyor-General had done. Here its channel separates into three principal branches, at half-a-mile apart, and, notwithstanding the promise it had given down to the point, at which he had now arrived, (latitude 24°52', and longitude 144°11',) having then travelled nearly 100 miles along its banks, Mr. Kennedy had great difficulty in finding water. In consequence indeed, of the unfavourable changes that had taken place in the river, he determined on leaving the party stationary, and proceeding down it with two men to the 26th parallel, whence, if he found that it still held to the south, he proposed returning with the intention of trying to find a practicable route to the Gulf of Carpentaria, in compliance with his instructions, and under an impression, I presume, that the fate of the Victoria would then have been fully determined.

In latitude 25°3', the river having changed its course to the W.S.W. was joined by a large creek from the "eastward." In latitude 25°7' it was turned by some low sandstone ranges on its left, and trended for thirty miles to the west, and even to the northward of that point, having almost connected ponds of water for that distance, varying in breadth, from 80 to 120 yards, and being bounded on either side by firm plains of white soil. About 25°9' and 143°16' the river was joined by a large tributary stream from the north-east, to which Mr. Kennedy gave the name of the "Thomson," and encouraged by the favourable changes which had now taken place, he returned for his party with the determination of following so fine a river to the last.

We shall now see how far his anticipations were confirmed, and how far his further investigation of the Victoria river, and his account of the country through which it flows, accords with the description I have given of the dreary region into which I penetrated.

On the 26th of September, Mr. Kennedy having brought down his party, resumed his journey, and crossing the Victoria, struck the N.E. tributary about three miles above its junction with the main stream, and fording at that point, kept on the proper right bank of the Victoria.

"At about a mile," says Mr. Kennedy, "it (the Victoria) there turns to the S.S.W. and south, spreading over a depressed and barren waste, void of trees or vegetation of any kind, its level surface being only broken by small doones of red sand, like islands upon the dry bed of an inland sea, which I am convinced at no distant period did exist there."

There cannot, I think, be any reasonable doubt, but that Mr. Kennedy had here reached the edge of the great central desert.

Both the river he was tracing, and the country were precisely similar in character to Cooper's Creek, and the country I had so long been wandering over. The former at one point having a fine deep channel, at another split into numberless small branches, and then spreading over some extensive level without the vestige of a water-course upon it. The country monotonous and sterile, its level only broken by low sandstone hills, or doones of sand, the whole bearing in its general appearance the stamp of a submarine origin.

Mr. Kennedy's last camp on the Victoria was in lat. 26°13'9"S. and in long. 142°20'E.; the most eastern point of Cooper's Creek gained by me was in lat. 27°46'S. and long. 141°51'E. This longitude, however, was by account, and I may have thrown it some few miles to the eastward; in like manner Mr. Kennedy's longitude being also by account, I believe he may have placed his camp a little to the west of its true position; but, as the two points are now laid down, there is a distance of 98 geographical miles between them, on a bearing of 13° to the east of north. Admitting the identity of the Victoria with Cooper's Creek, of which I do not think there is the slightest doubt, the course of the former in order to join the latter would be south, 13°W. the very course Mr. Kennedy states it had apparently taken up when he left it. "The lowest camp on the Victoria," he says, "was in lat. 26°13'9", and in long. 142°20', the river in several channels trending due south." If such is the case I must have misunderstood the signs of the natives, and been mistaken in my supposition that the vast basin into which I traced it, was the basin of Cooper's Creek, but I had so frequently remarked the rapid and almost instantaneous formation of such features in similar localities, that, I confess, I did not doubt the meaning the natives intended to convey.

There are several facts illustrative of the structure and lay, if I may use the expression, of the interior unfolded to us, in consequence of the farther knowledge Mr. Kennedy's exploration has given of that part through which the Victoria flows, which strike myself, who have so deep an interest in the subject, when they might, perhaps, escape the general reader; I have therefore thought it right to advert to them for a moment. He will not, however, have failed to observe, in the perusal of Mr. Kennedy's Report, that excepting where small sandstone ranges turned it to the westward, the tendency of the Victoria was to the south. The same fact struck me in reference to the Murray river, as I proceeded down it in 1830. I could not fail to observe its efforts to run away in a southerly direction when not impeded by cliffs or sand-hills. This would seem to indicate, that the dip of the continent is more directly to the south than to the west. There is a line of rocky hills, that turn Cooper's Creek to the latter point immediately to the south-west of the grassy plains on which I supposed it took its rise. From that point its general direction is to the westward for about eighty miles, when it splits into two branches, the one flowing to the north-west, and terminating in the extensive grassy plains described at page 39, Vol. II. xof the present work, the other passing to the westward and laying all the country under water during the rainy season, which Mr. Brown and I traversed on our journey to the north-west; the several creeks we discovered on that occasion, being nothing more than ramifications of Cooper's Creek, which thus, like all the other interior rivers of Australia, expends itself by overflowing extensive levels; but instead of forming marshes like the Lachlan, the Macquarie, and the Murrumbidgee, terminates in large grassy plains, which are as wheatfields to the natives, since the grass-seed they collect from them appears to constitute their principal food.

I have observed in the beginning of this work, that the impression on my mind, before I commenced my recent expedition, was, that a great current had passed southwards through the Gulf of Carpentaria which had been split in two by some intervening obstacle, that one branch of this current had taken the line of the Darling, the other having passed to the westward. Now, it would appear, that the sources of the Victoria are in long. 146°46', and we are aware, that the course of that river is to the W.S.W. as far as the 139th meridian; unless, therefore, there is a low and depressed country between the sources of the Victoria, and the coast ranges traversed by Dr. Leichhardt, through which the southerly current could have passed, my hypothesis, as regards it, is evidently wrong; and such, on an inspection of Sir Thomas Mitchell's map, appears to be the case, as he has marked a line of hills, connecting the basins of the Victoria with the higher ranges traversed by Doctor Leichhardt, nearer the coast. My object being to elicit truth, I have deemed it necessary to call the attention of the reader to this point, because it would appear to argue against the general conclusions I have drawn, since, if there is no apparent outlet, there could not have been any southerly current, as I have supposed; whereas, if the features of the country could have justified such a conclusion, the general ones I have formed would have been very considerably strengthened.

Mr. Kennedy's survey of the Victoria establishes the fact, that there is not a single stream or water-course falling into the main drainage of the continent, from the northward or westward, between the 24th and 34th parallels of latitude, a distance of more than 700 geographical miles—a fact which strongly proves the depressed nature of the north-west interior, and would appear to confirm the opinion already expressed, that the Stony Desert is the great channel into which such rivers as have a sufficiently prolonged course, are ultimately led, and towards which the northerly, and a great portion of the easterly drainage tends. How that singular feature may terminate, whether in an inland sea, or as an arid wilderness, stretching to the Great Australian Bight, it is impossible to say. From the general tendency of the rivers to fall to the south, it may be that the Stony Desert, as Mr. Arrowsmith supposes, has some connexion with Lake Torrens, but I think, for reasons already stated, that it passes far to the westward.

It may not be generally known, that Dr. Leichhardt is at this moment endeavouring to accomplish an undertaking, in which, if he should prove successful, he will stand the first of Australian explorers. It is to traverse the continent from east to west, nor will he be able to do this under a distance of more than 5,000 miles in a direct line. He had already started on this gigantic journey, but was obliged to return, as his party contracted the ague, and he lost all his animals; but undaunted by these reverses, he left Moreton Bay in December last, and has not since been heard of. One really cannot but admire such a spirit of enterprise and self-devotion, or be too earnest in our wishes for his prosperity. Dr. Leichhardt intends keeping on the outskirts of the Desert all the way round to Swan River, and the difficulties he may have to encounter as well as the distance he may have to travel, will greatly depend on its extent. We can hardly hope for intelligence of this dauntless explorer for two years; but if such a period should elapse without any intelligence of him, I trust there will not those be wanting to volunteer their services in the hope of rendering him assistance. Our best feelings have been raised to save the Wanderer at the Pole—should they not also be raised to carry relief to the Wanderer of the Desert? The present exploration of Dr. Leichhardt, if successful, will put an end to every theory, and complete the discovery of the internal features of the Australian continent, and when we look at the great blank in the map of that vast territory, we cannot but admit the service that intrepid traveller is doing to the cause of Geography and Natural History, by the undertaking in which he is at present engaged. It is doubtful to me, however, whether his investigations and labours will greatly extend the pastoral interests of the Australian colonies, for I am disposed to think that the climate of the region through which he will pass, is too warm for the successful growth of wool. As I stated in the body of my work, the fleece on the sheep we took into the interior, ceased to grow at the Depôt in lat. 29°40', as did our own hair and nails; but local circumstances may account for this effect upon the animal system, although it seems to me that the great dryness of the Australian atmosphere, where the heat is also excessive, as it must be in the interior and juxta-tropical parts of it, would prevent the growth of wool, by drying up the natural moisture of the skin. Nevertheless, if Dr. Leichhardt should discover mountains of any height or extent, their elevated plateaux, like that of the Darling Downs, which is one of the finest pastoral districts of New South Wales, and is in lat. $27^{1}/_{2}°$, would not be liable to the same objections; for I believe no better wool is produced than in that district, and that only there, and in Port Phillip, has the sheep farmer been able to clear his expenses this year. Were it not, therefore, for the almost boundless and still unoccupied tracts of land within the territory of New South Wales, we might look with greater anxiety, as regards the pastoral interests of Australia, to the result of Dr. Leichhardt's labours. At present, however, there seems to be no limit to the extent either of grazing or of agricultural land in New South Wales. The only thing to be regretted is, that the want of an industrious population, keeps it in a state of nature, and that the thousands who are here obtaining but a precarious subsistence, should not evince a more earnest desire to go to a country where most assuredly their condition would be changed for the better.

ANIMALS.

BUT FEW mammalia inhabit Central Australia. The nature of the country indeed is such, that we could hardly expect to find any remarkable variety. The greater part is only tenable after or during heavy rains, when the hollows in the flats between the sandy ridges contain water. On such occasions the natives move about the country, and subsist almost exclusively on the Hapalotis Mitchellii, and an animal they call the Talpero, a species of Perameles, which is spread over a great extent of country, being common in the sand hills on the banks of the Darling, to the S.E. of the Barrier Range, as well as to the sandy ridges in the N.W. interior, although none were met with to the north of the Stony Desert.

The Hapaloti feed on tender shoots of plants, and must live for many months together without water, the situation in which we found them precluding the possibility of their obtaining any for protracted intervals. They make burrows of great extent, from which the natives smoke them, and they sometimes procure as many as twelve or eighteen from one burrow. This animal is grey, the fur is exceedingly soft; although the animal is in some measure common, I could not procure any skins from the natives.

Very few kangaroos were seen, none indeed beyond the parallel of 28°. All that were seen were of the common kind, none of the minor description apparently inhabiting the interior, if I except some Rock Wallabi, noticed on the Barrier Range. The last beautiful little animal always escaped us in consequence of its extreme agility and watchfulness.

The Native Dog was not seen beyond lat. 28°. Nor was it found in a wild state beyond Fort Grey, to the best of my recollection; these miserable and melancholy animals would come to water where we were, unconscious of our presence, and would gain the very bank of the creek before they discovered us, rousing us by as melancholy a howl as jackal ever made; their emaciated bodies standing between us and the moon, were the most wretched objects of the brute creation.

The first Chœropus castanolus seen, was on the banks of the Darling, in the possession of the natives, but it was too much injured to be valuable as a specimen. A second was also killed there, but torn to pieces by the dogs. None were afterwards seen until after the Barrier Range had been crossed, when about lat. 27° several were captured alive, as detailed under the head Dipus. In like manner the first nest of the "Building Rats" I (Mus conditor, Gould) was found in the brushes on the Darling, where they were numerous. The last nest of these animals was on the bank of the muddy lagoon to the north of the Pine Forest, in which the party were so embarrassed, at the end of 1844.

The first Hapalotis, seen was in lat. 29½° on some plains to the eastward of the Depôt, where it was nearly captured by Mr. Browne. A second was taken by Mr. Stewart, at the tents, but in neither places were they found inhabiting the same kind of country as that in which they were subsequently found in such vast numbers. Mr. Gould thinks there were two species amongst those brought home, and it may be that these two were different from those inhabiting the sand hills: they only differed, however, in a darker shade in the fur, and a reddish mark on the back of the ears.

There were both rats and mice in the N.W. interior, numbers of which took up their abode in our underground room at the Depôt, but there was no apparent difference between them and the ordinary rat or mouse.

There was only one Opossum killed, or indeed seen to the westward of the Barrier Range, nor do they appear to inhabit the interior in any numbers. Since there were no signs of the trees having been ascended by the natives in search of them.

1. CANIS FAMILIARIS, var. AUSTRALASIÆ.—Dingo.

This animal was not very numerous in the interior, more especially towards the centre, for it was not noticed to the north of the Stony Desert. Wherever seen it was in the most miserable condition, and it is difficult to say on what they lived. This animal was of all colours. It appears to me that if these dogs are indigenous, nature has departed from her usual laws as regards wild beasts, in giving them such a variety of colours.

2. MACROPUS MAJOR.—Great Kangaroo.

This animal did not extend beyond 28°. Six or seven were there seen on a small stony range, but very few were observed to the westward of the Barrier Range.

3. MACROPUS LANIGER.—Red Kangaroo.

This fine animal did not extend beyond the neighbourhood and plains of the Murray, where it is not numerous. Several of the smaller kangaroos were taken during the progress of the Expedition up the Murray and Darling river; but as they have been frequently described, it is not thought necessary to insert them in this list.

4. CHŒROPUS CASTANOTUS, GRAY.

This animal was first killed on the Darling, but the specimen was destroyed by the dogs. Two or three were afterwards taken alive in latitude 26½°. They were found lying out in tufts of grass, and when roused betook themselves after a short run, to some hollow logs where they were easily cut out. The Chœropus is a beautiful animal, about eight inches long in the body, with a tail of considerable length, having a tuft at the end. The fur is a silvery grey, and very soft. When confined in a box they ate sparingly of grass and young leaves, but preferred meat and the offal of birds shot for them. The Chœropus is insectivorous, and I was therefore not surprised at their taking to animal food, which, however, not agreeing with them, they died one after the other. They squat like rabbits, laying their broad ears along their backs in the same kind of way.

5. HAPALOTIS MITCHELLII.

This beautiful little animal was, as I have observed in the introduction to this notice, first seen in the vicinity of the Depôt. It was subsequently found in vast numbers, inhabiting the sandy ridges from Fort Grey to Lake Torrens. Those immense banks of sand were in truth marked over with their footprints as if an army of mice or rats had been running over them. They are not much larger than a mouse, have a beautiful full black eye, long ears, and tail feathered towards the end. The colour of the fur is a light red, in rising they hop on their hind legs, and when tired go on all four, holding their tail perfectly horizontal. They breed in the flats on little mounds, burrowing inwards from the edge; various passages tending like the radii of a wheel to a common centre, to which a hole is made from the top of the mound, so that there is a communication from it to all the passages.

They are taken by the natives in hundreds, who avail themselves of a fall of rain to rove through the sandy ridges to hunt these little animals and the talpero, Perameles, as long as there shall be surface water. We had five of these little animals in a box, that thrived beautifully on oats, and I should have succeeded in getting them to Adelaide if it had not been for the carelessness of one of the men in fastening a tarpauline down over them one dreadful day, by which means they were smothered.

6. MUS CONDITOR, GOULD.—The Building Rat.

Inhabits the brushes in the Darling, in which it builds a nest of small sticks, varying in length from eight inches to three, and in thickness, from that of a quill to that of the thumb. The fabric is so firm and compact as almost to defy destruction except by fire. The animals live in communities, and have passages leading into apartments in the centre of the mound or pyramid, which might consist of three or four wheelbarrows full of the sticks, are about four feet in diameter, and three feet high. The animal itself is like an ordinary rat, only that it has longer ears and its hind feet are disproportioned to the fore feet. It was not found beyond latitude 30°.

7. ACROBATES PYOMÆA.—Flying Opossum Mouse.

This beautiful and delicate little animal was killed in a Box tree, whence it came out of a hole, and ran with several others along a branch, retreating again with great swiftness. It was so small that if the moon had not been very bright it could not have been seen. It is somewhat less than a mouse in size and has a tail like an emu's feather, its skin being of a dark brown.

8. LAGORCHESTES FASCIATUS (L. ALBIPILIS, GOULD?).—Fasciated Kangaroo. One only of this animal was seen on the plains of the interior. It is peculiar in its habits, in that it lies in open ground and springs from its form like a hare, running with extreme velocity, and doubling short round upon its pursuers to avoid them. The Lagorchestes is very common on the plains to the north of Gawler Town, but is so swift as generally to elude the dogs. It is marsupial, and about the size of a rabbit, but is greatly disproportioned, as all the Kangaroo tribe are, as regards the hind and fore quarters. In colour this animal is a silvery grey, crossed with dark coloured bars on the back.

9. PHALANGISTA VULPINA.—The Opossum.

Like the preceding, only one of these animals was seen or shot during the Expedition; it was in one of the gum-trees, taking its silent and lonely ramble amongst its branches, when the quick eye of Tampawang, my native boy, saw him. It does not appear generally to inhabit the N.W. interior. The present was a very large specimen, with a beautifully soft skin, and as it was the only one noticed during a residence of nearly six months at the same place, it was in all probability a stray animal.

10. VESPERTILIO—Little black Bat.

This diminutive little animal flew into my tent at the Depôt, attracted by the light. It is not common in that locality, or any other that we noticed. It was of a deep black in colour and had smaller ears than usual.

BIRDS.

I HAVE observed that a principal reason I had for supposing that there was either an inland sea, a desert country, or both in the interior, was from observations I had made—during several expeditions, and in South Australia, of the migration of certain of the feathered tribes to the same point—that is to say, that in lat. 30° and in long. 144°, I observed them passing to the N.W. and in lat. 35°, long. 138°, to the north. Seeing, on prolonging these two lines, that they would pass over a great portion of the interior before they met, about a degree beyond the tropic, I concluded that the nature of the intervening country was not such as they could inhabit, and that the first available land would be where the two lines thus met. It so happened that at the Depôt, in lat, 29 1/2° and in long. 142°, I was in the direct line of migration to the N.W., and that during our stay at that lonely post, we witnessed the migration of various birds to that quarter, though not of all. This was more particularly the case with the water-birds, as ducks, bitterns, pelicans, cormorants, and swans,—we saw few of the latter, but generally heard them at night passing over our heads from N.W. to S.E. or vice versâ; but we never afterwards found any waters which we could suppose those birds could frequent in the distant interior. On Strzelecki's Creek a small tern was shot, and on Cooper's Creek several seagulls were seen, but beyond these we had no reason to anticipate the existence of inland water from any thing we noticed as to the feathered races. On our first arrival at the Depôt there was a bittern, *Ardeeta flavicollis*, that frequented the creek in considerable numbers. This bird was black and white, with a speckled breast and neck. Every evening at dusk they would fly, making a hoarse noise, to the water at the bottom of the Red Hole Creek, and return in the morning, but as winter advanced they left us, and went to the N.W.

About February and the beginning of March, the *Epthianura tricolor* and *E. aurifrons,* and some of the Parrot tribe, collected in thousands on the creeks, preparatory to migrating to the same point to which the aquatic birds had gone. It was their wont to fly up and down the creeks, uttering loud cries, and collecting in vast numbers, but suddenly they would disappear, and leave the places which had rung with their wild notes as silent as the desert. The *Euphema elegans* then passed us, with several other kinds of birds, but some of them remained, as did also the *Euphema Bourkii,* which the reader will find more particularly noticed under its proper head.

The range of the Speckled Dove (*Geopelia cuneata*), so common on the Darling, extended to the Depôt, and two remained with us during the winter, and roosted two or three times on the tent ropes over my fire.

There were always an immense number of Raptores following the line of migration, and living on the smaller birds; nor was any thing more remarkable than the terror they caused amongst them. The poor things would hardly descend to water, and several of the Euphema came to the creek in the dark, when we could not see to fire at them, and several killed themselves by flying against our tent ropes.

The range of the Rose Cockatoo was right across the continent as far as we went—as well as that of the Crested Parroquet, which was, as I have observed, the last bird we saw, just before Mr. Browne and I turned homewards from our first going to the N.W. The *Cacatua sanguinea,* Gould, succeeded the Sulphur Crested Cockatoo to the westward of the Barrier Range, and was in flocks of thousands on Evelyn's Plains, near the Depôt, but I am not certain as to the point to which it migrated. It is remarkable, however, that the Sulphur Crested Cockatoo, though numerous along the whole line of the Darling, was never seen near the Depôt, or to the westward of the Barrier Range.

The *Amadina Lathami,* to which we always looked as the harbinger of good, was met with in every part of the interior—where there was water—and frequently at such vast distances from it, when migrating, I suppose, that vast numbers must have perished.

I have noticed the Pigeons in their proper place, and stated my opinion as to the point to which they went on leaving us; and I would refer my reader to my remarks on that head: he will find their habits and localities fully described there.

We fell in with the water-hen, *Tribonyx,* on one of the creeks on our journey to Lake Torrens, and again on Strzelecki's Creek, apparently migrating to the south. These birds ran alone, the banks like fowls, as they did in the located districts of Adelaide, as described by Mr. Gould, and that too in great numbers, and when disturbed took wing to the south. In like manner we observed the *Eudromias Australis,* migrating southwards in May. From these facts it would appear that the great line taken by the feathered tribes in migrating from the southern or south-eastern parts of the province is in a direction between the east and south points of the compass, and I cannot still help thinking that about a degree to the north of the Tropic, and about the meridian of 138, a more fertile country than any hitherto discovered will be found.

It may be necessary for me to observe that on our advance to Fort Grey, in August, we observed numerous *Calodercæ,* and other smaller birds in the brushes, apparently on the move whilst there was water for them, that had been left by the then recent rains. We did not again see these birds until we had passed the Stony Desert and entered the box-tree forest to the north of it, in which was the creek with the huge native well. There a variety of birds had congregated—the Rose Cockatoo, the piping Magpie, the Calodera, various parrots and parroquets, bronze-wing Pigeons, and numerous small birds.

At Cawndilla, Mr. Poole shot a *Euphema splendida,* Gould. It was in company with several others; but this bird was not again seen until we passed the 26th parallel, in September, when it was met by Mr. Browne and myself coming from the north. The following is a list of the birds seen during the expedition.

1. AQUILA FUCOSA, CUVIER—The Wedge-tailed Eagle.

Two of these birds frequented the Depôt Glen, in 29°40'00" and in longitude 142°, one of which was secured. They generally rested on a high pointed rock, whence their glance extended over the whole country, and it was only by accident that the above specimen was killed.

This powerful bird is common both on the Murray and the Darling, and is widely, perhaps universally distributed over the Australian continent, although the two birds in the Glen were the only ones seen in the interior to the N.W. of the Barrier, or Stanley's Range.

2. HALIASTUR SPHENURUS—The Whistling Eagle.

This species of Eagle is considerably smaller than the first and has much lighter plumage. It is a dull and stupid bird, and is easily approached. It was shot at the Depôt, in the month of April, 1845. Several others were seen during our stay there.

3. FALCO HYPOLEUCUS, GOULD.—The Grey Falcon.

This beautiful bird was shot at the Depôt, at which place, during our long stay, Mr. Piesse, my storekeeper, was very successful with my gun. A pair, male and female, were observed by him one Sunday in May, whilst the men were at prayers, hovering very high in the air, soon after which he succeeded in killing both. They came down from a great height and pitched in the trees on the banks of the creek, and on Mr. Piesse firing at and killing one the other flew away; but returning to look for its lost companion, shared its fate. Nothing could exceed the delicate beauty of these birds when first procured. Their large, full eyes, the vivid yellow of the ceres and legs, together with their slate-coloured plumage, every feather lightly marked at the end, was quite dazzling; but all soon faded from the living brightness they had at first. The two specimens were the only ones seen during an interval of seventeen months that the party was in the interior, and these, it appears probable to me, were on the flight, and were attracted down to us.

4. FALCO MELANOGENYS, GOULD—The Black-cheeked Falcon.

A single specimen of this bird was shot at the Depôt, when just stooping at a duck on some water in the glen. The strength of limb, and muscle of this fine species of falcon were extremely remarkable, and seemed to indicate that he despised weaker or smaller prey than that at which he was flying when shot. He had been seen several times before he was killed. His flight was rapid and resistless, and his stoop was always sure.

This must be a scarce bird, as the specimen was the only one seen.

5. FALCO SUBNIGER, G.R. GRAY—The Black Falcon.

The colour of this fine bird is a sooty black, but his shape is beautiful, and his flight, as his sharp pointed wings indicate, rapid. He was shot in some brushes behind the Depôt, where he had been spreading alarm amongst a flight of parroquets, (*Euphema Bourkii*).

This must also be a scarce bird, as he was the only one seen.

6. FALCO FRONTATUS—The White-fronted Falcon.

This is both a smaller and a more common bird; its range being very wide. This species followed the line of migration, and made sad havoc among the parroquets and smaller birds. He was generally hid in the trees, and would descend like an arrow when they came to water, frequently carrying off two of the little Amadina castanotis, a favourite bird of ours, one in each talon.

7. TINNUNCULUS CENCHROÏDES—Nankeen Kestril.

Like the last, small and swift of wing, following also the line of migration.
This bird is generally distributed over the continent and is known by the nankeen colour of his back.

8. ASTUR APPROXIMANS, VIG. AND HORSF—Australian Goshawk.

This bird was occasionally seen during the journey.

9. MILVUS AFFINIS, GOULD—Allied Kite.

This bird is common over the whole continent of Australia. They are sure to be in numbers at the camps of the natives, which they frequent to pick up what may be left when they go away. They are sure also to follow any party in the bush for the same purpose. About fifty of these birds remained at the Depôt, with about as many crows, when all the other birds had deserted us; and afforded great amusement to the men, who used to throw up pieces of meat for them to catch in falling. But although so tame that they would come round the tents on hearing a whistle, they would not eat any thing in captivity, and would have died if they had not been set at liberty again. It was this bird which descended upon Mr. Browne and myself in such numbers from the upper regions of the air, as we were riding on some extensive plains near the Depôt in the heat of summer. There can be no doubt but that in the most elevated positions where they are far out of the range of human sight, they mark what is passing on the plains below them.

10. ELANUS SCRIPTUS, GOULD—The Letter-winged Kite.

This beautiful bird was first seen on a creek to the eastward of the Barrier or Stanley's Range, and before the party had crossed that chain of bills. One was shot on the advance of the Expedition from the Darling in the early part of November 1844, in latitude 32°, and on the return of the party from the interior, in December 1845, several specimens were seen as low as Cawndilla, and ranging along the banks of the Darling. In the interval they were seen in flocks of from thirty to forty, either soaring in the air or congregated together in trees. They were never seen to stoop at any thing, nor could we detect on what they fed, but I am led to believe that it was mice. They are fond of hovering in the air, and in such a position look beautiful, the black bar across the wing, underneath them appearing like a W, and contrasting strongly with the otherwise delicate plumage of the bird. They left us for a time whilst we remained at the Depôt, and the first that were afterwards seen by us were on the return of Mr. Browne and myself from our first northern journey. These birds are widely distributed over that part of the interior traversed by the Expedition. Like *Elanus notatus*, it has a bright full eye, the iris inclined to a light pink. Its shoulders are black, and its back like a sea-gull, slate-coloured.

11. CIRCUS JARDINII, GOULD—Jardine's Harrier.

This bird, with its spotted plumage, was not common. A specimen was shot on the banks of the Darling, between Williorara and the junction with the Murray. None of the same bird were seen in the N.W. interior, or to the westward of the Barrier Range.

12. STRIX PERSONATA, VIG.—Masked Barn Owl

This fine night bird was very rare in the interior, and only one specimen was procured. Its plumage is characterised by that softness so peculiar to the genus to which it belongs, and in consequence of which its flight is so silent and stealthy that, like the foot-fall of the cat, it is unheard.

This owl was shot on the Darling, after having been startled out of a tree.

13. STRIX DELICATULUS, GOULD—Delicate Owl.

Nearly allied to the Strix flammea, or Barn Owl of England. This bird, widely spread over the continent of Australia, inhabits the interior in great numbers, wherever there are trees large enough for it to build in. Their young were just fledged when the Expedition descended into the western interior, and at sunset came out on the branches of the gum-trees, where they sat for several hours to be fed, making a most discordant noise every time the old birds came with a fresh supply of food, which was about every quarter of an hour. It was frequently impossible to sleep from the constant screeching of the young owls. Their food is principally mice, bats, and large moths.

14. ATHENE BOOBOOK—Boobook Owl.

So called from its whoop resembling that sound. Like others of its genus it comes from its hiding place at sunset, and its note in the distance is exactly like that of the cuckoo, but the sound changes as you approach it. This bird has a dark brown plumage, spotted white, and differs in many respects from the genus Strix, although very closely allied to it.

15. ÆGOTHELES NOVÆ-HOLLANDIÆ, VIG. AND HORSF—Owlet night Jar.

This small bird, although a night bird, is very frequently seen in the day time, sleeping on the branch of a Casuarina, to which they appear to be partial. It is very common in the brushes of the Murray belt, and when disturbed has an awkward flight, as if it knew not where to go. Its plumage is very downy and soft, and it weighs exceedingly light.

16. PODARGUS HUMERALIS. VIG. AND HORSF. Tawny-shouldered Podargus.

This singular bird is an inhabitant of the distant interior, and was seen on several occasions, but invariably near hills. The appearance of this uncouth bird is very absurd, with his enormous mouth that literally reaches from ear to ear, and his eyes half shut. Mr. Browne surprised five of these birds on a stone, on the summit of Mount Arrowsmith, about half a degree to the south of the Depôt. They were all sitting with their heads together, and all flew in different directions when roused.

17. EUROSTROPODUS GUTTATUS—Spotted Goat-sucker.

This rapid-winged night bird is widely distributed over South-eastern Australia, if not over every part of the Continent. I have often watched the motions of this light and airy bird round a pond of water close to which I have been lying, with the full bright moon above me, and been amazed at its rapid evolutions; and admired the wisdom of that Providence which had so adapted this little animal for the part it was to act on the great stage of the universe. So light, that it had no difficulty in maintaining a prolonged flight, with its noiseless wing, making its sweeps to greater or lesser distances, and seeming never to require rest. The habit of this Goat-sucker is to lie under any tree or brush during the day, from which it issues in great alarm on being roused.

18. CHELIDON ARIEL, GOULD.

The brown-headed Swallow, a common bird in the interior during the summer. Gregarious, and building clay nests, like bottles stuck against a tree, in rows one above the other. Instinct guides these little birds to select a tree that slopes and is concave, in which the nests will be protected from rain or storms. A white-headed swallow was also frequently seen, but it was always under circumstances that prevented our procuring a specimen.

19. MEROPS ORNATUS, LATH—Australian Bee-eater.

This beautiful little bird, with its varied plumage, is migratory, and visits the southern parts of the continent during summer, when its locality is near any river, or chain of ponds, although it is also found in other places. I first shot this pretty bird on the banks of the Macquarie in 1828, where it was in considerable numbers. It visits Adelaide, and we saw it in the interior almost to our extreme north.

20. HALCYON SANCTUS, VIG. AND HORSF—Sacred Halcyon.

This ill-proportioned bird in shape and general appearance is like the Kingfisher. Instead however of living on fish, he contents himself with lizards, beetles, grasshoppers, etc., and amongst these he makes a great havoc. The range of this bird did not extend beyond the lat. of the Depôt.

21. HALCYON PYRRHOPYGIA, GOULD—Red-backed Halcyon.

Similar in shape and figure to the last, but differing in plumage and in size having dull red feathers over the rump, the blue being also of a duller shade. It ranges far north.

22. ARTAMUS SORDIDUS—Wood Swallow.

The flight and habits of this bird are very like those of the swallow tribe. They huddle together to roost: selecting a flat round stump, round the edge of which they sit with their heads inwards, so presenting a singular appearance: or else they cling together to the number of thirty or forty on a branch like a swarm of bees. They were seen in every part of the interior over the whole of which they appear to range.

23. ARTAMUS PERSONATUS, GOULD—Masked wood Swallow.

So called because of a black mark on the throat and cheek resembling a mask in some measure. The plumage of this bird is light, the breast of the male almost approaching to a white, for size and shape there is little difference between this and the last. Both are equally common, and are seen together, ranging the brushes at a great distance from water.

24. ARTAMUS SUPERCILIOSUS, GOULD—White eye-browed wood Swallow.
A white line over the eye is the distinguishing mark of this bird. One or other species of Artamus was found when no other birds were to be seen. They generally sat on dead branches, and their flight extended no farther than from the one to the other.

25. PARDALOTUS STRIATUS TEMM—Striated Pardalote.
There are several species of this beautiful tribe of little birds, but the above was the only kind procured. The species under consideration occupies the higher branches of the gum-trees, and is so small that it is seen with difficulty.

26. GYMNORHINA LEUCONOTA, GOULD—The White-backed Crow Shrike,
This bird is somewhat larger than, and very much resembles a magpie, but the proportion of white is greater, and there is no metallic or varied tint on the black feathers as on the European bird. In South Australia it is a winter bird, and his clear fine note was always the most heard on the coldest morning, as if that temperature best suited him. All the species of this genus are easily domesticated, and learn to pipe tunes. They are mischievous birds about a house, but are useful in a garden. I had one that ranged the fields to a great distance round the house, but always returned to sleep in it.

27. CRACTICUS DESTRUCTOR.
This bird has the strong, straight, and hooked bill. He is an ugly brute in shape and plumage, but is a magnificent songster. His own notes ring through the wilds, and there is not a bird of the forest that he does not imitate. One of these birds regularly visited the camp at Flood Creek every morning to learn a tune one of the men used to whistle to him, and he always gave notice of his presence by a loud note of the most metallic sound. It breeds on the hills, and is generally found wherever there is shade and water.

28. GRALLINA AUSTRALS—Pied Grallina.
This harmless bird, somewhat larger than a field-fare, is found near water, where the banks are muddy. It is common on all the river flats, and lives on insects. Its pied plumage is very pretty, but its note is a melancholy one. Very few were seen to the westward of the Barrier Range, and those always close to lagoons.

29. GRAUCALUS MELANOPS—Black-faced Graucalus.
The colour of the plumage of this bird is that of slate, and it has a black throat. Its range is very extensive, but we did not see it in the distant north-west interior.

30. PTEROPODOCYS PHASIANELLA, GOULD—Ground Grauculus.
There were not more than six or seven of this bird seen during the progress of the Expedition, and that only at the Depôt. They were exceedingly wild and wary, keeping in the centre of open plains and feeding on locusts and grasshoppers. They always kept together, and flew straight from and to the trees on the banks of the creek. This bird is long in shape, and has a peculiar rise over the rump. It is elegantly formed. The head and back are slate-coloured; the rump white, with scollops, as also is the breast; the wings and tail being black and long. It was with great difficulty that we procured any specimen of this bird from its shyness. It apparently came from the N.E. and departed in the same direction when winter approached.

31. CAMPEPHAGA HUMERALIS, GOULD—White-shouldered Campephaga.
An insectivorous bird, frequenting the brushes of the interior, and of wide range; visiting the southern districts in summer, but evidently being a bird of a warm climate. A species very similar to the present inhabited Norfolk Island.

32. PACHYCEPHALUS GUTTURALIS—Guttural Pachycephala.
The strong bill of this bird indicates its character as living on insects. It is common, and has been so often described as to require no notice here.

33. PACHYCEPHALUS PECTORALIS, VIG. AND HORSF—Banded Thick-head.
Similar in habits to the last; and is abundant in all parts of South America.

34. COLLURICINCLA HARMONICA—Harmonious Colluricincla.
A bird of dull plumage, with the habits of a thrush, keeping in the bushes or young sapling gum-trees, near water, and living on insects of various kinds. Its note is sweet, and amongst Australian birds it may be considered a good songster. Its range is extensive. It was numerous on Cooper's Creek, in lat. 27½° and long. 142°.

35. OREOÏCA GUTTURALIS—Crested piping Thrush.
I found this bird common on the plains eastward of the Darling, and also in the western interior. It visits the south-eastern parts of the continent, and is common in South Australia; frequenting open forests, and betraying its presence by its monotonous notes. It is a strong built bird, with a dull plumage, but its crest adds much to its beauty, and it has a deep yellow iris.

36. ERYTHRODRYAS RHODINOGASTER—Pink-breasted wood Robin.
This pretty little bird is, like our own native Robin, fond of woodlands, and is generally found amongst thick brush, issuing from it to perch on dead branches.

Its breast is a fine bright pink; its plumage is otherwise black and white, and it has a spot of white over the nostrils. The range of this bird is extensive, and it is common to many localities.

37. PETROICA GOODENOVII—Red-capped Robin.

Similar in shape to the last, and essentially with the same plumage, with this exception, that the feathers over the nostril in this bird are a fine deep red, as well as its breast. It is found in South Australia, and was not uncommon in the interior.

38. PETROICA PHŒNICEA, GOULD—Flame-breasted Robin.

Similar in general appearance, but larger than either of the last; it is grey where it is black in the others, and is without any frontal mark. It has, like the others, a breast of red, approaching to a flame colour. This species is not common in the interior. None of the three described are songsters, and cannot therefore rival our own sweet bird in that respect.

39. DRYMODES BRUNNEOPYGIA GOULD—Scrub Robin.

This bird is considerably larger than the last described, and is an inhabitant of scrubs.

40. SPHENOSTOMA CRISTATA, GOULD—Crested Wedge Bill.

The note of this bird is generally heard when all the other birds are silent, during the heat of the day. Its range does not extend to the westward of the Barrier Range, or beyond 32½° of latitude.

41. MALURUS CYANEUS—Blue Wren.

This beautiful little warbler, so splendidly illustrated in the work of Mr. Gould, is common in South Australia. There are six or seven species of the genus, all equally beautiful.

42. MALURUS MELANOTUS.

This beautiful description of Malurus, common in the brushes of South Australia, was frequently met with, particularly in scrubby places.

43. MALURUS LEUCOPTERUS—White-winged.

The habits of this bird are exactly similar to those of a wren. It delights in being on the top of bushes, whence after singing for a minute or two it flies into the centre and secretes itself. The rich-coloured males of this family are generally followed by a number of small brown birds, their late offspring. This peculiarity has been mentioned fully by Mr. Gould in his splendid work on Australian birds.

44. EPTHIANURA AURIFRONS, GOULD—Orange-fronted Epthianura.

The general appearance of this beautiful little bird is very different from that of Australian birds in general. A few years ago a specimen came accidentally into my hands, and it was so unlike any bird I had seen that I doubted its having been shot in Australia, but concluded that it was a South American specimen. Two or three however were procured by the Expedition, in latitude 29°, longitude 141½°.

45. EPTHIANURA TRICOLOR, GOULD—Tricoloured Epthianura.

This beautiful little bird was procured, both on the summit of the Barrier Range, and on the plains to the westward of it, generally inhabiting open brush. It was conspicuous amongst the smaller birds on account of its bright red plumage, but it was by no means uncommon. This bird evidently migrates from the north-west, and the second time, when it was seen so far to the westward of the ranges, it was most likely on its return from that point.

46. PYRRHOLÆMUS BRUNNEUS, GOULD—Brown Red-throat.

A small and common brush bird, and a good warbler, more remarkable indeed for the sweetness of its song than for the beauty of its plumage.

47. CINCLORAMPHUS RUPESCENS.

A good songster, and generally distributed over the country.

48. AMADINA LATHAMI—Spotted-sided Finch.

This is, I believe, the largest of its genus, and is a beautiful little bird. It was not seen to the westward of Stanley's Barrier Range. Its range is, however, extensive, as it is found in most parts of New South Wales, as well as South Australia.

49. AMADINA CASTANOTUS, GOULD.

This pretty little bird is perhaps more numerous than any other in the interior of Australia. Never did its note fall on our ears there but as the harbinger of good, for never did we hear this little bird but we were sure to find water nigh at hand, and many a time has it raised my drooping spirits and those of my companions, when in almost hopeless search for that, to us, invaluable element.

The *Amadina castanotus* is gregarious, collecting together in hundreds on bushes never very far from water, to which they regularly go at sunset. They build in small trees, many nests being together in the same tree, and hatch their young in December. It was met with in every part of the interior wherever there was water, but hundreds must perish yearly from thirst, for the country must frequently dry up round them, to such a distance as to prevent the possibility of their flying to another place of safety. The hawks make sad havoc also amongst these harmless little birds, generally carrying off two at a time.

50. CINCLOSOMA CASTANOTUS, GOULD—Chestnut-backed Ground Thrush.
This is a bird of the great Murray belt, and was first shot by my very valued friend Mr. Gould, when in a bush excursion with me in South Australia. It is by no means a common bird, and is exceedingly wary.

51. CINCLOSOMA CINNAMONEUS, GOULD—Cinnamon-coloured Ground Thrush.
This third species of Cinclosoma appeared at the Depôt in latitude 29½° longitude 142°, during the winter months in considerable numbers, and a good many specimens were procured. Mr. Gould tells me this is the only new species procured during my recent Expedition, a proof, I think, of his indefatigable exertions in the prosecution of his researches. Indeed I can bear abundant testimony as to the perseverance and ability he displayed whilst with me, and the little regard he had to personal comfort, in his ardent pursuit of information as to the habits of the feathered tribes in the singular region where he was sojourning.

52. ZANTHOMYZA PHRYGIA—Warty-faced Honey-eater.
This Honey-eater, with alternate black and yellow plumage, frequented all the sand hills where Banksias grew, but as none of those trees are to be found to the westward of Stanley's Barrier Range, so these birds were confined to the country eastward of it.

They are found both in New South Wales and in South Australia; and most probably came to the latter place from the eastward.

50. ACANTHORHYNCHUS-RUFO-GULARIS, GOULD—Shiny Honey-eater.
A larger Honey-eater, with grey mottled plumage, generally found on the Banksia, and not very common.

53. ZOSTEROPS DORSALIS—Grey-backed white-eye.
Seen in many parts of the country through which the Expedition passed, but more common in the settled districts of the colony. It is exceedingly mischievous amongst the grapes, and frequents the gardens in such numbers as to be formidable.

54. CRYSOCOCCYX LUCIDUS—The shining Cuckoo.

This is the smallest of the Cuckoo tribe, and is known by the metallic lustre of its wings. It is beautifully figured in Mr. Gould's work. It was frequently seen in the interior.

55. CLIMACTERIS SCANDENS, TEMM—Brown Tree-Creeper.

This creeper was, with another *Climacteris Picumnus*, common in the pine forests and on the open box-tree flats all over the interior. It is not a showy bird in any way, but is very active and indefatigable in its search for insects. It is remarkable that no *Picus* has been found in Australia.

56. ACROCEPHALUS AUSTRALIS—The reed singing Bird.

This beautiful warbler is common in south-eastern Australia, wherever there are reeds by the banks of the rivers or creeks, but where they were wanting its voice was silent. On the banks of the Murray and the Darling its note was to be heard during the greater part of the night, almost equal to that of the nightingale, and like that delightful bird, its plumage is any thing but brilliant, it is however somewhat larger, and although its general shade is brown, it has a light shade of yellow in the breast that makes it brighter in its plumage than the European songster.

57. HYLACOLA PYRHOPYGIA.

A common species inhabiting scrubs.

58. HYLACOLA CAUTA, GOULD.

A small bush bird, common to the belts of the Murray and other similar localities.

59. CYSTICOLA EXILIS, GOULD—Exile Warbler.

This little bird has a varied note, indeed it is not a bad songster. It inhabits grass beds and scrubby lands, but its range does not extend beyond the 32° parallel. The Barrier Range appearing to form a limit to the wanderings of many of the smaller birds.

60. ACANTHIZA PYRRHOPYGIA—Red-rumped Acanthiza.

A small bush bird of brown plumage on the back, with a reddish spot over the rump.

61. ACANTHIZA CHRYSORRHÆA—Yellow-rumped Acanthiza.

This bird is similar to the last in every thing but the colour of the feathers over the rump, which in the present specimen is yellow. Very common on the plains and open glades of woods.

62. XEROPHILA LEUCOPSIS, GOULD—White-faced Xerophila.

It is singular, as Mr. Gould relates in his work, that this bird should not have been known or procured until he shot it, almost on the steps of Government house in Adelaide. It was occasionally seen in the interior, but not to the westward of the Barrier Range. It keeps generally on the ground. Mr. Gould has distinguished it in consequence its having a front of white. It is short and compact in form, and like the preceding bird keeps a good deal on the ground.

63. CALAMANTHUS CAMPESTRIS, GOULD—Field Reed Lark.

This bird is smaller than the regular lark, and differs from it in many respects: indeed it more resembles the tit lark than the sky lark, and altogether wants the melodious song of the latter. It is a very common bird all over such parts of Australia as I have visited; frequenting open ground.

64. CINCLORAMPHUS CANTILLANS, GOULD. Great singing Lark.

This bird, both in its habits and song, resembles the Bunting of Europe, rising like it from the top of one bush, with a fine full note, and descending with tremulous wing to another. Its range, as far as I can judge, is right across the continent, since we fell in with it at our most distant northern points. It is much larger than the above, has a stronger bill, and a dark breast. This bird is good eating.

65. CINCLORAMPHUS RUFESCENS—Singing Lark.

This is also a good songster.

66. CORCORX LEUCOPTERUS—While-winged Chough.

This bird has a dirty black plumage, excepting a white bar across the wings. It is generally seen in groups of six or seven, flying from tree to tree, and is widely distributed all over the continent.

67. CORVUS CORONÏADES, VIG. AND HORSF. White-eyed Crow.

This bird approaches somewhat to the raven. Its plumage is black and glossy, its neck feathers like a cock's hackle, and the iris white, the latter peculiarity giving it a singular appearance. Many of these birds remained with us at the Depôt after we had been deserted by most of the other kinds, and served to fatten an old native who had visited the camp, on whose condition they worked a perfect miracle. I suppose indeed that there never was such an instance of an individual becoming absolutely fat in so short a time, from a state of extreme emaciation, as in that old and singular savage, from eating the crows that were shot for him, and which constituted his chief, I might say, his only food.

68. POMATORHINUS SUPERCILIOSUS.

A bird that frequented the cypress and pine forests; running along the branches of the trees like rats, and chasing each other from one to the other. This bird is about the size of a thrush, but is very different in other respects. It has dark brown plumage, with a rufous breast.

69. POMATORHINUS TEMPORALIS.

A bird very similar in plumage and habits to the last, but smaller and quicker in its motions. I shot these birds on a former expedition to the eastward of the Darling, and both are figured in my former work, page 219, vol. II.

70. GLYCIPHILA FULVIFRONS—Fulvous-fronted Honey-eater.

A bird common amongst the honey-suckles (Banksias), in the sandy rises or mounds in the neighbourhood of the Darling. It appears in South Australia in similar localities, and has all the characters of its genus in the curved bill, pencilled tongue, and other points.

71. GLYCIPHILA ALBIFRONS, GOULD—White-fronted Glyciphila.

This bird is about the size of a chaffinch, and was first killed by me on the Darling.

72. PTILOTIS CRATITIUS, GOULD.

This Honey-eater is remarkable in having a narrow lilac skin on the cheek, with a light line of yellow feathers beneath it. It is long both in the body and tail, and is of graceful form. Its colour is grey, but the breast is of a lighter shade and is slightly mottled. First shot by Mr. Gould in South Australia, from whose searching eye, and persevering industry, few things escaped. It was not common in the interior, but was occasionally seen in favourable localities.

73. ANTHOCHÆRA CARUNCULATA—Wattle Bird.

Frequents Banksias, and is common wherever those trees are to be found. The Anthochæra carunculata is the largest of the wattle birds in South Australia. It has a grey plumage, mottled with white, and is by no means inelegant in its shape, being a long, slender, well proportioned bird. The whole of the Honey-suckers have curved bills and pencilled tongues.

74. ANTHOCHÆRA MELLIVORA—Brush Wattle Bird.

This Honey-eater is of very limited range, and was so seldom seen during the progress of the Expedition up the Darling, that it may almost be said to be confined to the located district of South Australia. Its range, however, is as far as to the parallel of 30°, beyond which point, as the majority of the honey-bearing trees cease, the larger Honey-suckers are not to be found, Like all the birds of the same genus, it is quick in its movements.

75. MELITHREPTUS GULARIS, GOULD—Black-throated Honey-eater.
This bird is distinguished by its black throat, and a white lunate mark on the nape of the neck. It is to be found in most places where honey-bearing flowers or trees are to be seen. The general plumage is a dull green.

76. MELITHREPTUS LUNULATUS—Lunulated Honey-eater.
This species partakes of all the characters of the genus, but is much smaller. The range of the Honey-Eaters does not extend beyond the 28th. Parallel—towards the N.W. interior, or Central Australia; as there are few honey-giving trees in that desert region. They are found all along the summits of the Barrier Range, however, in considerable numbers; and are always known by their loud wild note.

77. MYZANTHA GARRULA—The Old Soldier.
A very sociable and tame bird. Its range is over the whole of south-eastern Australia, and we saw nests of these noisy birds at Fort Grey, in 29°. The general colour is grey; their bill, and some portion of the head being yellow. They are fond of being near habitations, and frequent the trees round a stock station in great numbers.

78. SITTELLA PILEATA, GOULD—Black-capped Sittella.
A creeper, with a black head, and grey brown plumage. Not very common, though often seen in the interior. It is larger than the *S. Chrysoptera.*

79. CACATUA GALERITA—Sulphur-crested Cockatoo.
This Cockatoo, the most common in Australia, is snow-white, with the exception of its crest, which is of a bright sulphur. It is also the most mischievous of Australian birds, and not only plays sad havoc amongst the wheat when ripe, but soon clears a field that has been sown. They are in immense flocks, and when in mischief always have sentinels at some prominent point to prevent their being taken by surprise, and signify the approach of a foe by a loud scream. They build in the hollows of trees, and in vast numbers in the Murray cliffs, making them ring with their wild notes; and in that situation are out of reach of the natives. They are abundant along the line of the Darling as high as Fort Bourke, but do not pass to the westward of that river, nor do they inhabit the interior.

80. CACATUA LEADBEATERII—Leadbeater's Cockatoo.

This beautiful Cockatoo is, like the first, of white plumage, with a light red shade under the wings. He has a large sulphur and scarlet crest, which he erects to the best advantage when alarmed. This Cockatoo frequents the pine forests near Gawler Town, and is seen wherever that tree abounds; but he is not common, although widely distributed over the interior; his range extending to the latitude of Fort Grey, in 29°; far beyond where any pine trees were to be found.

81. CACATUA SANGUINEA, GOULD—Blood-stained Cockatoo.

This is a smaller bird than either of the preceding; it is also of white plumage, with a light red down under the feathers; and, although it has the power of erecting the feathers on its head, it may be said to be crestless. This bird succeeded *Cacatua galerita,* and was first seen in an immense flock on the grassy plains at the bottom of the Depôt Creek, feeding on the grassy plains or under the trees, where it greedily sought the seeds of the kidney bean. These cockatoos were very wild, and when they rose from the ground or the trees made a most discordant noise, their note being, if anything, still more disagreeable than that of either of the others. They left us in April, and must have migrated to the N.E., as they did not pass us to the N.W., nor were they any where seen so numerous as at this place.

82. CACATUA EOS—Rose Cockatoo.

This beautiful bird, seen in the depressed interior in such great numbers, has a slate-coloured back, wings and tail, whilst its breast and neck are of a beautiful rose-pink colour. It has a trifling crest, but not one like the two first described cockatoos. We carried this bird with us to the farthest north, as high up as the 25th parallel. There were several nests at Fort Grey, from which the men procured several young; one of which I brought alive to Adelaide. They hatch in the end of October, and build in the hollows of the box-trees. A flock of these cockatoos, turning their red breasts together to the sun in flying, look very beautiful.

83. LICMETIS NASICUS—Long-billed Licmetis.

This cockatoo is very like *Cacatua sanguinea* in colour and shape. It is white, with a dirty shade of yellow under the wing. The upper mandibula is much longer than the lower, overhanging it considerably. This it uses to grub up roots and other things on which it lives. These cockatoos were very numerous on the Murray, and are altogether distinct from the genus to which I have compared them; but their note is very similar, and, excepting to a naturalist, the difference is difficult to observe. The skin round the eye of both species is much larger than the cere round that of the common cockatoo.

84. CALYPTORHYNCHUS FUNEREUS?—Black Cockatoo
This fine bird was widely distributed over the brushy land of the interior, but was never seen in any considerable numbers. Its plumage is black, and the broad feathers in the tail are of a light yellow underneath. There is a supposition that when these cockatoos fly across the country uttering their hoarse note, it is a prelude to rain; but unfortunately I can bear testimony to the contrary, having often seen them so fly over my head when I would have given my right arm for water. I am not aware that the Black Cockatoo will survive captivity, I believe they always pine and die.

85. POLYTELIS MELANURA—Black-tailed Parroquet.
The Murray Parrot, with a bright yellow body and neck, the feathers at the back of the neck having a greener tinge. The long feathers of the wing are of a blue black, as also the tail, but in the wings there are three or four desultory red feathers. This bird visits the valley of the Murray in great numbers in the summer months, where its young are taken in great numbers, and easily tamed in cages. I was unable to make out where this bird comes from, or the point to which it migrates. Their place of abode during the winter is entirely unknown. It is a beautiful and a showy bird, making a noise something like the Green Leak, and was first shot by me on my return up the river, in 1836.

86. PLATYCERCUS BARNARDII, VIG. AND HORSF. Barnard's Parroquet.
This fine bird is found in the Murray Belt as well as in other localities, and is thence termed the parrot of the Murray Belt. It is one of the most beautiful of the parrot tribe, has a generally blue-green plumage on the back and neck, with a yellow crescent on the breast, and a purple below. This family are all distinguished by having long tails.

87. PLATYCERCUS ADELAIDIÆ, GOULD—The Adelaide Parroquet.
This fine and beautiful bird is common in South Australia, where it usurps the place of the Lory (*Platycercus penantii*) in New South Wales, and does equal mischief to the stack-yard. Its general plumage is yellow, but it has a dull red head, and blue cheeks. Its wings and tail, which is very long, are also blue, the longer feathers being almost black. Its back is marked with black scollops, and in size exceeds many of the Platycerci.

88. PSEPHOTUS HÆMATOGASTER, GOULD—The Crimson-bellied Parroquet.
This Parroquet is a bird of the interior, and was spread over the whole of it in greater or less numbers. Always numerous where box-trees were growing in the

vicinity of water. The *Psephotus hæmatogaster* is essentially a bird of the central parts of Australia, or else its range is confined between the 24th and 30th parallels of latitude. It is not a bird of bright plumage; it is distinguished by a bright crimson belly. It has likewise feathers of a peculiar bronze and yellow on the wings; the rest of the plumage being a dull blue green, excepting that over the bill it has some light blue feathers.

89. PSEPHOTUS HÆMATONOTUS, GOULD—Red-rumped Parroquet.

This is a bird of the interior, and was found on the most distant creeks, amongst the gum-trees. It was, however, fond of being on the ground, from whence it would rise and hide itself on being alarmed. It is a wild bird, and a noisy one. Its colours are generally dull.

90. EUPHEMA ELEGANS—Grass Parroquet.

This beautiful *Euphema*, is seen in great numbers on the sea-skirts of the plains of Adelaide, feeding on grass seeds. It was in course of migration when we were at the Depôt in lat. 29°40'; but after the other birds, and remained stationary for some time. It was never seen by us in the day time, but came regularly to water night and morning, when it was so dark that they could hardly be seen. The plumage of this bird is very beautiful. Its back and neck are green, as well as the crown of the head; its wings blue black; the breast and under tail feathers are of a bright yellow, with a blue and yellow band in the front.

91. EUPHEMA BOURKII—Bourke's Parroquet.

This elegant little bird was also a visitant at the Depôt, and remained throughout the winter; keeping in the day time in the barren brushes behind the camp, and coming only to water. The approach of this little bird was intimated by a sharp cutting noise in passing rapidly through the air, when it was so dark that no object could be seen distinctly; and they frequently struck against the tent cords in consequence. This *Euphema* has a general dark plumage, but with a beautifully delicate rose-pink shade over the breast and head, by which it will always be distinguished.

92. MELOPSITTACUS UNDULATUS—Warbling Grass Parroquet.

Called "Bidgerigung" by the natives. This beautiful little *Euphema* visits South Australia about the end of August or the beginning of September, and remains until some time after the breeding season. It is perhaps the most numerous of the summer birds. I remember, in 1838, being at the head of St. Vincents Gulf, early in September, and seeing flights of these birds, and *Nymphicus Novæ-Holl.* following each other in numbers of from 50 to 100 along the coast line, like starlings following a line of coast. They came directly from the north, and all kept the same straight line, or in each other's wake. Both birds subsequently disperse over the province. The plumage of this bird in a bright yellow, scolloped black, and three or four beautiful deep blue spots over each side the cheek.

93. NYMPHICUS NOVÆ-HOLLANDIÆ—The Crested Parroquet.

One of the most graceful of the parrot tribe, coming in, and I have stated above, with the *Melopsittacus,* and remaining during the summer. The general plumage is grey, with a white band across the wings. It has also a sulphur-yellow patch on the cheek, in the centre of which is one of scarlet. It has also a long, hairy crest, which it keeps generally erected. Both birds passed the Depôt in migrating, and Nymphicus was the last bird we saw to the north of the Stony Desert, in lat. 24½°and long. 138° on its return to the province in September.

94. TRICHOLOSSUS PORPHYROCEPHALUS, DIET—Porphyry-crowned Parroquet.

This pretty bird has a green plumage, but is distinguished by a deep blue patch on the crown of the head; from which it derives its name.

95. PEZOPORUS FORMOSUS—The Ground Parrot.

This bird was only twice seen in the interior, but on both occasions in the same scrubby and salty country it is known to frequent in New South Wales and other places. A specimen was shot by Mr. Stuart, in the bed of a salt lagoon in 26½° of latitude, and 141° of longitude, but none of these birds were seen to the west of that point. It has dark green plumage mottled with black, and has a patch of dull red over the bill.

96. PHAPS CHALCOPTERA—Common Bronze-wing.

This fine pigeon, so well known in the located parts of the continent, was also generally spread over the interior. Its habits are peculiar, insomuch that it goes to water at so late an hour that it is almost impossible to see them. They were rather numerous at the Depôt, but very few were shot there. In the more distant interior, when we should frequently have been glad of one of these birds to give a relish to our monotonous diet, they were equally as difficult to be shot, and although we sat at the edge of any pond near which we happened to be, and watched with noiseless anxiety, they would get to the water, and the sharp flap of their wings in rising, alone told us we had missed our game. The natives of the Murray set nets across any gully down which they fly to water on the banks of the Murray, and so catch them in great numbers. The Bronze-wing is strong in his flight, and is a plump bird, and capital to eat. Its general colour is brown lightly mottled, it has a dirty-white crown, and the wing feathers are a beautiful bronze.

97. PHAPS ELEGANS—Small Brush Bronze-wing.

This is much smaller than the above, and not so common. It inhabits close brushes, and is flushed like a woodcock, there seldom being more than two together. Its plumage is darker than Phaps chalcoptera, nor is there any white about it except on the crown of the head, the secondary wing feathers being of a bronze colour, without any shade of blue and green, so prominent in the first described of these birds.

98. PHAPS HISTRIONICA, GOULD—The Harlequin Bronze-wing.

This beautiful pigeon is an inhabitant of the interior. Its range was between the parallels of 31½° and 26°, but it was never seen to the south of Stanley's Barrier Range, if I except a solitary wanderer on the banks of the Murray. These birds lay their eggs in February, depositing them under any low bush in the middle of open plains. In the end of March and the beginning of April, they collect in large flats and live on the seed of the rice-grass, which the natives also collect for food. During the short period this harvest lasts, the flavour of these pigeons is most delicious, but at other times it is indifferent. They feed on the open plains, and come to water at sunset, but like the Bronze-wing only wet the bill. It is astonishing indeed that so small a quantity as a bare mouthful should be sufficient to quench their thirst in the burning deserts they inhabit. They left us in the beginning of May, and I think migrated to the N.E., for the farther we went to the westward the fewer did we see of them. This bird has a white and black head, the crown being white, and its back is a rusty brown, the long feathers of the wings of a slate colour, with a white spot at the end of each as well as at the end of the tail feathers; the belly being a beautiful deep slate colour.

99. GEOPHAPS PLUMIFERA, GOULD.

It was on the return of the party from the eastern extremity of Cooper's Creek, that we first saw and procured specimens of this beautiful little bird. Its locality was entirely confined to about thirty miles along the banks of that creek, and it was generally noticed perched on some rock fully exposed to the sun's rays, and evidently taking a pleasure in basking in the tremendous heat. It was very wild and took wing on hearing the least noise, but its flight was short and rapid like that of a quail, which bird it resembles in many of its habits. In the afternoon this little pigeon was seen running in the grass on the creek side, and could hardly be distinguished from a quail. It never perched on the trees, but when it dropped after rising from the ground, could seldom be flushed again, but ran with such speed through the grass as to elude our search. The *Geophaps plumifera* was found, I believe, in considerable numbers on the Lind and the Burdekin by Doctor Leichhardt, during his journey from Moreton Bay to Port Essington.

100. OCYPHAPS LOPHOTES—Crested Pigeon of the Marshes.

The locality of this beautiful pigeon is always near water. It is a bird of the depressed interior, never ascending to higher land where there are extensive marshes covered with the polygonum geranium. In river valleys, on the flats of which the same bramble grows, the Ocyphaps lophotes is sure to be found. It was first seen by me on the banks of the Macquarie, in lat. 31° during my expedition to the Darling, but there is no part of the interior over which I have subsequently travelled where it is not, and it is very evident that its range is right across the continent from north to south. The general colour of this bird is a light purple or slate colour, and its form and plumage are both much more delicate than that of the Bronze-wing, but it is by no means so fine a bird, its flesh being neither tender nor well-flavoured.

This bird is figured in my former work, page 79, vol. I. It has a crest, and is marked on the back and wings very similar to *Geophaps plumiferus*. This bird builds in low shrubs in exposed situations, and lays two eggs on so few twigs that it is only surprising how they remain together.

101. GEOPELIA CUNEATA—Speckled Dove.

All that we read or imagine of the softness and innocence of the dove is realised in this beautiful and delicate little bird. It is very small and has a general purple plumage approaching to lilac. It has a bright red skin round the eyes, the iris being also red, and its wings are speckled over with delicate white spots. This sweet bird is common on the Murray and the Darling, and was met with in various parts of the interior, but I do not think that it migrates to the N.W. Two remained with us at the Depôt in latitude 39°40', longitude 142°, during a greater part of the winter, and on one occasion roosted on my tent ropes near a fire. The note of this dove is exceedingly plaintive, and is softer, but much resembles the coo of the turtle-dove.

102. GEOPELIA TRANQUILLA, GOULD—Ventriloquist Dove.

This bird, somewhat larger than the preceding, is not by any means so delicate in appearance. The colour of its plumage is similar in some respects, but has close black scollops on the breast and neck without any spots on the wings. This bird also frequents the banks of the Darling and the Murray, but is not so common as *Geopelia cuneata*. I first heard it on the marshes of the Macquarie, but could not see it. The fact is that it has the power of throwing its voice to a distance, and I mistook it for some time for the note of a large bird on the plains, and sent a man more than once with a gun to shoot it, without success. At last, as Mr. Hume and I were one day sitting under a tree on the Bogan creek, between the Macquarie and the Darling, we heard the note, and I sent my man Fraser to try once more if he could discover what bird it was, when on looking up into the tree under which we were sitting we saw one of these little doves, and ascertained from the movement of its throat that the sound proceeded from it, although it still fell on our ears as if it had been some large bird upon the plain. I have therefore taken upon me to call it the "Ventriloquist."

103. PEDIONOMUS TORQUATUS, GOULD—The plain Wanderer.

This singular bird, in plumage and habit so like the Quail, was first discovered on the plains of Adelaide by Mr. Gould, where it appeared in considerable numbers in the year 1839-40. It was afterwards procured by a persevering collector in that colony, Mr. Strange, who is now in Sydney. Although in many respects resembling a Quail, this bird has long legs like a Bustard, but has a hind toe which that bird has not. We fell in with several in the N.W. interior, but they were all solitary birds. How far therefore we might conclude that they migrate northwards may be doubtful, although, it is impossible to suppose they would proceed in any other direction. The *Pedionomus* is a stupid little bird, and is more frequently caught by the dog than shot. Its general colour is a light brown, speckled with black like a quail. Its neck is white, spotted thickly with black, and has a white iris.

104. HEMIPODIUS VARIUS—Varied Quail.

This bird is the prettiest of its tribe, and is very common in many of the located parts of south-eastern Australia, but is not a bird of the interior, and was not observed beyond the flats of the Darling, where it was occasionally flushed from amongst the long grass.

105. COTURNIX PECTORALIS—Quail.

This bird is very common on the better description of plains in South Australia, and two or three specimens were shot during the early progress of the Expedition, but it was not seen to the north of Stanley's Range. It is to be observed, indeed, that few quails of any kind were seen in the interior. This variety is a very pretty bird, with bright brown plumage, mottled like that of the ordinary quail, and is characterized by a black spot on the breast.

106. SYNOÏCUS AUSTRALIS—Swamp Quail, or Partridge.

Synoïcus Australis is a smaller bird than those just described, but the colour of the plumage is much the same. It is generally found in marshes, or marshy ground, and frequently in bevies.

107. SYNOÏCUS CHINENSIS.

This beautiful little quail is generally found in marshes, or in high rushy ground. It is not a common bird. In size this quail is not larger than a young guinea fowl that has just broken the shell. It has dark plumage on the back and head—a deep purple breast and belly, and a white horse-shoe on the upper part of the neck. The female has general dark plumage, speckled black.

108. DROMAIUS NOVÆ-HOLLANDIÆ—The Emu.

This noble bird ranges over the whole of the continent, although we did not see any to the north of the Stony Desert. A good many were killed by the dogs at Fort Grey. They travel many miles during a single night to water, as was proved by a pack of thirteen coming down to the Depôt Creek to drink, that we had seen the evening before more than 12 miles to the north. Those we saw in the distant interior did not differ from the common emu.

109. OTIS AUSTRALASIANUS—The Bustard.

This fine and erectly walking bird is also common over the whole of the interior, migrating from the north in September and October. Several flights of these birds were seen by us thus migrating southwards in August, passing over our heads at a considerable elevation, as if they intended to belong on the wing. I have known this Otis weigh 28lbs. Its flesh is dark and varied in shade. The flavour is game and the meat is tender.

110. LOBIVANELLUS LOBATUS—The wattled Peewit.

This bird is most abundant over all south-east Australia, on plains, marshes, and rivers, its cry and flight are very like that of our Peewit at home, and it adopts the same stratagem to draw the fowler from its young. It is a pretty bird, with bright yellow eye and a singular wattle coming from the bill along the cheek. It is also remarkable for a spur on the shoulder which it uses with much force in fighting with any crow or hawk.

111. EDICNEMUS GRALLARIUS—The southern stone Plover.

There are few parts in the located districts of Australia in which this bird is not to be found. Its peculiar and melancholy cry, ran through the silence of the desert itself, and wherever rocks occurred near water they were also seen but not in any number. We caught a fine young bird at Flood's Creek, but as it was impossible to keep it, we let it go. This bird very much resemble the stone Plover of England, but there are some slight differences of plumage.

112. SARCIOPHORUS PECTORALIS—Black-breasted Dottrel.

This bird is remarkable for a small red wattle protruding from the bill, with a grey back and wings. It takes its name from its black breast.

113. EUDROMIAS AUSTRALIS, GOULD—Aust. Dottrell.

This singular bird like several others of different genera, made its appearance in 1841 suddenly on the plains of Adelaide, seeming to have come from the north. It occupied the sand hills at the edge of the Mangrove swamps and fed round the puddles of water on the plains. This bird afforded my friend Mr. Torrens, an abundant harvest, as they were numerous round his house, but although some few have visited South Australia every year, they have never appeared in such numbers as on the first occasion. The plumage is a reddish brown, with a dark horseshoe on the breast. It has a full eye, and runs very fast along the ground, Mr. Browne and I met or rather crossed several flights of these birds in August of 1845, going south. They were in very large open plains and were very wild.

114. HIATICULA NIGRIFRONS—Black-fronted Dottrell.

Much smaller than the preceding. A pretty little bird with a plaintive note, generally seen in pairs on the edge of muddy lagoons. Its plumage is a mixture of black, white, and brown, the first colour predominating on the head and breast. It runs with great swiftness, but delights more in flying from one side of a pond to the other.

115. CHLADORHYNCHUS PECTORALIS—The Banded Stilt.

This singular bird, with legs so admirably adapted by their length for wading into the shallow lakes and sheets of water, near which it is found, is in large flocks in the interior. It was in great numbers on Lepson's Lake to the northward of Cooper's Creek, and on Strzelecki's Creek was sitting on the water with other wild fowl making a singular plaintive whistle. It is semi-palmated, has black wings, and a band of brown on the breast, but it is otherwise white. Its bill is long, straight and slender, and its legs are naked for more than an inch and half above the knee.

116. HIMANTOPUS LEUCOCEPHALUS, GOULD—The white-headed Stilt.

The present bird is about the size of *Chladorhynchus pectoralis*, and in plumage is nearly the same. This bird was not found in the distant interior but in the shallow basin and round the salt lagoons of Lake Torrens.

117. SCHŒNICLUS AUSTRALIS—Australian Sand-piper.

A bird very much resembling the British Dunlin. General plumage, grey with a white breast. A quick runner, and fond of low damp situations as well as open plains. Common on the banks of all rivers and lagoons.

118. SCOLOPAX AUSTRALIS, LATH.—Snipe.

Considerably larger than the Snipe of England. Common in South Australia but very scarce in the interior. In the valley of the Mypunga there are great numbers of snipe which build there, but it is only in such localities, where the ground is constantly soft that they are to be found. Their flesh is delicate and their flavour good.

119. RHYNCÆA AUSTRALIS, GOULD—Painted Snipe.

This beautiful bird was also very scarce in the interior, having been seen only on one occasion. It is not a common bird indeed anywhere. Some three or four couple visit my residence at Grange yearly, and remain in the high reeds at the bottom of the creek. As they are with us during the summer they doubtless build, but we never found one of their nests. They lay basking in the shade of a tree on the sand hills during the day, and separate when alarmed. It is full as large as *Scolopax Australis,* but its plumage is black-banded on the back with a general shade of green. Its head is black and brown. It has a black horse-shoe on the breast, the belly being white, and the quill feathers are grey with a small brown spot on each.

120. GRUS AUSTRALASIANUS—Crane, or Native Companion.

This large sized Crane is common near the waters of the interior, but he is a wary bird, and seldom lets the fowler within shot. When seen in companies they often stand in a row, as they fly in a line like wild fowl. Their general plumage is slate colour, but they have a red ceres or skin on the head. One of these birds was tame in the Government domain at Paramatta in 1829, and a goose used daily to visit it and remain with it for many hours. I have frequently seen them together, and the goose has allowed me to approach quite close before he flew. At last I suppose the poor bird was shot, as he suddenly ceased to visit his friend, and the Native Companion died some little time afterwards.

121. HERODIAS SERMATOPHORUS, GOULD—White Heron.

This beautiful Heron is common all over the inhabited parts of the Australian Continent, and is seen at a great distance in consequence of its snow-white plumage. It was not however seen in the interior, although it was frequently seen on the line both of the Darling and the Murray.

122. NYCTICORAX CALEDONICUS—Nankeen Bird.

A Night Heron with a nankeen-coloured back and wings, and white breast, with a black crown to the head from which three long fine white feathers project. It is altogether a bird frequenting water, building in trees as the Heron does. It is about the size of a well grown young fowl, but is not good eating.

123. BOTAURUS AUSTRALIS—The Bittern.

Is well known with its dark brown mottled plumage and hoarse croaking note. These birds are very numerous in the reedy flats of the Murray, whence they call to one another like bull frogs. It is a higher bird than the above, with a ruff down the neck, which behind is naked. He has a fine bright eye, and darts with his bill with astonishing rapidity and force.

124. BOTAURUS FLAVICOLLIS—Spotted Bittern.

This bird was very numerous at the Depôt Creek, remaining during the day in the trees in the glen. There was, as the reader may recollect, a long sheet of water at the termination of the Depôt Creek distant about thirty miles. It was the habit of these birds to fly from the glen across the plains to this lower water, where they remained until dawn, when they announced their return to us by a croaking note as they approached the trees. They collected in the glen about the end of April, and left us, but, I am not certain to what quarter they passed, although I believe it was to the north-west, the direction taken by all the aquatic birds. This bird had a black body, and white neck with a light shade of yellow, and speckled black.

125. PORPHYRIO MELANOTUS, TEMM—The black-backed Porphyrio.

This bird is very common on the Murray, where birds of the same kinds have such extensive patches of reeds in which to hide themselves. Although dark on the back their general plumage is a fine blue, and their bills and legs are a deep red as well as the fleshy patch on the front of the crown. It was not seen by us to the westward of the Barrier Range, nor is it an inhabitant of any of the creeks we passed to the N.W. This Coote is of tolerable size, but is not fit to eat, its flesh being hard, and the taste strong.

126. TRIBONYX VENTRALIS, GOULD—The black-tailed Tribonyx.

This bird, like the *Eudromias Australis* or Australian Dottrel appeared suddenly in South Australia in 1840. It came by the successive creeks from the north, fresh flights coming up to push those which had preceded them on. It was moreover evident that they had been unaccustomed to the sight of man, as they dropped in great numbers in the streets and gardens of Adelaide, and ran about like fowls. At last they increased so much in numbers as to swarm on all the waters and creeks, doing an infinity of damage to the crops in the neighbourhood. They took the entire possession of the creek near my house, and broke down and wholly destroyed about an acre and a quarter of wheat as if cattle had bedded on it. These birds made their first appearance in November, and left us in the beginning of March, gradually retiring northwards as they had advanced.

The plumage of this bird is a dark dusky green, and it has a short black tail which it cocks up in running. Its bill is green and red, and it has all the motions and habits of a water rail, and although it has visited the province annually, since its first visit, it has never appeared in such vast numbers as on the first occasion.

The line on which this bird migrates seems to be due north. It was never seen at the Depôt or on any of the creeks to the west excepting Strzelecki's Creek, and a creek we crossed on our way to Lake Torrens, when on both occasions they were migrating southwards.

127. RALLUS PECTORALIS, CUVIEA—Water Rail.

This bird could hardly be distinguished from the English rail in shape and plumage. It is admirably adapted for making its way through reeds or grass, from its sharp breast. There are numbers of this rail on the Murray, but not many on the Darling; the natives can easily run it down. It was seen on two or three ponds in the interior and must have considerable powers of flight to wing its way from the one to the other as they successively dry up.

128. BERNICLA JUBATA—Mained Goose, wood Duck.

There are two varieties of this beautiful goose, one bird being considerably larger than the other, but precisely the same in plumage. In the colony they are called the wood duck, as they rest on logs and branches of trees, and are often in the depth of the forest. They have an exceedingly small bill characteristic of their genus, and a beautifully mottled neck and breast, the head and neck being a light brown. The smaller species is very common all over South-eastern Australia, but the larger bird is more rare. Three only were shot during the progress of the Expedition. Their range did not extend beyond 28°.

129. CYGNUS ATRATUS—The black Swan.

A description of this bird is here unnecessary. I may merely observe that the only swan seen on the waters of the interior was a solitary one on Cooper's Creek. They frequently passed over us at night during our stay at the Depôt, coming from and going to the N.W being more frequently on the wing when the moon was shining bright than at any other time.

130. CASARCA TADORNOÏDES—Chesnut-coloured Sheldrake.

This beautiful duck, the pride of Australian waters, is a bird of the finest plumage. He is called the Mountain Duck by the settlers, and may be more common in the hills than the low country, since he is seldom found in the latter district. This bird builds in a tree, and when the young are hatched, the male bird carries them in his bill down to the ground. Strange, whose name I have already mentioned, had an opportunity to watch two birds that had a brood of young in the hollow of a lofty tree on the Gawler; and after the male bird had deposited his charge, he went and secured the young, five in number, which he brought to me at Adelaide, but I could not, with every care, keep them alive more than a month. This bird is very large as a duck; his head and neck are a fine green colour, and he has a white ring round his neck, as also a white band across his wings. It is not a good eating bird, however, as is often the case with the birds of finer plumage.

131. ANAS SUPERCILIOSA, GMEL—The Wild Duck.

Unlike the preceding, this bird is one of the finest eating birds of Australia, being the wild duck of that continent. It is a fine bird in point of size, but cannot boast the plumage of our mallard. It is a bird of dark, almost black plumage, with a few glossy, green, secondary feathers, characteristic of the genus. It is spread over the whole of the interior, even to the north of the Stony Desert, but was there very wild, and kept out of our reach.

132. SPATULA RHYNCHOTIS—Australian Shoveller.

Not quite so large as the wild duck, but extremely good eating. This bird is not common in the interior., and was only seen once or twice amongst other ducks. Its plumage is a dark brown, and it has a light dull blue band across the wing. It takes its name from its peculiar bill, and may be termed the Shoveller of Australia. The specimens we procured in the interior are precisely the same as those of the southern coast of the continent.

133. MALACORHYNCHUS MEMBRANACEUS—Membrane Duck.

A beautiful duck, of delicate plumage, but little fit for the table. It is very common on most of the Australian creeks and streams, and is called the Whistling Duck. This duck is rather larger than our teal. It has a grey head, with a brown tinge, and is mottled in the breast something like the woodcock. Its eye is dark and clear, and it has a line of rose-pink running longitudinally behind it.

134. ANAS PUNCTATA, CUV.—Common Teal

Somewhat larger than the English Teal, and equally good for the table. The plumage of this little bird is dark, like that of the wild duck, from which, in this respect, it hardly differs. It is the most numerous of the water birds of the interior, and was sure to be in greater or less numbers on any extensive waters we found. A pair had a brood on one of the ponds in the Depôt Glen; but the whole were taken off by a kite, Milvus affinis, that watched them land and then flew at them.

So long as they kept in the water they were safe, but on land soon fell a prey to the kite.

135. LEPTOTARSIS EYTONI, GOULD—Eyton's Duck.

This new and fine bird was first shot on Strzelecki's Creek by Mr. Browne; and was subsequently seen by me in considerable numbers on Cooper's Creek. Its range was not to the westward, nor was it seen north of the Stony Desert. I believe I am wrong in stating that the first was killed at the place above mentioned; for, if my memory does not deceive me, we had already secured a specimen at the Depôt. In its general plumage it is of a light brown, with a mottled breast and neck. It has long white feathers crossing the thighs, with a fine black line along them, and altogether it is a handsome bird. Under ordinary circumstances we might have fared well on this duck at Cooper's Creek; but it was so wild as to keep out of our reach, being evidently hunted by the natives of the creek.

136. BIZIURA LOBATA—Musk Duck.

This ugly bird was common on the Murray, and was seen by me in hundreds on Lake Victoria; but it is seldom seen on the Darling—never to the westward of Stanley's Range. It is an Oxford grey in colour, with a light shade of brown; he flaps only, not being able to do more than skull along the top of the water. It trusts therefore for its safety to diving; and is so quick as to be shot with difficulty. The peculiarities of this bird are twofold: first its strong, musky smell, and secondly the large appendage the male bird has attached to the under part of the bill.

137. XEMA JAMESONII—Jameson's Gull.

This bird was seen only on Cooper's Creek in lat. 27° long. 102°; where three or four were sitting on some rocks in the middle of the water, and far out of gun's reach. They appeared to be similar to the English gull, with a slate-coloured back and wings, and white breast. On firing a shot, they rose and followed the ducks which rose at the same time up the creek, and when flying they seemed exactly to resemble the common gull. The only swan we saw was on this sheet of water, with eight or ten cormorants.

138. HYDROCHEDIDON FLUVIATILIS, GOULD—The Marsh Tern.

The only specimen seen during the Expedition, was shot by Mr. Stuart on Strzelecki's Creek. It was flying up and down the creek, plunging into the water every now and then. This light and airy bird had a slate-coloured back, with black neck and breast; the crown of the head was black, delicately spotted white.

139. PHALACROCORAX SULCIROSTRUS—Groove-billed Cormorant.

Of a fine dark glossy green plumage; common on all the creeks and rivers of the interior. These birds were very numerous at the Depôt, and were constantly coming in from, and flying to the N.W. But although we afterwards penetrated some hundreds of miles in that direction, we never discovered any waters to which they might have gone.

140. PELECANUS CONSPICILLATUS, TEMM—The Pelican.

Like the swans, these birds frequently passed over us, coming from, and going to that point to which all the aquatic, as well as many of the ordinary birds winged their way. We sometimes saw them low down, sweeping over the ground in circles, as if they had just risen from the water; but in neither instance could such have been the case. On several occasions we might have shot them, but they were useless, and would have encumbered us much.

141. PODICEPS GULARIS—Grebe.

The common Diver; frequenting the pools and rivers of the interior: of dark brown plumage and silver-white belly. There are two or three varieties of this bird, that I have seen on other occasions; but none, with the exception of the present specimen, during the recent Expedition.

No. I.
LIST OF SPECIMENS, AND THE NAMES OF THE VARIOUS ROCKS,

Collected during the Expedition.

1-4: Tertiary Fossil, or limestone, (opalescent) from above the fossil cliffs.
5 Ferruginous sandstone.
6 Soapstone, apparently a recent deposit.
7 Gneiss.
8 Hornstone, a variety Of.
9 Specular iron ore, lamellar with quartz.
10 Granite, with mammillary hematite—hornstone.
11 Specular iron ore, and iron ore highly magnetic.
12 Granite, white, a variety of.
13 Soapstone or clay, schorl, and slate with mica and chlorite.
14 Gneiss, a variety.
15 Granite, grey, both fine and coarse.
16 Granite, white, fine grained.
17 Hornstone, and mica slate (waved).
18 Clay.
19 Magnesian limestone, and limestone slaty and impure.
20 White conglomerate rock, appearing a binary granite.
21 Indurated clay.
22 Silicious pebbles.
23 Silicious rock, with veins of quartz.
24 Silicious rock.
25 Rock composed principally of silica and alumen forming sandstone.
26 Milky quartz.
27 Rounded balls, composed of sand and clay, cemented
28 by oxide of iron; hollow, but without crystals; rounded by the action of water.
29 Hornstone.
30 Granite, grey, a variety.
31 Ferruginous sandstone.
32 Silicious rock, with veins of quartz.
33 Mica slate.
34 Quartz, indurated with red veins.
35 Silicious rock, dusky.
36 Silicious rock, white.
37 Gypsum, or sulphate of lime.

38 Quartz veins from slate; trap rock, containing horn-blende and feldspar; limestone, recent, with clay and slate imbedded.
39 Impure and slaty limestone; hornslate, a variety.
40 Hemætite, a silicious oxide of iron; quartz veins in slate; silicious rock; chalcedony; sandy clay.
41 Indurated and dusky quartz.
42 Quartz, a hard, fine-grained dusky variety.
43 Ditto.
44 Silicious rock, appearing a knob, from a slate formation.
45 Limestone (fibrous).
46 Silicious rock.
47 Horn slate.
48 Silicicrus rock; iron-stone pebbles.
49 Hornstone.
50 Quartz.
51 Quartz.
52 Trap rock.
53 Quartz.
54 Hornstone.
55 White rock.
56 White sandstone.
57 Sandstone.
58 Sandstone.
59 Silicious oxide of iron.
60 Gypsum.

It will be seen, by an inspection of the map, that there is a large interval of low depressed country, between Stanley's and Grey's Ranges. The rock formation on the latter being almost exclusively of one kind. Beyond Grey's Range, no elevation in the interior, on the N.W. line traversed by the Expedition, was seen; but on the Stony Desert the fragments of rock, with which it was covered, were composed of indurated quartz, rounded by attrition, and coated with oxide of iron. North of the Stony Desert, sandstone occurred in the bed of Eyre's Creek, and milky quartz cropped out of the ground, in lat. 25°35' and in lon. 138°39'. The valley of Cooper's Creek was., however, bounded in by low quartzose hills, covered with sand. The general level of the interior was otherwise ferruginous clay, on which the long sandy doones or ridges rested, excepting where their regularity was broken by flooded plains. The clay rested on sandstone, which, with a few exceptions, where fossil tertiary limestone occurred, similar to that of the Murray cliffs, was ferruginous sandstone, at the depth of two feet and a half or three feet.

No. II.
LOCALITIES
OF THE DIFFERENT GEOLOGICAL SPECIMENS,

Collected by the Central Australian Expedition.

1-4: From the cliffs of the Murray River, both above and below the great north-west bend, bounding the valley of that river, with an average height of 150 or 200 feet.
5 From the sandstone hills on the Murray.
6 From Carnapaga, on the first creek to the N.W. of the Darling River.
7 From station No.3, on the Barrier or Stanley's Range, Mount Darling.
8 From the Glen of Yancowinna.
9 From the Iron Ridge, south of the Glen of Yancowinna.
10 From Mount Bourke, on Stanley's Range, No.1 station.
11 From the Iron Stone Hill on the Range (Piesse's Knob).
12 From a central hill on the Range.
13 From a central hill.
14 From Lewis's hill.
15 From the Black Hill Mount Robe.
16 From a valley in the Range.
17 From the bed of the Creek.
18 From the Rocky Glen.
19 From the outer Range to the westward of the Barrier, Station No.1.
20 From the same, Station No.2.
21 From the Stony Creek.
22 Gathered from the plains between the creeks to the west of the Ranges.
23 From a distant hill in Stanley's Range—the base.
24 From the summit of the same.
25 From a rugged detached hill.
26 From a small hill near the Range.
27 From the nearer plains.
28 Ditto.
29 From a water-worn hill near Flood's Creek,
30 From Station No.38, Mount Wood.
31 From the summit of the Range, Station No.39.
32 From Station No.40. Mount Lyell, fifty miles east.
33 From some low hills, near Flood's Creek.
34 From the last hill on Stanley's Barrier Range.
35 From the Magnetic Hill, Mount Arrowsmith.

36 From the Table Hill, Mount Browne.
37 From the White Hill.
38 From the Depôt Glen.
39 From the Black Hill, Mount Robe.
40 Ditto.
41 From the summit of Grey's Range.
42 From the last hill to the north, lat. 28°26'.
43 From the most distant hill to the north-east.
44-46 From the Depôt Glen.
47,48 From the Plains to the north of the Red Hill, Mount Poole.
49-54 From various parts of the Depôt Glen, and the Range with which it is connected. This Range is separated from the main ranges, but still occupies the eastern side of the high land, running between the eastern and western waters.
55 From the summit of the Red Hill, Mount Poole.
56,57 From the base of the same hill.
58 From the summit (2nd specimen).
59 From the plains north of the Depôt
60 from the plains.

BOTANICAL APPENDIX,

BY

ROBERT BROWN, Esq., D.C.L., F.R.S., F.L.S., &c.

MY FRIEND, Captain Sturt, having placed at my disposal the Collection of Plants formed in his recent Expedition into the Southern Interior of Australia, I am desirous of giving some account of the principal novelties it contains.

The collection consists of about one hundred species, to which might be added, if they could be accurately determined, many other plants, chiefly trees, slightly mentioned in the interesting narrative, which is about to appear, and to which the present account will form an Appendix. I may also observe, in reference to the limited number of species, that Captain Sturt and his companion, Mr. Brown, seem to have collected chiefly those plants that appeared to them new or striking, and of such the collection contains a considerable proportion.

In regard too to such forms as appear to constitute genera hitherto undescribed, it greatly exceeds the much more extensive herbarium, collected by Sir Thomas Mitchell in his last expedition, in which the only two plants proposed as in this respect new, belong to genera already well established, namely, Delabechia to Brachychiton, and Linschotenia to Dampiera.

In Captain Sturt's collection, I have been obliged, from the incomplete state of the specimens, to omit several species, probably new, from the following account, in which the plants noticed, chiefly new genera and species, are arranged according to the order of families in the Prodromus of De Candolle.

BLENNODIA.

Cruciferarum genus, prope Matthiolam.

CHAR. GEN.—Calyx clausus, foliolis lateralibus basi saccatis. Petala æqualia, laminis obovatis. Stamina: fillamentis edentulis. Ovarium lineare. Stylus brevissimus. Stigma bilobum dilatatum. Siliqua linearis valvis convexitisculls, stigmate coronata, polysperma, Semina aptera pube fibroso-mucosa tecta! Cotyledones incumbentes!

Herba (v. Suffrutex) erecta ramosa canescens, pube ramosa; foliis latolinearibus remotè dentatis; racemis termina. libus.

1. BLENNODIA canescens.

Loc. In arenosis depressis.

Desc. Sufruticosa, sesquipedalis, caule ramisque teretibus. Folia vix pollicaria paucidentata. Racemi multiflori, erecti, ebracteati. Flores albicantes. Calyx incanopubescens. Petalorum ungues calyce paulo longiores. Stainina 6, tetradynama, filamentis linearibus membranaceis apice sensim angustato.

Obs. This plant has entirely the habit, and in many important points the structure of Mattbiola, near which in a strictly natural method it must be placed; differing, however, in having incumbent cotyledons, and in the mucous covering, of its seeds. The mucus proceeds from short tubes covering the whole surface of the testa, each containing a spiral fibre, which seems to be distinct from the membrane of the tube. A structure essentially similar is known to occur generally in several families: to what extent or in what genera of Cruciferæ it may exist, I have not ascertained; it is not found, however, in those species of Matthiola which I have examined.

STURTIA.

Malvacearum genus, proximum Gossypio, affine etiam Senræ.

CHAR. Gen;.—Involucrum triphyllum integerrimum. Calyx 5-dentatus, sinubus rotundatis. Petala cuneato-obovata, basi inæquilatera. Columna staminum. polyandra. Ovaria 5, polysperma. Styli cohærentes. Stigmata distincta linearia. Pericarpia. . . Semina. . .

Suffrutex orgyalis glaber; foliis petiolatis obovatis integerrimis; floribus pedunculatis solitariis.

2. STURTIA Gossypioides.

Loc. In the beds of the creeks on the Barrier Range.—D. Sturt.

Desc. Suffirutex orgyalis glaber. Folia ramorum alterna, diametro unciali, trinervia; petiolo folium subæquanti, basi in stipulam subscariosam aduatam dilatato. Pedunculi vel. potius rami floriferi suboppositifolii nee verè axillares uniflori, juxta apicem folio nano petiolato stipulis 2 distinctis stipato instructi. Involucrum foliaceum venosurn, foliolis distinctis, cordatis, punctis nigricantibus, glandulosis conspersis. Calyx dentibus acutis, sinubus rotundatis. Petala sesquipollicaria, uti calycis tabus glanduloso-punctata glandulis nigricantibus semi-immersis, purpurea basibus atrò purpureis margine barbatis. Columna staminum e basi nuda super ad apicem usque antherifera: antheris reniformibus, loculis apice, confluentibus. Pollen hispidurn.

Obs. Sturtia is no doubt very nearly related to Gossypium, from which it differs in the entire and distinct leaves of its foliaceous involucrum, in the sharp teeth and broad rounded sinuses of the calyx, and possibly also in its fruit and seeds, which are, however, at present unknown. They agree in the texture and remarkable glands of the calyx, and in the structure of the columna staminum. Senra, which like Sturtia, has the foliola of its three-leaved involucrum distinct and entire, differs from it in having its calyx 5-fid with sharp sinuses, in the absence of glands, in the reduced number of stamina, and in its dispermous ovaria.

3. TRIBULUS (Hystriæ) lanatus, foliis 8-10-jugis, fructibus undique tectis spinis subulatis longitudine inæqualibus: majoribus sparsis longitudinem cocci superantibus.
Loc. In collinis arenosis. Lat. 26° D. Sturt.
Desc. Herba diffusa, sericea, incana. Folium majus cuiusque paris 8-10 jugum, foliolis ovatis. Flores magni, Calyx æstivatione leviter imbricatà. Petala calyce duplo longiora. Stamina decem, antheris linearibus.
Obs. I. A species nearly related to T. Hystrix, found on the west coast of Australia, or on some of its islands, in the voyage of the Beagle, may be distinguished by the following character. Tribulus (occidentalis) sericeo-lanatus, foliis suboctojugis, coceis undique densè armatis: spinis omnibus conico-subulatis longitudine invicem æqualibus. These two species differ from all others in the uniform shape of the spines, which equally cover the whole external surface of the fruit.
Obs. II. The American species of the Linnean genus Tribulus are distinguishable from the rest of the published species, by having ten monospermous cocci, by their persistent calyx, and the absence of glands subtending the 5 filaments opposite to the sepals.

This tribe was originally separated as a genus by Scopoli, under the name of Kallstrœmia, which has been recently adopted by Endlicher.

Another tribe exists in the intratropical part of the Australian continent, to which, nearly 40 years ago, in the Banksian Herbarium, I gave the generic name of Tribulopis, and which may readily be distinguished by the following characters.

TRIBULOPIS.

Calyx 5-partitus deciduus. Petala 5. Stamina decem (nunc 5.) Filamenta quinque, sepalis opposita, basi glandula stipata. Ovaria 5, monosperma. Cocci, præter tubercula 2 v. 4 baseos, læves.
Herbæ annuœ prostratœ; foliis omnibus alternis!
TRIBULOPIS (Solandri.) folfis bi-trijugis, foliolis subovatis inæquilateris, coecis basi quadrituberculatis.
Loc. In ora orientali intratropica Novæ Hollandiæ prope Endeavour River, anno 1770. D.D. Banks et Solander.
TRIBULOPIS (angustifolia), foliis 3-4 jugis (raro bijugis), foliolis linearibus, tuberculis baseos coccorum abbreviatis.
Loc. Ad fundum sinus Carpentarim annis 1802 et 3. R. Brown.
TRIBULOPIS (pentandra), folfis bijugis, foliolis oblongo-lanceolatis pari superiore

duplo majore, floribus pentandris, petalis lanceolatis.
Loc. In insulis juxta fundum sinus Carpentariæ anno 1803. R. Brown.

4. CROTALARIA (Sturtii) tomentosa, foliis simplicibus ovalibus utrinque sericeo-tomentosis, petiolis apice geniculatis, racemis terminalibus multifloris.
Loc. On the top of the ridges in pure sand, from S. Lat. 28° to 26° D. Sturt.
Desc. Frutex 2-3-pedalis (D. Sturt). Folia alterna, ovata passim ovalia, obtusa, sesquipollicem longa, utrinque velutina; petiolus, teres, basi vix crassiore apice curvato. Racemus terminalis; pedicellis approximatis calycem vix æquantibus apice bibracteatis. Flores sesquipollicares. Calyx 5-fidus; laciniis lanceato-linearibus acutis subæqualibus tubum paulo superantibus. Corolla sordè flava, calyce plus duplo major. Vexillum magnum, basi simplici nec auriculata, late ovatum, acutum. Alæ vexillo fere dimidio breviores, basi semicordata. Carina longitudine vexilli, acuminata, basi gibbosa, ibique aperta marginibus tomentosis. Stamina 10 diadelpha, simplex et novemfidum. Antheræ quinque majores lineares, juxta basin affixæ; quinque reliquæ ovatæ, linearibus triplo breviores, incumbentes. Ovarium lineare, multi-ovulatum. Stylus extra medium et præsertim latere interiore barbatum. Stigma obtusum. Legumen desideratur.
Obs. A species very nearly related to C. Sturtii, having flowers of nearly equal size, and of the same colour and proportion of parts, found in 1818, by Mr. Cunningham, on the north-west coast of Australia, and since in Captains Wickham and Stokes Voyage of the Beagle; may be distinguished by the following character:—Crotalaria (Cunninghamii) tomentosa, foliis simplicibus ovali-obovatis utrinque sericeo-tomentosis, petiolis apice curvatis, pedunculis axillaribus unifloris.

5. CLIANTHUS (Dampieri) herbaceus prostratus sericeo-villosissimus, foliolis oppositis (rarissime alternis) oblongis passim lineari-oblongis obovatisve, pedunculis erectis scapiformibus, floribus subumbellatis, calycibus 5-fidis sinubus acutis, ovariis (leguminibusque immaturis) sericeis.
Clianthus Oxleyi A. Cunningham in Hort. Soc. Transac. II. series, vol. 1. p. 522.
Donia speciosa Don, Gen. Syst. vol. 2. p. 468.
Clianthus Dampieri Cunningham, loc. cit.
Colutea Novæ Hollandiæ, &c. Woodward in Dampier's Voy. vol. 3. p. 111. tab. 4. f. 2.
Loc. In ascending the Barrier Range near the Darling, about 500 feet above the river. D. Sturt.
Obs. In July, 1817, Mr. Allan Cunningham, who accompanied Mr. Oxley in his first expedition into the Western Interior of New South Wales, found his Clianthus Oxleyi on the eastern shore of Regent's Lake, on the River Lachlan. The same plant

was observed on the Gawler Range, not far from the head of Spencer's Gulf by Mr. Eyre in 1839, and more recently by Captain Sturt, on his Barrier Range near the Darling. I have examined specimens from all these localities, and am satisfied that they belong to one and the same species.

In March (not May) 1818, Mr. Cunningham, who accompanied Captain King in his voyages of survey of the coasts of New Holland, found on one or the islands of Dampier's Archipelago, a plant which he then regarded as identical with that of Regent's Lake. This appears from the following passage of his MS. Journal:—
"I was not a little surprised to find Kennedya speciosa, (his original name for Clianthus Oxleyi), a plant discovered in July 1817, on sterile bleak open flats, near Regent's Lake, on the River Lachlan, in lat. 33°13'S. and long. 146°40'E. It is not common, I could see only three Plants, of which one was in flower. This island is the Isle Malus of the French." Mr. Cunningham was not then aware of the figure and description in Dampier above referred to, which, however, in his communication to the Horticultural Society in 1834, he quotes for the plant of the Isle Malus, then regarded by him as a distinct species from his Clianthus Oxleyi of the River Lachlan. To this opinion he was probably in part led by the article Donia or Clianthus, in Don's System of Gardening and Botany, vol. 2. p. 468, in which a third species of the genus is introduced, founded on a specimen in Mr. Lambert's Herbarium, said to have been discovered at Curlew River, by Captain King. This species, named Clianthus Dampieri by Cunningham, he characterises as having leaves of a slightly different form, but its principal distinction is in its having racemes instead of umbels; at the same time he confidently refers to Dampier's figure and description, both of which prove the flowers to be umbellate, as he describes those of his Clianthus Oxleyi to be. But as the flowers in this last plant are never strictly umbellate, and as I have met with specimens in which they are rather corymbose, I have no hesitation in referring Dampier's specimen, which many years ago I examined at Oxford, as well as Cunningham's, to Clianthus Dampieri. This specimen, however, cannot now be found in his Herbarium, as Mr. Heward, to whom he bequeathed his collections, informs me: nor can I trace Mr. Lambert's plant, his Herbarium having been dispersed.

Since the preceding observations were written, I have seen in Sir William Hooker's Herbarium, two specimens of a Clianthus, found by Mr. Bynoe, on the North-west coast of Australia, in the voyage of the Beagle. These specimens, I have no doubt, are identical with Dampier's plant, and they agree both in the form of leaves and in their subumbellate inflorescence with the plant of the Lachlan, Darling, and the Gawler Range. From the form of the half-ripe pods of one of these specimens, I am inclined to believe that this plant, at present referred to Clianthus will, when its ripe pods are known, prove to be sufficiently different from the original New Zealand species to form a distinct genus, to which, if such should be the case, the generic name Eremocharis may be given, as it is one of the greatest ornaments of the desert regions of the interior of Australia, as well as of the sterile islands of the North west coast.

CLIDANTHERA.

CHAR. GEN.-Calyx 5-fidus. Petala longitudine subæqualia. Stamina diadelpha: antheræ uniformes; loculis apice confluentibus, valvula contraria ab apice ad basin separanti dehiseentes! Ovarium monospermum. Stylus subulatus. Stigma obtusum. Legumen ovatum, lenticulari-compressum, echinatum.
Herba, v. Suffrutex, glabra, glandulosa; ramulis angulatis. Folia cum impari pinnata; foliolis oppositis, subtus glandulosis. Stipulm parvtv, basi petioli adnata.. Flores spicati, parvi, albicantes.
Obs. Subgenus forean Psoralete, cui habitu simile, foliis calycibusque pariter glandulosis; diversum dehiscentia ineolita antherarum!

6. CLIDANTHERA psoralioides.
Loc. Suffrutex bipedalis in paludosis. D. Sturt.
Desc. Herba, vel suffrutex, erecta, bipedalis, glabrinscula. Ramuli angulati. Folia mim impari pinnata, 4-5-juga; foliola opposita, lanceolata, subtus glandulis crebris parvis manifestis, marginibus acabris. Spicæ densæ, multifloræ. Calyx 5-fidus, parum inæqualis, acutus, extus glandulis dense conspersus. Corolla., Vexillum lamina oblonga subconduplicata nec explanata, basi simplici absque auriculie; ungue abbreviato. Alæ vexillo paulo breviores, carinam sequantes, laminis oblongis, auriculo baseos brevi. Caripetala alis conformes. Stamina diadelpha, simplex et novemlidum; antheræ subrotundæ v. reniformes, valvula ventrali antheræ dimidio minore subrotunda. Ovarium hispidum ovulo reniformi. Legumen basi calyce subemarcido cinctum, echinatum. Semen reniforme, absque strophiola; integumento duplici. Embryo viridis; cotyledones obovatæ, accumbentes.
Obs. This plant, which in some respects resembles certain species of Glycyrrhiza, appears to be not unfrequent in the southern interior. It was found in one of the early expeditions of Sir Thomas Mitchell, and Mrs. (Capt.) Grey, observed it on the flats of the Murray.

7. SWAINSONA (grandiflora) suffruticosa pubescens, folfis 8-10-jugis inexpansis incano-tomentosis; foliolis oblongis obtusis retusisve: adultis semiglabratis: rachi subincana, racemo, multifloro folium superante, bracteolis lanceato-linearibus acutis wquantibus tubum. calycis albo-lanati quinque fidi: laciniis acutissimis longitudine feré tubi, vexillo bicalloso.

Loc. Common on the rich alluvial flats of the Murray and Darling. D. Sturt.
Obs. This plant is, perhaps, not specifically distinct from S. Greyana Lindl. Bot. Regist. 1846, tab. 66, of which the figure is a good representation of S. grandiflora in every respect, except in the form and proportions of the teeth of the calyx and lateral bracteæ. In these points it exactly agrees with complete specimens, for which I am indebted to Mrs. Grey, from the banks of the Murray, and Mr. Eyre's station (Moorundi), about 98 miles from Adelaide, where it was first found in November, 1841. The following characters, if constant, will sufficiently distinguish it from S. grandiflora.

SWAINSONA (Greyana) suffruticosa pubescens, foliis 5-9-jugis inexpansis incano-tomentosis; foliolis oblongis obtusis retusisve: adultis semiglabratis: rachi subincana, racemis multifloris folio longioribus, bracteis lateralibus lanceato-linearibus brevioribus tubo calycis albo-lanati quinque-dentati: dentibus obtusiusculis tubo dimidio brevioribue, vexillo bicalloso.
In the second edition of Hortus Kewensis, (vol. 4. p. 326), I excluded from the generic character of Swainsona the calli of the vexillum, having observed two Australian species where they were wanting, but which in every other respect appeared to me referable to this genus; for the same reason I continue to introduce the calli, where they exist, into the specific characters, as was done in Hortus Kewensis, l. c. In the generic character of Swainsona, given in De Candolle's Prodromus, (vol. 2. p. 271), the calli of vexillum are transferred to the calyx; this can only be regarded as an oversight, which perhaps has been corrected by the author himself, and which, so far as I know, has never been adopted in any more recent work in which the generic character of Swainsona is given.

8. SWAINSONA? (laxa) glabra, caule ramoso, folfis 6-7-jugis; foliolis oblongo-ovalibus obtusis, racemis elongatis laxis, pedicellis calyce glabro quinquedentato brevioribus, bracteolis subulatis, vexillo ecalloso.
Loc. Statio nulla indicata, in Herb. D. Sturt.
Obs. There is something in the aspect of this plant not entirely agreeing with the other species of the genus; and as the fruit is unknown, and the flowers yellow, I refer it with a doubt to Swainsona.

PENTADYNAMIS.

CHAR. GEN.—Calyx 5-fidus subwqualis. Vexillum explanatum, callo baseos laminw in unguern decurrenti. Carina obtusa, basin versus gibba, longitudine alarum. Stamina diadelpha; aritheris 5 majoribus linearibus, reliquis ovatis. Ovarium polyspermum. Stylus e basi arcuata porreetus, postice barbatus. Legumen compressum.

Herba (Suffrutex sec. D. Sturt), bipedalis sericeo-incana; caule angulato erecto. Folia ternata; foliolis sessilibus, linearibus, obtusis. Flores racemosi, flavi.

9. PENTADYNAMIS incana.
Loc. On sand-hills with Crotalaria Sturtii. D. Sturt.
Desc. Herba ereeta, ramosa, sericeo-incana. Folia alterna, ternata; petiolo elongato, teretiusculo, foliolo terminali longiore viz unciali. Racemi multiflori, erecti; pedicelli subæquantes calycem. Bracteolæ subulatæ, infra apicem pedicelli, basin calycia attingentee. Calyx 5-fidus; lacinfis acutis tubum sequantibua. Corolla flava, calyce plus duplo longior. Vexillum explanatum, basi abisque auriculis sed callo in unguera decurrenti ibique barbato auctum. Carina infra medium gibba proreceptione baseos styli. Staminum antheræ majores lineares, basi vel juxta basin affixæ; 5 minoreo ovatæ, incumbentes. Ovanum lineare, pubescens. Stigma terminale, obtusum, Legumen immaturum incanum, stylo e basi arcuata porrecto terminatum, calyce subemarcido subtensum.
Obs. In the collection of the plants of his last expedition, presented to the British Museum by Sir Thomas Mitchell, there is a plant which seems to belong to the genus Pentadynamis, which is probably, therefore, one of the species of Vigna, described by Mr. Bentham.

10. CASSIA (Sturtii), tomentoso-incana, foliis 4-jugis foliolis lanceolato-linearibus planis: glandula depressa inter par infimum, racemo corymboso paucifloro cum pedunculo suo folium paulo superante v. æquante, calyce tomentoso.
Loc. In sandy brushes of the Western interior. D. Sturt.
Obs. Species proximo. C. artemisiæfoliæ De Cand. Prodr. quæ Cassia glaucescens Cunningh. MSS. 1817, cui foliola teretiuscula, et racemus corymbosus cum pedunculo suo folio brevior.

11. CASSIA (canaliculata), cinerascens pube tenuissima, folfis 2-jugis (raro 1-jugis) foliolis angustato-linearibus canaliculatis: glandula inter par inferius et dum unijuga inter terminale, calycibus glabriusculis, racemis corymbosis paucifloris folio brevioribus.
Loc. In the bed of the creeks of the Barrier Range, about thirty-six miles from the Darling, in lat. 32°S. D. Sturt.
Obs. Proximo C. eremophilæ Cunningh. MSS. quæ sequentibus notis a Cassia phyllodinea et C. zygophylla, Benth. facile distinguenda.
CASSIA (eremophila), glabra, folfis unijugio rarb passim bijugis foliolis linearibus canaliculatis latitudine racbeoa linearis aversw, corymbis paucifloris folio brevioribus.

Loc. In desertis prope fluvium Lachlan, anno 1817, detexit. D. Cunningham.
CASSIA (zygophylla), glabra foliis unijugis; foliolis linearibus planis rachi duplo latioribus, corymbis paucifloris folio brevioribus.
Cassia zygophylla, Benth. in Mitch. trop. Austr. p. 288.
Another species nearly related to C. zygophylla is readily distinguished by the following character.
CASSIA, (platypoda), glabra, folfis unijugis; foliolis linearibus apiculo recurvo duplo augustioribus rachi aversa lanceolato-lineari.
Loc. Juxta fluvium Murray, anno 1841, detexit Domina Grey.

12. CASSIA. (phyllodinea), canescens pube arctissimè adpressa, phyllodiis aphyllis linearibus planis falcatis aversis, calycibus glabris, legumine plano-compresso.
Loc. In Herbario D. Sturt specimen exstat nulla stationis aut loci indicatione, sed candem speciem ad fundum sinus Spencer's gulf dicti in sterilibus apricis anno 1802 legi.
Desc. Frutex quadripedalis, ramosissimus. Phyllodia semper aphylla, averse, linearia, acuta, basi attenuata, plus minusvè falcato-incurva, biuncialia, 1/6 circiter unciæ lata, exstipulata, paginis pube arctissime adpressa canescentibus, margine superiore glandula unica depressa obsolete. Flores flavi, in umbella axillari 2-3 flora.
Obs. Cassia phyllodinea is one of the very few species of the genus, which, like the far greater part of New Holland Acaciæ lose their compound leaves, and are reduced to the footstalk, or phyllodium, as it is then called, and which generally becomes foliaceous by vertical compression and dilatation. A manifest vertical compression takes place in this species of Cassia.

A second species, Cassia circinata of Benth. in Mitch. trop. Austr. p. 384, is equally reduced to its footstalk, but which is without manifest vertical compression. To this species may perhaps be referred Cassia linearis of Cunningham MS., discovered by him in 1817, but which appears to differ in having a single prominent gland about the middle of its phyllodium: Bentham's plant being entirely eglandular.

These two, or possibly three species, belong to the desert tracts of the South Australian interior. In the same regions we have another tribe of Cassiæ closely allied to the aphyllous species; they have only one pair of foliola which are caducous, and whose persistent foot-stalk is more or less vertically compressed. Along with these, and nearly related to them, are found several species of Cassia, having from two to four or five pairs of foliola which are narrow, but their footstalks are without vertical compression, and their foliola are caducous, chiefly in those, however, which have only two pairs.

PETALOSTYLIS.

Cæsalpinearum genus, Labicheæ proximum.

CHAR. GEN.—Calyx 5-phyllus, æqualis. Petala 5 subæqualia, patentia. Stamina: Filamenta quinque sepalis opposita, quorum tria antherifera, antheris basifixis linearibus, duo reliqua castrata. Ovarium oligospermum. Stylus maximus petaloideus, trilobus, lobo medio longiore axi incrassata desinente in stigma obtusum simplex!

Frutex glaber, erectus. Folia allerna, pinnata cum impari, foliolis alternis. Racemi axillares, paucflori. Flores flavi.

13. PETALOSTYLIS Labicheoides.

Loc. In the bed of a creek along with Sturtia. D. Sturt.

Obs. Eadem omnino species exstat inter plantas in Insulis Archipelagi Dampieri juxta oram, septentrio-occidentalem Novæ Hollandiæ in itinere navis Beagle dictæ lectas.

Desc. Frutex facie fere Cassiæ et Labicheæ. Folia alterna, cum impari pinnata, foliolis alternis brevissimè petiolatis oblongo-lanceolatis cum mucronulo terminali paulo majore. Stipulæ parvæ caduæ. Racemi pauciflori, axillares, folio breviores. Alabastrum, ovali-oblongum acutiusculum. Calyx viridis, sepalis subæqualibus oblongis acutis, æstivatione imbricatis. Petala quinque subacqualia, oblonga, flava æstivatione imbricata, sepalis sesquilongiora. Stamina 3 antherifera æqualia, filamentis abbreviatis, antheris acutis bilocularibus, loculis sulco longitudinali insculptis; 2 reliqua rudimenta parva subfiliformia. Ovarium sessile, lineare, 3-4-spermum. Stylus lobo medio, triplo longiore, oblongo-lanceolato, lobis lateralibus auriculiformibus semiovatis obtusis. Stigma imberbe.

Obs. The structure of the style, which forms the only important character of this genus, so far as the specimens enable me to judge, is so remarkable and peculiar, as to render it necessary to state, that I have found it quite uniform in all the flowers I have examined; namely, in four immediately before, and in three after expansion.

PODOCOMA.

CHAR. GEN.—Involucrum imbricatum, foliolis angustis acutis. Ligulæ pluriseriales, angustissimæ, femineæ. Flosculi pauciores hermaphrodito-masculi. Ligularum pappo capillari, stipitato, denticulato. Receptaculum epaleatum.

Herba humilis, setosa; caule densé, foliato; folia petiolata, cuneata, incisa, setis albis conspersa.

14. PODOCOMA cuneifolia.
Loc. In Herbario D. Sturt abeque ulla indicatione loci vel stationis.
Obs. This plant appears to be generically distinct from Erigeron, particularly in its stipitate pappus. The specimens, however, are so incomplete, that I am unable to determine whether what I have considered stem, may not be a branch only.

LEICHHARDTIA.

CHAR. GEN Calyx 5—partitus. Corolla urceolata; tubo intus imberbi, fauce annulo integerrimo inerassata. Corona staminea 5-phylla, foliolis antheris oppositis, iisque brevioribus, indivisis. Antheræ membrana (brevi) terminatæ. MassæPollinis ercetæ basi affixæ. Stigma vix divisurn.
Suffrutex volubilis; foliis linearibus, fascicularibus, extra-alaribus; falliculis ventricosis ovato-oblongis.

15. LEICHHARDTIA australis.
Doubah Mitchell, trop. Austr. p. 85.
Loc. Common on the Murray and in the interior. D. Sturt.
Desc. Suffrutex pubescens, subeinereus; ramis striationec omnino teretibus. Folia sesquipollicaria, linearia, acuta. Fasciculi multiflori. Calycia foliola obtusa, pube tenui cinerascentia. Corolla glabra; tubo absque squamulis denticuliave, ventricoso; limbo vix longitudine tubi, laciniis conniventibus sinistrorum imbricatis. Coronæ foliola e basi dilatata adnata linearia, indivisa. Massæ Pollinis (Pollinia) lineares.
Obs. Doubah was originally found by Sir T. Mitchell, but with fruit only, in one of his journeys, and also in his last expedition; and, according to him, the natives eat the seed-vessel entire, preferring it roasted. Captain Sturt, on the other hand, observes, that the natives of the districts where he found it, eat only the pulpy seed-vessel, rejecting the seeds.

16. JASMINUM lineare. Br. prodr. 1. p. 521.
Jasminum Mitchellii. Lindl. in Mitch. trop. Anstr. p. 365.
Obs. In Captain Sturt's collection there are perfect specimens of this plant, on which a few remarks may be here introduced, chiefly referring to its very general existence in the sterile regions of the interior of Southern Australia, and even extending to the north-west coast.

The species was established on specimens which I collected in 1802, in the sterile exposed tract at the head of Spencer's Gulf. With these I have compared and found identical Mr. A. Cunningham's specimens gathered in the vicinity of the Lachlan, in 1817; Captain Sturt's, in his earlier expeditions, from the Darling; those of Sir Thomas Mitchell, in his different journeys; and specimens collected in one of the islands of Dampier's Archipelago. In this great extent of range, it exactly agrees with a still more remarkable plant, and one much less likely to belong to a desert country, namely, Clianthus Dampieri.

I have considered Jasminum Mitchellii as hardly a variety of J. lineare, the character of this supposed species depending on its smooth leaves, and its axillary nearly sessile corymbi or fasciculi, which are much shorter than their subtending leaves; but even in the specimen contained in the collection presented to the British Museum by Sir Thomas Mitchell, the young branches, as well as the pedunculus and pedicelli, are covered with similar pubescence, and in the same degree as that of J. lineare; the specimens from Dampier's Archipelago have leaves equally smooth, but have the inflorescence of J. lineare and I have specimens of J. lineare in which, with the usual pubescence of that species, the inflorescence is that of Mitchellii. Among Sir Thos. Mitchell's collection at the Museum, there is a Jasminum not noticed by Professor Lindley, which, though very nearly related to J. lineare, and possibly a variety only, may be distinguished by the following character.

Jasminum (micranthum) cinereo-pubescens, foliis ternatis; foliolis lanceato-linearibus, pedunculis axillaribus 1-3 floris, corollæ laciniis obtusis dimidio tubi brevioribus.

17. GOODENIA (cycloptera) ramosissima pubescens, foliis radicalibus serrato-incisis; caulinis lanceolato-ellipticia obsoletè serratis in petiolam, attenuatis, pedunculis axillaribus unifloris folia subwquantibus, seminibus orbiculatis membrana angusta cinctis.

Loc. Indicatio nulla stationis in Herb. D. Sturt.

18. SCÆVOLA (depauperata), erecta ramosissima, ramis alternis; ultimis oppositis divaricatis, falfis minimis sublinearibus: ramorum alternis ramulorum oppositis, pedunculis e dichotomiis ramulorum solitariis unifloris.

Loc. In salt ground, in lat. 26°S. D. Sturt.

Desc. Herbacea, vix suffruticosa, adulta glabriuscula, erecta, ramosissima. Rami ramulique angulati; ultimi oppositi, indivish divaricati, apice diphylli, folfis minimis et rudimento minuto floris abortivi. Folia sessilia, linearia, scuts, brevissima, ramos subtendentia alterna, ramulos ultimos bmebiatos opposita. Pedunculi e dichotomiis ramulorum ultimorum penultimo-rumque eolitarii, uniflori, ebracteati. Calyx: limbo supero quinquepartito; laciniis lineari-lanceatis, æqualibus, pubescentibus. Corolla: tubo hine ad basin usque fisso, limbo unilabiato, 5-partito; laciniis lanceolatis, æqualibus, marginibus angustis induplicatis, extus uti tubus pubescentibus, intus glabris trinerviis, nervo medio venoso. Stamina: filamenta distincta, anguste linearia, glabra, axi incrassata; antheræ liberæ, lineares, imberbes, basi affiæ, loculis longitudinaliter dehiscentibus. Ovarium biloculare? loculis monospermis, ovulis erectis. Stylus cylindraceus, glaber. Stigmatis indusium margine ciliatum et extus pilis copiosis longis strictis acutis albis tectum v. cinctum.

19. EREMOPHILA (Cunninghamii) arborescens, foliis alternis linearibus mucronulo recurvo, sepalis fructûs unguiculatis eglandulosis, corolla extus glabra.
Eremophila? arborescens, Cunningh. MSS. 1817.
Eremodendron Cunninghami, De Cand. prodr. xi. p. 713. Delessert ic. select. vol. v. p. 43. tab. 100. (ubi error in num. ovulorum.)
Loc. In the sandy brushes of the low western interior, not beyond lat. 29°S. D. Sturt.
Obs. The genus Eremophila was founded on very unsatisfactory materials, namely, on two species, E. oppositifolia and alternifolia which I found growing in the same sandy desert at the head of Spencer's Gulf in 1802, the only combining character being the scariose calyx, which I inferred must have been enlarged after flowering. This, however, proves not to be the case in E. alternifolia, which Mrs. Grey has found in flower towards the head of St. Vincent's Gulf: and from analogy with other species since discovered, it probably takes place only in a slight degree in E. oppositifolia, whose expanded flowers have not yet been seen.

In 1817, Mr. Cunningham, in Oxley's first expedition, discovered a third and very remarkable species in flower and unripe fruit, which he referred, with a doubt, to Eremophila, and which M. Alphonse De Candolle has recently separated, but as it seems to me on very insufficient grounds, with the generic name of Eremodendron, established entirely on Mr. Cunningham's specimens. A fourth species has lately been described by Mr. Bentham, in Sir Thos. Mitchell's narrative of his Journey into Tropical Australia; and some account of a fifth is given in the following article.

These five species may be arranged in four sections, distinguished by the following characters:

α. Folia opposita; sepala unguiculata.
Eremophila oppositifolia. Br. prodr. 1. p. 5 18.
β. Folia alterna; sepala unguiculata, eglandulosa; antheræ exsertæ.
E. Cunninghamii.
δ. Folia alterna; sepala brevè unguiculata, eglandulosa; stamina inclusa.
Eremophila Mitchelli. Benth. in Mitch. trop. Austr. p. 31.
Eremophila Sturtii.
γ. Folia alterna sepala glanduloso-tuberculata, sepala cuneato-obovata, sessilia, glandulosa.
E. alternifolia. Br.prodr. i. p. 518.

This last species might he separated from Eremophila; it is not however referable to Stenochilus, with some of whose specles it nearly agrees in corolla, but from all of which it differs in its landular scariose calyx.

20. EREMOPHILA. (Sturtii), pubescens, foliis angustè linearibus apiculo recurvo, corollis extus pubescentibus limbo intus barbato, staminibus inclusis.
Loc. On the Darling; flowers purplish, sweet-scented. D. Sturt.
Desc. Frutex orgyalis (D. Sturt). Calyx 5-partitus, æqualis sepalis obovato-oblongis, basi angustioribits sed in unguem vix attenuatis, membranaceis, uninerviis, venosis. Corolla bilabiata, tubo amplo recto, labiis obtusis, extus pubescens, intus hine (inferius) barbata. Labium superius tripartitum; lobo medio bifido (e duobus conflato); laciniis omnibus obtusis; inferius obcordatum bilobum lobis rotundatis, densius barbatum. Stamina quatuor didynama, omnino inclusa. Filaments, glabra. Antheræ reniformes, loculis apice confluentibus. Ovarium densè lanatum. Stylus glaber. Stigma indivisum, apice styli vix crassius.
Obs. Species proxima E. Mitchelli Benth. in Mitch. Trop. Autr. p. 31.

21. STENOCHILUS longifolius. Br. prodr. i. p. 517;
Stenochilus pubifiorus. Benth. in Mitch. trop. Aust. p. 273.
Stenochilus salicinua. Benth. in Mitch. trop. Austr. p. 25 1.
Loc. Nulla stationis indicatio.

22. STENOCHILUS maculatus, Ker in Bot. Regist. tab. 647.
Cunningh. MSS. 1847.
β Stenochilus curvipes. Benth. in Mitch. trop. Austr. p. 221. Varietas S. maculati, sepalorum acumine paulo breviore.
Obs. M. Alphonse De Candolle, in Prodr. xi. p. 715. refers S. ochroleucus of Cunningh. MSS. 1817, as a variety to S. maculatus; it is however very distinct, having a short erect peduncule like that of S. glaber, to which it is much more nearly related, differing chiefly in its being slightly pubescent.

23. GREVILLEA (EUGREVILLEA) Sturtii, foliis indivisis (non-nullis raròbifidis) augustè linearibus elongatis uninerviis: marginibus aretè revolutis, racemis oblongis cylindraceisve: rachi pedicellis perianthiisque inexpansis glutinoso-pubescentibus, ovario sessili, stylo glabro.
Loc. On sand-hills in lat. 27°S. D. Sturt.
Desc. Arbor 15-pedalis (Sturt.) Rami teretes, pube arete adpressa persistenti incani. Folia 6-10-pollices longa, vix tres lineas lata, subter pubescentia incana, super tandem glabrata. Thyrsus terminalis, 2-4 uncialis, rachi pedicellisque pube erecta nec appressa secretione glutinosa intermista. Flores aurantiaci.

Obs. In the collection presented to the British Museum by Sir Thomas Mitchell, of the plants of his last expedition, there is a very perfect specimen, in flower, of Grevillea Sturtii.

The following observations respecting the Grevilleæ of the same collection may not be without interest.

Grevillea Mitchellii, Hooker, in Mitch. Trop. Austr. p. 265, proves to be Gr. Chrysodendron, prodr. fl. Nov. Holl. p. 379, the specific name of which was not derived from the colour of the under surface of the leaves, which is, indeed, nearly white, but from the numerous orange-coloured racemes, rendering this tree conspicuous at a great distance.

Grevillea longistyla and G. juncea of the same narrative, both belong to that section of the genus which I have named Plagiopoda.

A single specimen, in most respects resembling Gr. longistyla, of which possibly it may be a variety, but which at least deserves notice, has all its leaves pinnatifid, instead of being undivided. It may be distinguished by the following character:—
Grevillea (Plagiopoda) neglecta, foliis pinnatifidis subtus niveis; laciniis linearlobus, stylis glabris.

A single specimen also exists of Grevillea (or Hakea) lorea, prodr. flor. Nov. Holl. p. 380, but without fructification.

24. GREVILLEA (CYCLOPTERA ?) lineata, foliis indivisis lineari-ens formibus enerviis subter striis decem paucioribus elevatis uniformibus interstitia bis-terve latitudine superantibus, cicatrice insertionis latiore quam longa utrinque obtusa, racemis terminalibus alternis, pistillis semuncia brevioribus stigmate conico.

Loc. It takes the place of the gum-tree (Eucalyptus) in the creeks about lat. 29°30'S. D. Sturt.

Obs. It is difficult to distinguish this species, which, according to Captain Sturt, forms a tree about 20 feet in height, from Grevillea striata. I have endeavoured to do so in the above specific difference, contrasted with which the leaves of G. striata have always more than 10 striæ, which are hardly twice the breadth of the pubescent interstices, and the cicatrices of whose leaves are longer than broad, and more or less acute, both above and below. This is a source of character which in the supplement to the Prodr. Floræ Novæ Hollandiæ I have employed in a few cases both in Grevillea and Hakea, but which I believe to be important, as it not only expresses a difference of form, but also in general of vascular arrangement.

25. PTILOTUS (latifolius) capitalis globosii; bracteis propriis calycem superantibus, foliis ovatis petiolatis.

Loc. In lat. 26°S. D. Sturt.

Desc. Herba diffusa, ramosa, incana. Folia alterna, petiolata, latè ovata, integerrima. Capitula ramos terminantia, solitaria vel duo approximata. Bracteæ laterales scariosæ, sessiles, latè ovatæ, enerviæ. Perianthium; foliolis subæqualibus, lana implexa alba basi tectis, ante expansionem ungue nervoiso tune brevissimo, post anthesin laminam scariosam enervem fere æquante. Stamina 5 antherifera; filamenta basi in cyathulum edentulum connata. Antheræ biloculares, loculis utrinque distinctis medio solum conjunctis. Ovarium monospermum, glabrum. Stylus filiformis, glaber. Stigma capitatum, parvum. Utriculus evalvis, ruptilis.
Obs. I was at first inclined to consider this plant as a genus distinct from Ptilotus, more, however, from the remarkable difference in habit than from any important distinction in the flower, for its character would have chiefly consisted in the great size, of its lateral branches, and in the form of its antheræ.

In a small collection formed during the voyage of Captains Wickham and Stokes, there is a plant very nearly related to, and perhaps not specifically distinct from Ptilotus latifolias, but having narrower leaves. It was found on one of the islands of Dampier's Archipelago.

26. NEURACHNE (paradoxa) glaberrima, culmo dichotomo, foliis rameis abbreviatis, fasciculis paucifloris, glumis perianthiisque imberbibus valvula exteriore cujusve floris septemnervia.
Loc. Nulla indicatio loci v. stationis, in Herbario. D. Sturt.
Desc. Gramen junceum, facie potlus Cyperaceæ cujusdam. Folia radicalia in specimine unico, viso defuere; ramos subtendentia abbreviata, vagina aperta ipsum folium superante; floralia subspathiformia sed foliacea nec membranacea. Fasciculi pauciflori: spiculæ cum, pedunculo brevissimo articulatæ et solubiles, et subtensæ bractea nervosa carinata ejusdem circiter longitudinis, Gluma bivalvis biflora, nervosa, acuta, mutica; valvulæ subæquales septemnervice; exterioris nervis tribus axin occupantibus sed distinctis reliquis per paria a marginibus et axilibus subæquidistantibus; interioris nervis æquidistantibus, externis margine approximatis. Perianthium inferius (exterius), bivalvis, neuter; valvula exterior septemnervis, exteriori glumæ similis textura forma et longitudine; valvula interior (superior) angustior pauloque brevior, dinervis, nervis alatis marginibus veris latis induplicatis. Perianthium superius hermaphroditam, paulo brevius, pergamineo-membranaceum, nervis dilutè viridibus; valvula exterior quinquenervis, acuta, coneava; interior ejusdem. fere longitudinis, dinervis. Stamina 3, filamentis linearibus. Ovariam oblongum, imberbe. Styli duo. Stigmata plumosa, pallida ?
Obs. Neurachne paradoxa, founded on a single specimen, imperfect in its leaves and stem, but sufficiently complete in its parts of fructification, differs materially in habit from the original species, N. alopeuroides, as well as from N. Mitchelliana of

Nees, while these two species differ widely from each other in several important points of structure.

In undertaking to give some account of the more remarkable plants of Captain Sturt's collection, it was my intention to have entered in some detail into the general character of the vegetation of the interior of Australia, south of the Tropic. I am now obliged to relinquish my original intention, so far as relates to detail, but shall still offer a few general remarks on the subject.

These remarks will probably be better understood, if I refer, in the first place, to some observations published in 1814, in the Botanical Appendix to Captain Flinders's Voyage.

From the knowledge I then had of New Holland, or Australian vegetation, I stated that its chief peculiarities existed in the greatest degree in a parallel, included between 33° and 35°S. lat. which I therefore called the principal parallel, but that these peculiarities or characteristic tribes, were found chiefly at its western and eastern extremities., being remarkably diminished in that intermediate portion, included between 133° and 138°E. These observations related entirely to the shores of Australia, its interior, being at that period altogether unknown; and the species of Australian plants, with which I was then acquainted, did not exceed 4,200. Since that time great additions have been made to the number, chiefly by Mr. Allan Cunningham, in his various journeys from Port Jackson, and on the shores of the North and North-west coasts during the voyages of Captain King whom he accompanied; by Messrs. William Baxter, James Drummond, and M. Preiss, at the western extremity of the principal parallel, and by Mr. Ronald Gunn in Van Diemen's Land. It is probable that I may be considered as underrating these additions, when I venture to state them as only between two and three thousand; and that the whole number of Australian plants at present known, does not exceed, but rather falls short of 7,000 species.

These additions, whatever their amount may be, confirm my original statement respecting the distribution of the characteristic tribes of the New Holland Flora; some additional breadth might perhaps be given, to the principal parallel, and the extent of the peculiar families may now be stated as much, greater at or near its western, than at its eastern extremity.

With the vegetation of the extra-tropical interior of Australia, we are now in some degree acquainted, chiefly from the collections formed by the late Mr. Allan Cunningham, and Charles Fraser, in Oxley's two expeditions from Port Jackson into the western interior, in 1817 and 1818; from Captain Sturt's early expeditions, in which the rivers Darling, Murrumbidgee, and Murray, were discovered; from those of Sir Thomas Mitchell, who never failed to form extensive collections of plants of the regions he visited; and lastly, from Captain Sturt's present collection.

The whole number of plants collected in these various expeditions, may be estimated at about 700 or 750 species; and the general character of the vegetation, especially of the extensive sterile regions, very nearly resembles that of the heads of the two great inlets of the south coast, particularly that of Spencer's Gulf; the same or a still greater diminution of the characteristic tribes of the general Australian Flora being observable. Of these characteristic tribes, hardly any considerable proportion is found, except of Eucalyptus, and even that genus seems to be much reduced in the number of species; of the leafless Acaciæ, which appear to exist in nearly their usual proportion; and of Callitris and Casuarina. The extensive families of Epacrideæ, Stylideæ, Restiaceæ, and the tribe of Decandrous Papilionaceæ, hardly exist, and the still more characteristic and extensive family of Proteaceæ is reduced to a few species of Grevillea, Hakea, and Persoonia.

Nor are there any extensive families peculiar to these regions; the only characteristic tribes being that small section of aphyllous, or nearly aphyllous Cassiæ, which I have particularly adverted to in my account of some of the species belonging to Captain Sturt's collection; and several genera of Myoporinæ, particularly Eremophila and Stenochilus. Both these tribes appear to be confined to the interior, or to the two great gulfs of the South coast, which may be termed the outlets or direct continuation of the southern interior; several of the species observed at the head of Spencer's Gulf, also existing in nearly the same meridian, several degrees to the northward. It is not a little remarkable that nearly the same general character of vegetation appears to exist in the sterile islands of Dampier's Archipelago, on the North-west coast, where even some of the species which probably exist through the whole of the southern interior are found; of these the most striking instances are, Clianthus Dampieri, and Jasiminum lineare, and to establish this extensive range of these two species was my object in entering so minutely into their history in the preceding account.

A still greater reduction of the peculiarities of New Holland vegetation, takes place in the islands of the South coast.